New Challenges of Parkinson's Disease

New Challenges of Parkinson's Disease

Guest Editors

Cristina Agliardi
Franca Rosa Guerini

MDPI

Basel • Beijing • Wuhan • Barcelona • Belgrade • Novi Sad • Cluj • Manchester

Guest Editors
Cristina Agliardi
Laboratory of Molecular
Medicine and Biotechnology
(LAMMB)
IRCCS Fondazione
Don Carlo Gnocchi
Milano
Italy

Franca Rosa Guerini
Laboratory of Molecular
Medicine and Biotechnology
(LAMMB)
IRCCS Fondazione
Don Carlo Gnocchi
Milano
Italy

Editorial Office
MDPI AG
Grosspeteranlage 5
4052 Basel, Switzerland

This is a reprint of the Special Issue, published open access by the journal *International Journal of Molecular Sciences* (ISSN 1422-0067), freely accessible at: https://www.mdpi.com/journal/ijms/special_issues/T30IA485L6.

For citation purposes, cite each article independently as indicated on the article page online and as indicated below:

Lastname, A.A.; Lastname, B.B. Article Title. *Journal Name* **Year**, *Volume Number*, Page Range.

ISBN 978-3-7258-6274-0 (Hbk)
ISBN 978-3-7258-6275-7 (PDF)
https://doi.org/10.3390/books978-3-7258-6275-7

Contents

About the Editors

Cristina Agliardi

Cristina Agliardi is a researcher at the IRCCS Don Carlo Gnocchi Foundation in Milan, Italy, since 2004, where she has conducted her scientific work throughout her career. Her research focuses on the identification of genetic factors involved in the etiology, prognosis, and rehabilitative outcomes of major neurological disorders, including Parkinson's disease, Alzheimer's disease, and Multiple sclerosis. In recent years, she has also investigated neural-derived extracellular vesicles as a promising source of novel biomarkers for the diagnosis and monitoring of neurodegenerative diseases. She has published extensively in peer-reviewed journals, and she has participated in multiple research projects. She serves as Guest Editor for this Special Issue.

Franca Rosa Guerini

Franca Rosa Guerini is a senior scientist and Head of the Genetics Section at the Laboratory of Molecular Medicine and Biotechnologies, IRCCS S. Maria Nascente, Don Carlo Gnocchi Foundation. Since 1990, she has conducted research in the fields of genetics and immunogenetics of neurodegenerative and neurodevelopmental disorders. Dr. Guerini is actively involved in the development and writing of scientific projects in the field of neuroscience and has published more than 150 peer-reviewed scientific papers on the neurogenetics and immunogenetics of Alzheimer's disease, multiple sclerosis, Parkinson's disease, and Neurodevelopmental disorders. She has also participated in multiple research projects. She serves as Guest Editor for this Special Issue.

Preface

This Reprint brings together original research articles and reviews that provide a comprehensive overview of recent advances in Parkinson's disease research. The contributions cover a wide range of topics, including α-synuclein pathology, intracellular peptides, gut–brain interactions, regulatory T cells, cerebrospinal fluid biomarkers, and epigenetic changes, including microRNAs and extracellular vesicles. Several studies explore novel molecular and cellular mechanisms underlying disease onset and progression, while others focus on innovative approaches for early diagnosis and therapeutic intervention. Collectively, these works highlight the importance of integrative strategies combining genetics, molecular biology, and functional studies to better understand Parkinson's disease, identify reliable biomarkers, and develop targeted treatments, ultimately improving patient care and informing future research directions.

Cristina Agliardi and Franca Rosa Guerini
Guest Editors

International Journal of
Molecular Sciences

Editorial

Special Issue: "New Challenges of Parkinson's Disease"

Cristina Agliardi * and Franca Rosa Guerini

IRCCS Fondazione Don Carlo Gnocchi ONLUS, via Capecelatro 66, 20148 Milan, Italy; fguerini@dongnocchi.it
* Correspondence: cagliardi@dongnocchi.it

Parkinson's disease (PD) is the second most common neurodegenerative disorder affecting more than 10 million people worldwide [1]. It is marked by progressive dopaminergic neuronal loss in the *Substantia nigra* and pathological misfolded α-synuclein aggregation that leads to typical motor symptoms (tremors, rigidity, bradykinesia, postural instability) [2]. While dopamine depletion is closely linked to motor deficits, the emergence of certain non-motor symptoms, such as depression, behavior change, psychosis, hallucinations (in particular visual), is complex and likely multifactorial [1]. For instance, visual hallucinations and psychosis can arise not only from the underlying pathology, but also as side effects of long term dopaminergic treatment, particularly with L DOPA [3,4]. Moreover, cognitive decline and dementia are common in advanced PD and are strongly associated with widespread α-synuclein pathology beyond the dopaminergic system [5]. As a consequence, all of these symptoms strongly impact the patients' quality of life.

Early and accurate diagnosis is crucial not only for improving prognostic outcomes, but also for enabling timely therapeutic interventions. However, early diagnosis remains challenging, as the hallmark symptoms of PD often overlap with those of other neurodegenerative diseases [6]. Moreover, current diagnostic practices primarily rely on clinical assessments [7], highlighting the need for more refined and objective diagnostic tools. In this context, the identification and validation of reliable biomarkers are of growing importance, as they can support differential diagnosis, monitor disease progression, and guide the development of personalized treatment strategies, ultimately improving patient care and quality of life [8–10].

This Special Issue of the *International Journal of Molecular Sciences* features four reviews and six original research articles that explore new challenges in understanding, diagnosing, and treating PD.

The contributions in this issue cover a broad spectrum of topics, ranging from the identification of biochemical, genetic, and epigenetic biomarkers to the application of advanced approaches such as multi-omics integration, artificial intelligence, and functional imaging. Particular attention is also given to the innovative use of extracellular vesicles (EVs) and to emerging therapeutic strategies. The review articles provide an overview of several recent developments. Arya et al. in Contribution 1 categorize PD biomarkers into biochemical, genetic, and neuroimaging domains, highlighting the diagnostic and prognostic potential of α-synuclein assays, oxidative stress markers, and mitochondrial-derived peptides. Notably, the α-synuclein Seed Amplification Assay (α-synuclein-SAA) represents a paradigm shift, enabling the detection of misfolded α-synuclein aggregates with high sensitivity and specificity. Advances in genetic studies, including GWAS and artificial intelligence (AI)-based approaches, have identified numerous PD-associated loci and revealed key pathogenic mechanisms such as lysosomal dysfunction and impaired stress responses. Emerging evidence also underscores the role of epigenetic regulation, particularly miRNAs, which show strong discriminative power and potential for personalized therapeutic strategies.

Hwang et al. in Contribution 2 demonstrate that multi-omic integration, particularly combining proteomic and miRNome data, can uncover key pathogenic mechanisms in PD including oxidative stress, mitochondrial dysfunction, and impaired neurotransmission. Artificial intelligence enhances this approach by extracting predictive patterns from complex datasets, thereby facilitating the discovery of novel therapeutic targets. Complementarily, Gualerzi et al. in Contribution 3 highlight the diagnostic and therapeutic potential of EVs, which carry disease-relevant proteins and miRNAs, enable differential diagnosis, and correlate strongly with clinical severity.

Klokkaris and Migdalska-Richards in Contribution 4, together with Gualerzi et al. in Contribution 3, review emerging therapeutic strategies for PD, focusing on epigenetic interventions and EV-based drug delivery. Epigenetic approaches—including HDAC inhibition, RNA interference, CRISPR/Cas9 editing, and miRNA modulation—show promise in reducing α-synuclein pathology and improving neuronal viability, although their efficacy requires careful modulation to avoid off-target effects. EVs, with their ability to cross the blood–brain barrier and deliver therapeutic molecules such as siRNA, dopamine, or anti-inflammatory compounds, represent a highly innovative but still experimental platform facing challenges in large-scale production and standardization.

The original research articles of the present issue are based on both animal models and clinical experimental research; they illustrate the diversity of mechanisms implicated in PD pathogenesis and highlight novel therapeutic strategies ranging from microbiota modulation to immune cell therapy.

The pathogenic role of α-synuclein remains a cornerstone of PD research. Barnes et al. in Contribution 5 demonstrated that both soluble and insoluble α-synuclein fractions from a transgenic mouse model induce pathology independent of injection site, suggesting that neuronal vulnerability rather than the site of entry dictates disease progression. Complementing this, Seo et al. in Contribution 6 reported that decreased serum/glucocorticoid-related kinase 1 expression in the intestine correlates with increased α-synuclein levels, providing evidence of gut–brain axis involvement in PD pathology. Gastrointestinal disorders are some of the major prodromal symptoms of PD. Together, these studies underscore the systemic dimension of α-synuclein aggregation and propagation.

Mounting evidence highlights the gut microbiota as a modulator of levodopa pharmacokinetics. Ai et al. in Contribution 7 revealed that targeted microbial modulation enhances levodopa bioavailability and motor recovery in PD models, suggesting that microbial composition explains interindividual therapeutic heterogeneity. Parallel to microbial factors, intracellular peptides may represent another underexplored contributor. Fiametti et al. in Contribution 8 identified five peptides that ameliorated motor deficits in a zebrafish PD model without cytotoxicity, pointing to peptides as potential therapeutic agents with neuroprotective roles.

Additionally, inflammation and mitochondrial dysfunction have long been suspected and explored in PD. Gevezova et al. in Contribution 9 provided evidence that YKL-40, an inflammatory glycoprotein, correlates with altered mitochondrial bioenergetics in PD patients. Elevated YKL-40 levels were associated with increased respiratory reserve and ATP production, suggesting that YKL-40 could serve as both a biomarker and a mechanistic link between neuroinflammation and bioenergetic dysfunction.

An innovative immunological strategy was proposed by Park et al. in Contribution 10, who developed α-synuclein-specific regulatory T cells (Tregs). These cells ameliorated PD progression in mice by reducing motor deficits, dopaminergic neuronal loss, and microglial activation. This approach illustrates how harnessing adaptive immunity could provide disease-modifying effects beyond symptomatic treatment.

Collectively, these studies exemplify the multifactorial nature of PD pathogenesis and therapeutic development. α-Synuclein remains central, but the microbiota, peptides, metabolic regulators, and immune responses are increasingly recognized as integral players. Future research should focus on integrating these domains to design combination therapies, possibly involving microbiota-targeted interventions, peptide-based neuroprotective agents, and immune modulation. Moreover, novel biomarkers may enable patient stratification for precision medicine approaches for both pharmacological and rehabilitative therapies.

The reviewed studies reinforce PD as a systemic disorder involving complex cross-talk between neuronal, immune, microbial, and metabolic networks. Advancing from bench to bedside will require multidisciplinary strategies that target not only the central nervous system, but also peripheral contributors such as the gut and immune system. Together, these findings herald a new era of integrative therapeutics for PD.

Conflicts of Interest: The authors declare no conflicts of interest.

List of Contributions:

1. Arya, R.; Haque, A.K.M.A.; Shakya, H.; Billah, M.M.; Parvin, A.; Rahman, M.M.; Sakib, K.M.; Faruquee, H.M.; Kumar, V.; Kim, J.J. Parkinson's Disease: Biomarkers for Diagnosis and Disease Progression. *Int. J. Mol. Sci.* **2024**, *25*, 12379. https://doi.org/10.3390/ijms252212379.
2. Hwang, J.S.; Kim, S.G.; George, N.P.; Kwon, M.; Jang, Y.E.; Lee, S.S.; Lee, G. Biological Function Analysis of MicroRNAs and Proteins in the Cerebrospinal Fluid of Patients with Parkinson's Disease. *Int. J. Mol. Sci.* **2024**, *25*, 13260. https://doi.org/10.3390/ijms252413260.
3. Gualerzi, A.; Picciolini, S.; Bedoni, M.; Guerini, F.R.; Clerici, M.; Agliardi, C. Extracellular Vesicles as Biomarkers for Parkinson's Disease: How Far from Clinical Translation? *Int. J. Mol. Sci.* **2024**, *25*, 1136. https://doi.org/10.3390/ijms25021136.
4. Klokkaris, A.; Migdalska-Richards, A. An Overview of Epigenetic Changes in the Parkinson's Disease Brain. *Int. J. Mol. Sci.* **2024**, *25*, 6168. https://doi.org/10.3390/ijms25116168.
5. Barnes, J.; Vermilyea, S.C.; Meints, J.; Martell-Martinez, H.; Lee, M.K. Soluble and Insoluble Lysates from the Human A53T Mutant α-Synuclein Transgenic Mouse Model Induces α-Synucleinopathy Independent of Injection Site. *Int. J. Mol. Sci.* **2025**, *26*, 6254. https://doi.org/10.3390/ijms26136254.
6. Seo, M.H.; Kwon, D.; Kim, S.H.; Yeo, S. Association between Decreased SGK1 and Increased Intestinal α-Synuclein in an MPTP Mouse Model of Parkinson's Disease. *Int. J. Mol. Sci.* **2023**, *24*, 16408. https://doi.org/10.3390/ijms242216408.
7. Ai, P.; Xu, S.; Yuan, Y.; Xu, Z.; He, X.; Mo, C.; Zhang, Y.; Yang, X.; Xiao, Q. Targeted Gut Microbiota Modulation Enhances Levodopa Bioavailability and Motor Recovery in MPTP Parkinson's Disease Models. *Int. J. Mol. Sci.* **2025**, *26*, 5282. https://doi.org/10.3390/ijms26115282.
8. Fiametti, L.O.; Franco, C.A.; Nunes, L.O.C.; de Castro, L.M.; Santos-Filho, N.A. Study of Intracellular Peptides of the Central Nervous System of Zebrafish (*Danio rerio*) in a Parkinson's Disease Model. *Int. J. Mol. Sci.* **2025**, *26*, 2017. https://doi.org/10.3390/ijms26052017.
9. Gevezova, M.; Kazakova, M.; Trenova, A.; Sarafian, V. YKL-40 and the Cellular Metabolic Profile in Parkinson's Disease. *Int. J. Mol. Sci.* **2023**, *24*, 16297. https://doi.org/10.3390/ijms242216297.
10. Park, S.Y.; Yang, H.; Kim, S.; Yang, J.; Go, H.; Bae, H. Alpha-Synuclein-Specific Regulatory T Cells Ameliorate Parkinson's Disease Progression in Mice. *Int. J. Mol. Sci.* **2023**, *24*, 15237. https://doi.org/10.3390/ijms242015237.

References

1. van der Meer, F.; Jorgensen, J.; Hiligsmann, M. Burden of non-motor symptoms of Parkinson's disease: Cost-of-illness and quality-of-life estimates through a scoping review. *Expert Rev. Pharmacoecon. Outcomes Res.* **2025**, *25*, 17–27. [CrossRef] [PubMed]
2. Jankovic, J. Parkinson's disease: Clinical features and diagnosis. *J. Neurol. Neurosurg. Psychiatry* **2008**, *79*, 368–376. [CrossRef] [PubMed]
3. Factor, S.A.; Feustel, P.J.; Friedman, J.H.; Comella, C.L.; Goetz, C.G.; Kurlan, R.; Parsa, M.; Pfeiffer, R.; Parkinson Study Group. Longitudinal outcome of Parkinson's disease patients with psychosis. *Neurology* **2003**, *60*, 1756–1761. [CrossRef] [PubMed]
4. Cummings, J.L. Behavioral complications of drug treatment of Parkinson's disease. *J. Am. Geriatr. Soc.* **1991**, *39*, 708–716. [CrossRef] [PubMed]
5. Halliday, G.; Hely, M.; Reid, W.; Morris, J. The progression of pathology in longitudinally followed patients with Parkinson's disease. *Acta Neuropathol.* **2008**, *115*, 409–415. [CrossRef] [PubMed]
6. Xie, L.; Hu, L. Research progress in the early diagnosis of Parkinson's disease. *Neurol. Sci.* **2022**, *43*, 6225–6231. [CrossRef]
7. Alster, P.; Madetko, N.; Koziorowski, D.; Friedman, A. Diagnostic challenges in Parkinson's disease. *Neurol. I Neurochir. Pol.* **2021**, *55*, 121–129. [CrossRef]
8. Delenclos, M.; Ross, O.A.; Wszolek, Z.K. Biomarkers in Parkinson's disease: Advances and strategies. *Mayo Clin. Proc.* **2022**, *97*, 304–316. [CrossRef] [PubMed]
9. Mazzetti, S.; Contaldi, E.; Basellini, M.J.; Novello, C.; Calogero, A.M.; Straniero, L.; Garrì, F.; Ferri, V.; Calandrella, D.; Del Sorbo, F.; et al. Comparing GBA1-Parkinson's disease and idiopathic Parkinson's disease: α-Synuclein oligomers and synaptic density as biomarkers in the skin biopsy. *Brain Pathol.* **2024**, *34*, e13284. [CrossRef] [PubMed]
10. Müller, B.; Mollenhauer, B. Status of current biofluid biomarkers in Parkinson's disease. *Mov. Disord. Clin. Pract.* **2023**, *10*, 18–26. [CrossRef]

International Journal of
Molecular Sciences

MDPI

Article

Soluble and Insoluble Lysates from the Human A53T Mutant α-Synuclein Transgenic Mouse Model Induces α-Synucleinopathy Independent of Injection Site

Justin Barnes [1], Scott C. Vermilyea [2], Joyce Meints [1], Héctor Martell-Martinez [1] and Michael K. Lee [1,2,*]

1 Department of Neuroscience, Institute for Translational Neuroscience, University of Minnesota, Minneapolis, MN 55455, USA; bubberman13@gmail.com (J.B.); meint002@umn.edu (J.M.); martineh@umn.edu (H.M.-M.)
2 Aligning Science Across Parkinson's (ASAP) Collaborative Research Network, Chevy Chase, MD 20815, USA; svermily@umn.edu
* Correspondence: mklee@umn.edu

Abstract: Pathological aggregation of α-synuclein (αS) is implicated in the pathogenesis of Parkinson's disease (PD) and other α-synucleinopathies. The current view is that neuron-to-neuron spreading of αS pathology contributes to the progression of α-synucleinopathy. We used an A53T mutant human αS transgenic mouse model (*TgA53T*) to examine whether the site of pathogenic αS inoculation affects the pattern of neuropathology and whether soluble and insoluble fractions derived from crude pathogenic tissue lysates exhibit differential capacities to initiate αS pathology. To test whether the inoculation site impacts the ultimate spatial/temporal patterns of αS pathology, αS preformed fibrils (PFFs), or brain homogenates from *TgA53T* mice with α-synucleinopathy, were injected into the cortex/striatum, brainstem, or skeletal muscle. In all cases, inoculation of pathogenic αS induced end-stage motor dysfunction within ~100 days post-inoculation (dpi). Significantly, irrespective of the inoculation sites, the ultimate distribution of the αS pathology was like that seen in normally aged *TgA53T* mice at end-stage, indicating that the intrinsic neuronal vulnerability is a significant determinant in the induction of αS pathology, even when initiated by inoculation of pathogenic αS. Temporal analysis of brainstem-injected *TgA53T* mice show that initial αS pathology was seen by 30 days post-inoculation and inflammatory changes occur at later stages. In addition, we show that both highly soluble (S150) and insoluble (P150) fractions from end-stage *TgA53T* mice can seed de novo αS pathology in vivo. Moreover, the endoplasmic reticulum (ER)-enriched fraction from the TgA53T mice were highly pathogenic as the ER fraction induced αS pathology faster than other fractions when injected unilaterally into *TgA53T* mice. Our results suggest that multiple αS species from the brain can initiate the development of progressive αS pathology.

Keywords: alpha-synuclein; alpha-synucleinopathy; alpha-synuclein spreading; endoplasmic reticulum; oligomers; transgenic mouse

1. Introduction

Parkinson's disease (PD) is a common late-onset, progressive neurodegenerative disorder characterized by the presence of intracellular aggregates of α-synuclein (αS) termed Lewy bodies (LBs) and neurites (LNs) [1]. While the etiology of PD is unknown in most cases, degenerating neuronal populations in PD exhibit αS abnormalities, indicating that the αS abnormalities are mechanistically linked to the pathogenesis of PD and other α-synucleinopathies [2,3].

The temporal analysis of αS pathology in human cases [4–6] indicates that αS pathology in the PD brain originates in several focal sites, particularly the dorsal motor nucleus of vagus (DMV) and the anterior olfactory nucleus. With the preclinical progression of the disease, αS pathology moves anteriorly in a topographically defined manner until reaching the cortex during the symptomatic stage. This staging led to the hypothesis that αS pathology moves via cell-to-cell transmission of amyloidogenic αS species that serves as a conformational template to seed pathology [4,5]. This hypothesis was also further supported by the presence of LBs in fetal dopamine neurons grafted into PD patients [7,8]. Cell-to-cell transmission of α-synucleinopathy has received formal experimental support from brain injection studies establishing αS spreads trans-synaptically in mice [9–11]. Currently, direct inoculation of αS preformed fibrils (PFFs) or tissue lysates/fractions containing αS aggregates can induce αS pathology when directly injected into the central nervous system (CNS) or peripherally [12]. However, most current studies use αS PFF derived from recombinant αS. Thus, it is unknown if the spatial–temporal patterns of αS pathology are comparable following the inoculation of various pathogenic αS species to various brain regions of αS Tg models.

To better understand the spreading of α-synucleinopathy, we directly compared spatial–temporal patterns of αS pathology following the inoculation of pathogenic αS species into different brain regions (cortex/striatum and dorsal brainstem) and skeletal muscle of A53T mutant human αS (*TgA53T*) transgenic mice. We show that the resulting pattern of pathology is independent of injections sites, indicating that specific neuronal populations are predisposed to the effects of toxic αS species in vivo. We also show that both the soluble and insoluble fractions from end-stage *TgA53T* mice can initiate pathology in younger *TgA53T* mice and endoplasmic reticulum (ER)-associated αS is more pathogenic than other fractions.

2. Results

2.1. Lysates from Symptomatic TgA53T Mice and αS PFF Induce α-Synucleinopathy with Similar End-Stage Distribution

Although the intracortical/intrastriatal (IC/IS) transmission model shows the induction and spread of αS pathology in TgA53T mice [9], studies show that injections of pathogenic αS lysates or preformed αS fibrils (PFFs) in multiple areas, including the skeletal muscle, can induce widespread αS pathology over time [13]. Thus, we examined whether the inoculation site impacts the onset and spread of αS pathology in TgA53T mice. Because brainstem (BrSt) neurons are affected early by αS pathology in both human PD [14] and in the TgA53T mouse model [15], we compared the disease produced by the IC/IS inoculation with the BrSt injections of αS PFF. We injected young, disease-free 3–6-month-old TgA53T mice (all from line G2-3 unless noted) with end-stage lysate (ESL or S3000, a crude 3000× *g* pooled lysate) from BrSt/Spinal Cord (SpC) of end-stage (ES)-affected TgA53T mice or lysates from disease-free (asymptomatic lysate, ASL) 5–6 months-old TgA53T mice (Figure 1a–c). Immunoblot analysis (Figure 1b) shows that both ASL and ESL contain a similar amount of total αS but the ESL contains a greater amount of pS129αS, consistent with pathology in ES TgA53T mice. Like that seen by Luk et al. [9], the mean of survival for bilateral IC/IS-injected mice was 98 ± 8.6 days post-inoculation (dpi) (Mean ± SD) (Figures 1d and S1). Unilateral inoculation of ESL into the BrSt resulted in earlier onset of motor dysfunction with a mean survival of 71.3 ± 18.7 dpi (Figures 1d and S1a), a significantly shorter disease time course than with IC/IS dual injections. Inoculation of αS PFF into BrSt leads to a mean survival time (77.4 ± 13.16 dpi, $n = 5$) that is like the ESL inoculation (Figures 1d and S1a). All the mice injected with ASL or saline remained disease free at ~150 dpi (Figure 1d) except one ASL-inoculated animal who succumbed to disease and prematurely died at ~9 months of age, a time point where normal-aged TgA53T mice start to show disease [15]. Thus, it is likely that this animal developed disease independently of the injected lysate.

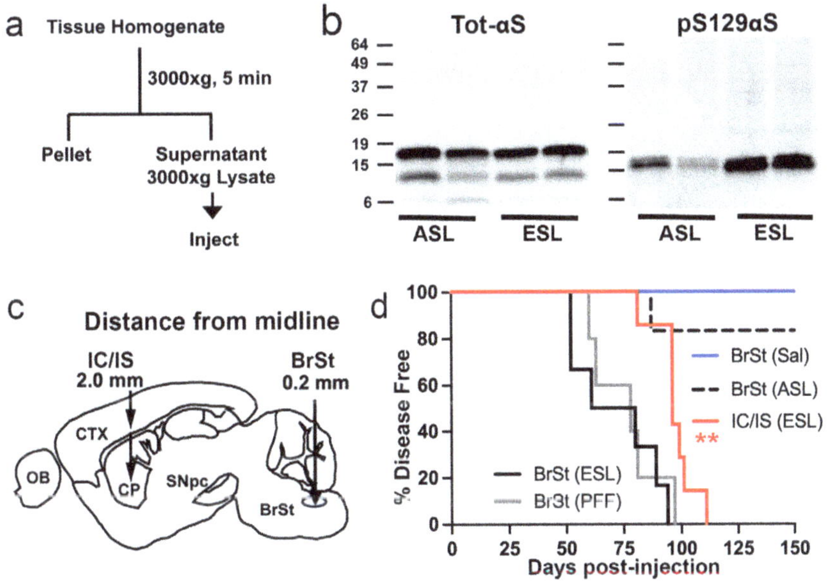

Figure 1. Injection of *TgA53T* brain-derived lysate into cortex/striatum or brainstem induces α-synucleinopathy and premature death in young *TgA53T* mice. (**a**) Schematic representation of the protocol used to generate the injected ESL and asymptomatic ASL. (**b**) Analysis of ASL and ESL for total αS (top) and pS129 αS levels. While similar levels of total αS are present in ASL and ESL, ESL contains more pS129 αS than ASL. (**c**) Diagrams depicting the stereotaxic injection sites for the IC/IS and BrSt injection models. Stereotaxic coordinates: IC/IS 2.0 mm lateral from the midline, +0.2 mm relative to bregma, and 0.8 and 2.6 mm deep from the dura; BrSt: 0.2 mm lateral from the midline, −7.34 mm from bregma, 3.75 mm deep from the dura. (**d**) Kaplan–Meier survival curve of mice IC/IS injected with ESL and BrSt-injected animals with αS PFF, ESL, ASL, and saline. Control animals receiving saline and ASL all survived disease free to 200 days post-injection (dpi), except one animal injected with ASL that developed disease and died 136 dpi. ** $p < 0.01$, IC/IS ESL vs BrSt ESL or BrSt PFF, Mantel–Cox log-rank test. All $n = 6$ except BrSt (PFF), $n = 5$. Abbreviations: α-synuclein, αS; asymptomatic lysate, ASL; end-stage lysate, ESL; intracortical/intrastriatal, IC/IS; brainstem, BrSt; preformed fibril, PFF.

Biochemical analysis of the ESL-inoculated animals confirms that the ESL, but not ASL, leads to αS pathology (Figure 2a). Analysis of Triton X100-detergent soluble and insoluble fractions reveals the accumulation of insoluble high molecular weight (HMW) species of αS (Figure 2a, asterisks) in the SpC of ESL-injected mice. In the ESL-injected mice, insoluble pS129αS accumulates in BrSt and SpC, corresponding to the regions most affected by αS pathology (Figure 2a). Consistent with lack of αS pathology, pS129αS does not accumulate in ASL- or saline-injected mice (Figure 2a). This pattern of αS pathology is equivalent to that seen with the ES TgA53T mice that normally develop α-synucleinopathy from aging [15].

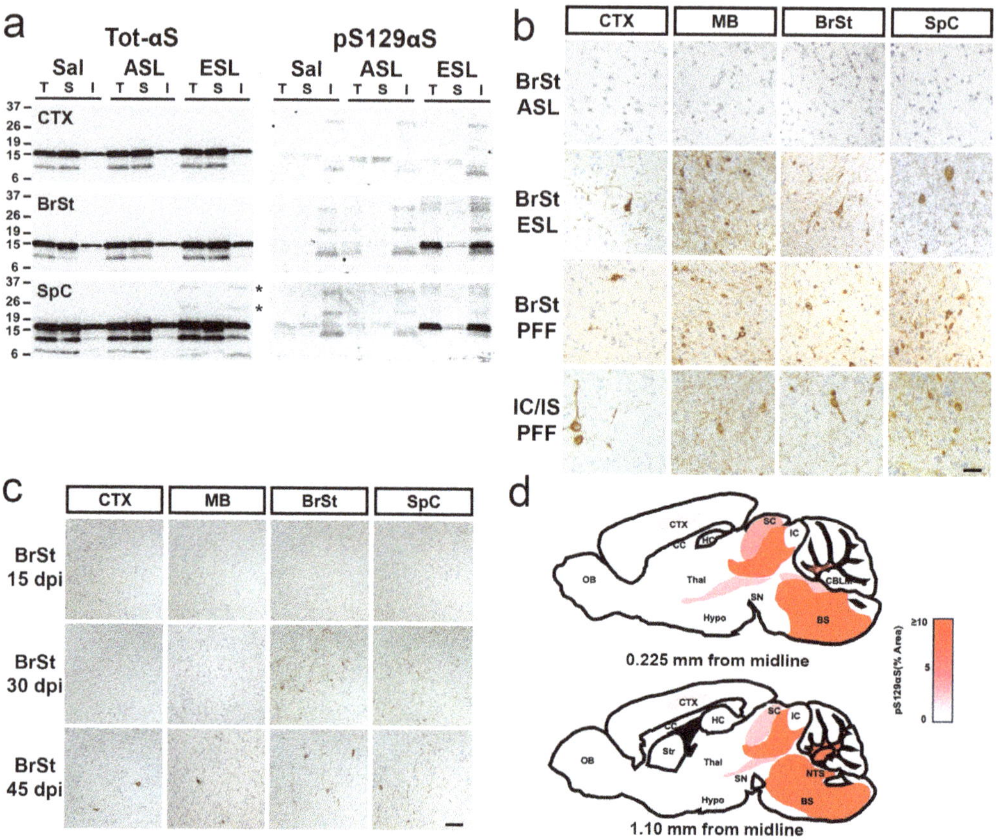

Figure 2. Aggregation of αS is induced by ESL in *TgA53T* mice with pathological distribution that is independent of injections site. (**a**) Accumulation of pathological HMW αS species (*) and pS129 αS in affected brain regions in detergent-insoluble fraction. Total lysates (T) from unaffected CTX and affected BrSt and SpC from saline, ASL, and ESL of BrSt-injected animals were separated into TX-soluble (S) and TX-insoluble (I) fractions. Detection of total αS (left) demonstrates the presence of HMW-insoluble αS species in SpC (*) ESL-inoculated mice. Immunoblots for pS129 αS clearly show a complete lack of pathological αS in ASL-injected animals and CTX of ESL-inoculated mice. Virtually all of the HMW αS species (*) and pS129 αS partition with the TX-insoluble fraction in BrSt and SpC. (**b**) Aggregated αS, marked by pS129 αS immunoreactivity, is abundant in animals injected with ESL or αS PFF. (**c**) Immunoreactivity for pS129 αS occurs between 15 and 30 days post-BrSt injection of ESL. (**d**) Diagrams depicting the distribution of αS pathology as observed by immunohistochemical analysis. Shades of red denote the relative abundance of αS inclusions (% Area covered by pS129 αS). Note that the sections are unilateral to the injection site even though pathology is detected bilaterally. Abbreviations: α-synuclein, αS; asymptomatic lysate, ASL; end-stage lysate, ESL; high molecular weight, HMW; cortex, CTX; brainstem, BrSt, BS; cerebellum, CBLM; corpus callosum, cc; hippocampus, HC; hypothalamus, Hypo; inferior and superior colliculus, IC, SC; nucleus tractus solitarius, NTS; olfactory bulb, OB; spinal cord, SpC; striatum, Str; substantia nigra, SN; thalamus, Thal; triton-X, TX; preformed fibril, PFF; saline, Sal. Scale bars: 50 μm.

Immunohistochemical detection of pS129αS demonstrates robust staining in multiple brain tissues from IC/IS- and BrSt-injected end-stage animals (Figure 2b). Consistent with the lack of motor phenotype, no pathological alterations are noted in ASL-inoculated mice or saline-injected animals (Figure 2b). Significantly, despite the differences in the site of inoculation (IC/IS vs BrSt), the overall patterns of pathology in the ES mice were comparable to each other with the most abundant pathology in the BrSt and SpC (Figure 2b), a pattern very similar to what is seen in the TgA53T mice that naturally develop αS pathology with aging [15]. However, the IC/IS-injected animals exhibit more pS129αS pathology in CTX compared to BrSt-injected animals (Figure S1b). To determine if components of lysates other than pathogenic αS could be responsible for the distribution of the pathology, we also examined animals that were BrSt-injected with αS PFF. Our results show that the pattern of pathology seen in the ESL-inoculated animals is very similar to animals injected with αS PFF into BrSt (Figures 2b and S1c). Moreover, the pattern of αS pathology induced by BrSt injections is also very similar to the pathology achieved in this line of mice by IC/IS and peripheral, intramuscular (IM) inoculation of αS PFF [16,17]. Analysis of spinal cords from animals unilaterally injected with ESL (IC/IS) or αS-PFF (IM) confirm that αS pathology progresses to bilateral pathology (Figure S1d).

To determine if the prominent subcortical pathology is unique to the TgA53T(G2-3) line or occurs in another line of TgA53T, we performed IC/IS injections on TgA53T(H5), which express 50% less αS than TgA53T(G2-3) [18]. Because of lower transgene expression, injected TgA53T(H5) animals developed a progressive motor phenotype later (~150 dpi) than the TgA53T(G2-3) (~100 dpi) following IC/IS injections of αS PFF. Neuropathological analysis shows that, as with TgA53T(G2-3), αS pathology in ES mice was most abundant in the BrSt/SpC region (Figure S2). Significantly, the TgA53T(H5) animals also showed a more widespread distribution of αS pathology, including αS pathology in the hippocampus (Figure S2a). In summary, all animals inoculated with ESL or αS PFF showed the most severe BrSt and SpC αS pathology, regardless of the initial sites of injections.

Collectively, these results indicate that in TgA53T mice, the ES distribution of αS pathology is independent of the inoculation site and likely determined by factors other than local seeding by pathogenic αS. Our Tg models show the highest levels of total αS in the CTX [17] (Figure S2c). Thus, the relative lack of forebrain pathology is not because of insufficient αS expression in this area. However, because endogenous αS or βS could impact αS aggregation [19–23], we examined the expression of human αS along with levels of endogenous mouse αS and βS in various brain regions of TgA53T mice (Figure S2c). The results confirm that the transgene-derived HuαS is widely expressed in all brain regions. However, endogenous mouse αS and βS are completely absent from BrSt/SpC regions (Figure S2c), suggesting that the lack of endogenous αS and/or βS might contribute to the vulnerability of neurons to developing αS pathology in a TgA53T model.

While previous studies with the IC/IS injection model show that cortical and striatal αS pathology occurs prior to BrSt or SpC [9], we observe more prominent early subcortical pathology with the BrSt inoculation. Thus, following the BrSt inoculation, the appearance of pS129αS over time shows that initial αS pathology develops proximal to the inoculation site where αS pathology is seen in the pons by 30 dpi (Figure 2c). By 45 dpi, regions rostral and caudal to the BrSt inoculation sites exhibit αS pathology (Figure 2c). Thus, we propose that following BrSt injection of pathogenic αS, the pathology spreads in both rostral and caudal directions but the most prominent pathology occurs in subcortical regions (Figure 2d). Analysis of IC/IS-injected TgA53T(H5) animals at the intermediate stage (90 dpi) shows cortical pS129αS but the BrSt and SpC pS129αS pathology was more prominent (Figure S2b). Overall, the current results collectively show that regardless of the initial injection site, subcortical areas (e.g., BrSt, SpC) exhibit the earliest αS pathology and culminate in the most severe αS pathology in ES TgA53T mice.

2.2. Neuroinflammation Follows Onset of α-Synucleinopathy

It has been proposed that inoculation with pathogenic αS could lead to early neuroinflammation that may promote further propagation of α-synucleinopathy [24]. Thus, we examined the spatial–temporal relationship between αS pathology and neuroinflammation. First, we confirmed that at ES, animals following BrSt injection exhibit coincident αS pathology and neuroinflammation. Tissue sections from BrSt-injected animals were stained for microglia (Iba1; Figure 3a) and astrocytes (GFAP; Figure 3b). Based on the morphology of the cells, no obvious neuroinflammatory changes are seen in the absence of significant αS pathology, such as in CTX (Figure 3a,b) and in animals injected with ASL (Figure 3a,b). However, in ES animals following ESL injection, areas with significant αS pathology (BrSt, SpC) exhibit abundant microglia and astrocytes with highly activated morphology. Immunoblotting for Iba1 and GFAP reveals no change in Ctx, but a significant increase in BrSt and SpC of ESL-injected mice at ES disease (Figures 3c,d and S3).

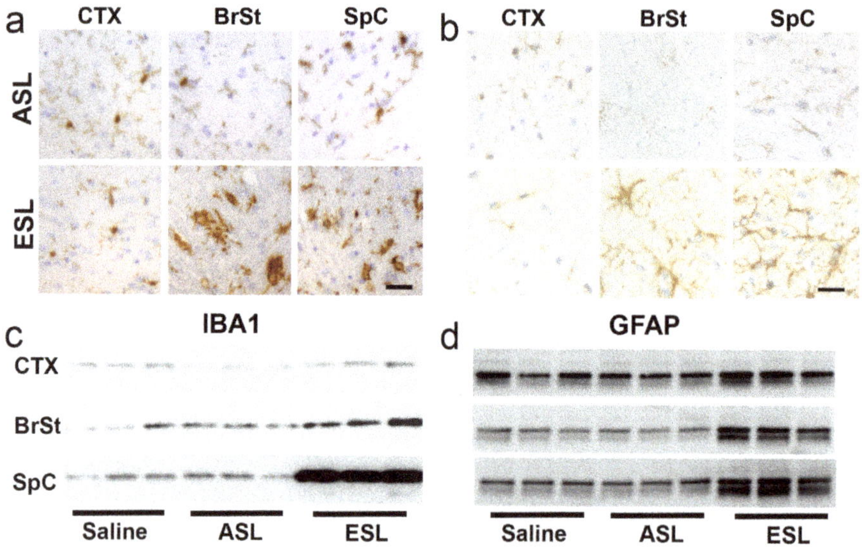

Figure 3. Neuroinflammation is increased in affected brain regions. (**a**,**b**) Tissue sections stained for Iba1 (**a**) and GFAP (**b**). BrSt and SpC from ESL-injected TgA53T mouse reveal abundant activated microglia and astrocytes. Areas lacking αS pathology, such as the CTX, as well as animals injected with ASL, are free of glial activation. (**c**,**d**) Biochemical analysis of whole-tissue lysates from CTX, BrSt, and SpC of saline-, ASL-, and ESL-injected mice show robust increases in Iba1 (**c**) and GFAP (**d**) in BrSt and SpC of ESL-injected mice. Quantitation of the blots is shown in Figure S3. Abbreviations: alpha synuclein, αS; brainstem, BrSt; spinal cord, SpC; end-stage lysate, ESL; cortex, CTX; asymptomatic lysate, ASL. Scale bars: 25 μm.

To better determine the spatial–temporal relationship between the onset of microglial activation and αS pathology, the sections from BrSt-inoculated mice harvested at 15-, 30-, and 45-dpi that were evaluated for αS pathology (Figure 2c) were stained for microglia (Iba-1) and astrocytes (GFAP) (Figure S4). With BrSt inoculation, αS pathology is seen in the pons by 30 dpi (Figure 2c). However, there is no obvious signs of astrocytic or microglial activation at 30- or 45-dpi (Figure S4). Because αS pathology is still sparse at 45 dpi, we performed double immunofluorescence analysis of pS129αS with Iba1 or GFAP (Figure 4) to determine if the presence of αS pathology could be associated with local inflammatory changes. Double immunofluorescence analyses show that αS pathology in ES mice is associated with the nearby reactive astrocytes and microglia. However, such glial responses were not coincident with αS pathology at 30- or 45-dpi (Figure 4b,c). Similar analysis of early-stage mice following IM αS PFF injections shows that the initial onset of pS129αS pathology in SpC, occurring as early as 15–30 dpi, is not accompanied by obvious glial activation (Figure S5). While other A53T Tg models show αS pathology in astrocytes [25,26], we do not observe significant pS129αS in astrocytes in ES mice (Figure 4).

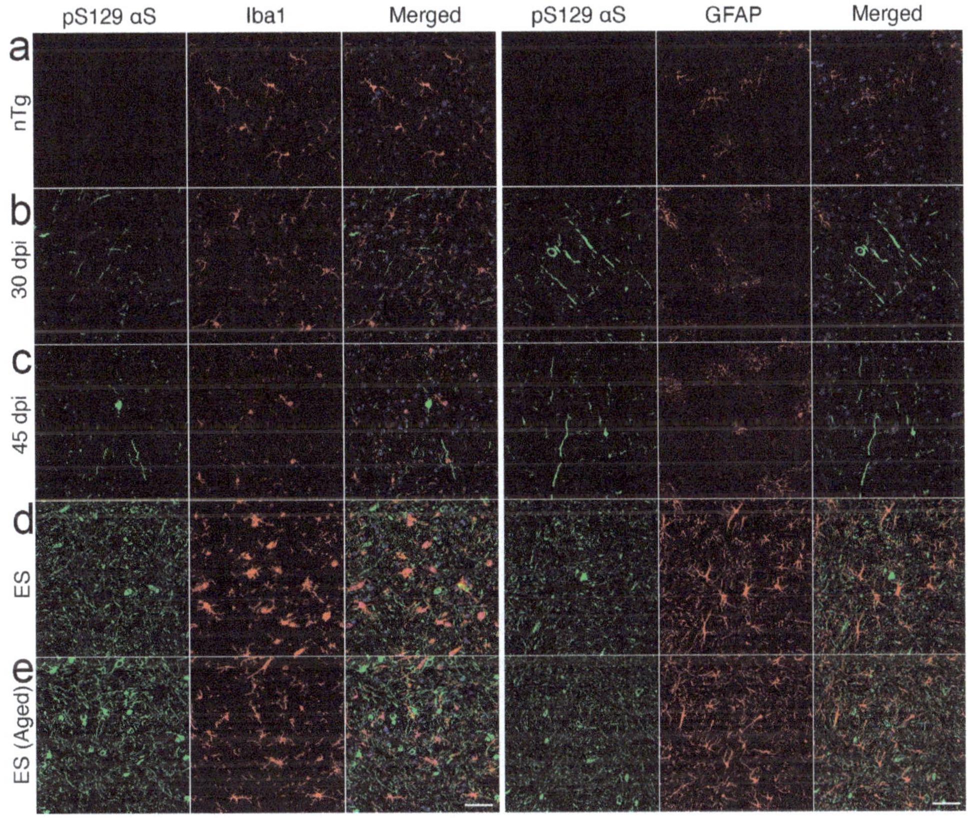

Figure 4. Neuroinflammation occurs after αS pathology in *TgA53T* mice. BrSt sections from control mice (**a**, ntg) and *TgA53T* mice injected with ESL into BrSt were analyzed at 30 dpi (**b**) 45 dpi (**c**), and ES (**d**, injected); (**e**, aged). The sections were stained for pS129 αS and Iba1 or GFAP. Aggregated αS (pS129 αS) is clearly seen in 30- and 45-dpi samples (**b**,**c**) but the overall morphology and density of Iba-1 and GFAP staining are comparable to nTg mice (**a**), indicating that the presence of pS129 αS is not accompanied by obvious glial activation. In contrast, there is a clear increase in Iba1 and GFAP staining with well-established pS129 αS pathology in ES animals (**d**,**e**). Abbreviations: alpha synuclein, αS; brainstem, BrSt; end-stage lysate, ESL; end-stage, ES; non transgenic, nTg. Scare bar: 50 μm.

Collectively, our results indicate that glial activation, particularly activation of microglia, is only evident after substantial αS pathology is established. A similar pattern of initial αS pathology followed by neuroinflammation was reported with IM inoculation of the M83 line [27]. Thus, while microglia and astrocytes may modulate neuronal spreading of αS pathology or neuronal survival, our results indicate that it is unlikely that neuroinflammation is a significant factor in the initial onset of αS pathology in a TgA53T model of α-synucleinopathy.

2.3. Both Highly Soluble and Insoluble Fractions Induce αS Pathology

Because αS fibrils and disease-associated αS aggregates in TgA53T mice are detergent-insoluble [9,13,16,27], we tested whether the induction of αS pathology by the ESL is mediated by the insoluble αS species. The pathogenic S3000 ESL was centrifuged at 150,000× *g* to obtain highly soluble (supernatant, S150) and insoluble (pellet, P150) fractions (Figure 5a). Biochemical analyses of the fractions show that very little αS is found in P150 from asymptomatic animals (Figure 5b). Moreover, most of the pS129αS, representing the overt αS pathology, is highly enriched in the P150 fractions from the symptomatic mice (Figure 5b). In contrast, the amount of total αS is the same in the S150 fractions, regardless of the disease state of the animal (Figure 5b). Despite the high levels of total αS, the levels of pS129αS in S150 are lower than in the P150 with no differences between the disease state of the animal (Figure 5b). Side-by-side comparison and quantitative analysis of αS species in ESL (S3000), S150, and P150 show that, compared to ESL, pS129αS is highly enriched in the P150 fraction and the S150 fraction is low in pS129αS (Figure S6). Thus, we expected that αS pathology would be selectively induced by the P150-ESL fraction.

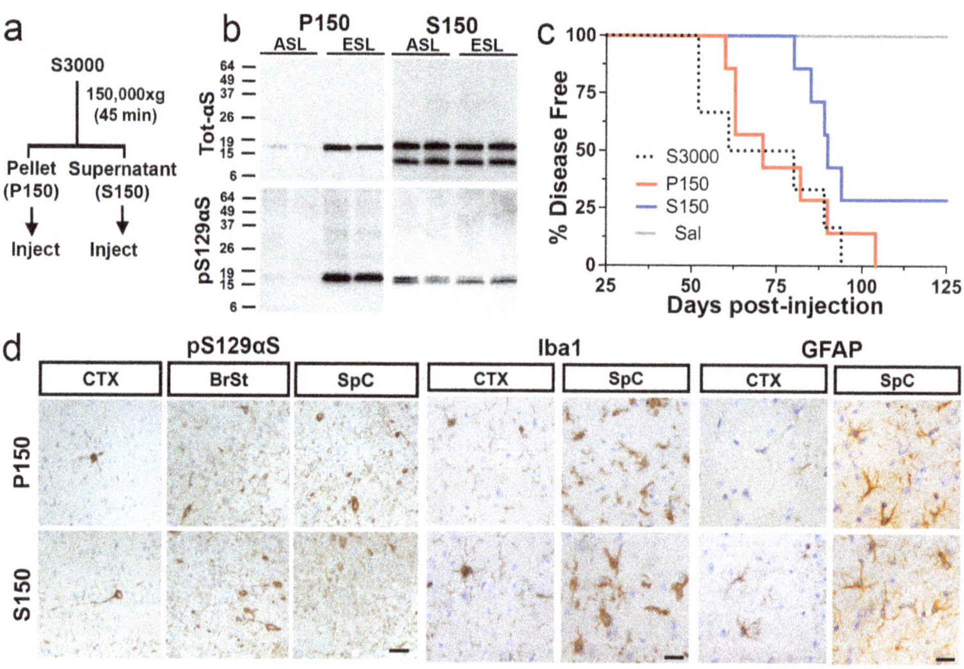

Figure 5. Both soluble and insoluble fractions from ESL induce α-synucleinopathy in *TgA53T* mice. (**a**) Flow diagram showing the separation of S3000 into soluble (S150) and insoluble (P150) fractions. (**b**) Biochemical analysis of S150 and P150 fractions derived from ASL and ESL. Equal amount of total protein was analyzed. P150 fractions from ASL contain very little αS while total αS and pS129 αS are abundant in P150 fractions from ESL. Despite the abundant total αS in S150, much lower pS129 αS is present in the S150. (**c**) Kaplan–Meier survival curve of mice injected with S150 and P150 into BrSt. Also shown in Figure 1d is the curve for the S3000-ESL-injected mice. Both S150 and P150 injections lead to motor dysfunction and reduced life span. S3000, $n = 6$; P150, $n = 7$; S150, $n = 7$; Sal, $n = 6$. (**d**) Neuropathological analysis of affected S150- and P150-injected animals shows expected distribution of αS pathology (pS129 αS) and corresponding neuroinflammation (Iba1 and GFAP). Abbreviations: end-stage lysate, ESL; asymptomatic lysate, ASL; α-synuclein, αS. Scale bars: 50 μm for pS129 αS, 25 μm for Iba1 and GFAP.

Following BrSt injections of P150 and S150 fractions, we were surprised to find that, regardless of the solubility of the inoculated material, inoculated TgA53T mice developed motor dysfunction leading to premature death with average lifespans of 75 $\pm$ 7 dpi for the P150 fraction and 88 $\pm$ 3 dpi for the S150 fraction (Figure 5c). While two out of seven subjects injected with S150 did not develop a disease phenotype by 150 dpi, there was no differences in the average lifespan (Figure 5c, p = 0.1397) of symptomatic animals injected with either S150 or P150.

Histological (Figure 5d) and biochemical (Figure S6) analyses of ES P150- and S150-injected animals demonstrate virtually indistinguishable αS pathology both in the spatial pattern and severity of αS aggregation (compare Figures 2, 5 and S6). Aggregation of αS occurred throughout the CNS but was most robust in the BrSt and SpC (Figure 5d). Biochemical analysis of brain tissue also reveals an accumulation of detergent-insoluble αS species in the insoluble fraction from BrSt and SpC (Figure S6a) and dramatic increases in pS129αS in both BrSt and SpC (Figure S6b), a pattern very similar to ESL-injected mice (Figure S7). Finally, based on immunostaining for Iba1 and GFAP, the distribution and intensity of neuroinflammation are similar in S150- and P150-BrSt-injected animals (Figure 5d), and are nearly identical to those observed in ESL-injected mice (see Figure 3). Thus, by three broad metrics- survival, immunohistochemical staining, and biochemistry- the α-synucleinopathy and disease induced by injection of all three end-stage lysates (S3000, S150, P150) are essentially identical.

Since S3000 ASL does not induce disease and there are no obvious differences between S150 from ASL and ESL by immunoblot analysis (Figure 5b), S150 from ESL may contain soluble pathogenic αS conformers that are SDS-labile. Thus, the incomplete penetrance of end-stage S150 to cause disease could be due to the possibility that toxic αS assemblies in S150 fractions are more prone to degradation following cellular uptake compared to more mature insoluble toxic αS species in P150 (Figure 5b). To test this hypothesis, we performed dot blot analysis of P150 and S150 for αS oligomers, using FILA1 [28] and the OC antibody [29] (Figure S8a,b). We previously showed that FILA1 can recognize both soluble and insoluble αS aggregates derived from the TgA53T model [30] and the OC antibody selectively recognizes mature fibrils [29]. Our results show that while levels of FILA1+ oligomers are similar between P150 and S150, the levels of OC+ oligomers are significantly more abundant in P150 (Figure S8a,b). We also subjected S150 and P150 to proteinase-K treatment as more compact mature aggregates should be more resistant to proteinase-K treatment. The results show that αS in P150 is more resistant to proteinase K proteolysis than αS in S150 (Figure S8c). Thus, we conclude that the presence of FILA1+ oligomers/aggregates in both S150 and P150 is pathogenic. Further, more labile FILA1+ oligomers in S150 may be responsible for the partial penetrance of S150 in inducing disease.

2.4. Microsomes from Symptomatic TgA53T Mice Induce Rapid α-Synucleinopathy

In the TgA53T model, α-synucleinopathy is associated with the accumulation of αS oligomers/aggregates in the endoplasmic reticulum (ER) and chronic ER stress that contributes to neurodegeneration [30,31]. Analyses of TgA53T mice inoculated with S3000 ESL show that αS pathology in these mice is also associated with signs of chronic ER stress (Figure S9), suggesting that ER stress occurs even when α-synucleinopathy is induced by exogenous inoculation of pathogenic αS. We recently showed that ER-enriched microsomes from TgA53T mice, containing αS aggregates, are highly toxic to cultured neurons and aggressively induce αS aggregates and cell death in cultured neurons [32]. Thus, we examined whether the microsome fraction from symptomatic TgA53T mice can induce pathology following BrSt injection.

Microsomes from pooled BrSt/SpC or CTX from the symptomatic TgA53T mice were used for BrSt injections of 3–4-month-old TgA53T mice. Analysis of the fractions shows that pS129αS is enriched in microsomes (P100 or ER) from BrSt/SpC, where very little pS129αS is seen in P100 from CTX (Figure 6a). Comparative analysis of total αS and pS129αS in various fractions from end-stage TgA53T mice indicates that the P100 fraction contains similar levels of total αS and pS129αS as the P150 fraction (Figure S9b). Analysis of organelle markers shows that the P100 fraction is also enriched in the ER marker (Grp78/BiP). Analysis of injected animals shows that P100 from symptomatic TgA53T mice develop progressive motor abnormalities by ~50 dpi while the animals inoculated with P100 from CTX did not show any disease phenotype (Figure 6b). Significantly, microsome inoculation leads to onset of motor symptoms much faster than any of the other fractions tested in this study (compare Figures 1d, 5c and 6b), indicating that microsome-associated αS oligomer/aggregates are highly pathogenic in vivo. Neuropathological analysis shows that the early motor deficits were indeed associated with significant αS pathology (Figure 6c) with the pattern seen with other αS fractions. Collectively, the results show that ER-associated αS oligomers/aggregates are highly pathogenic in vivo [30,32].

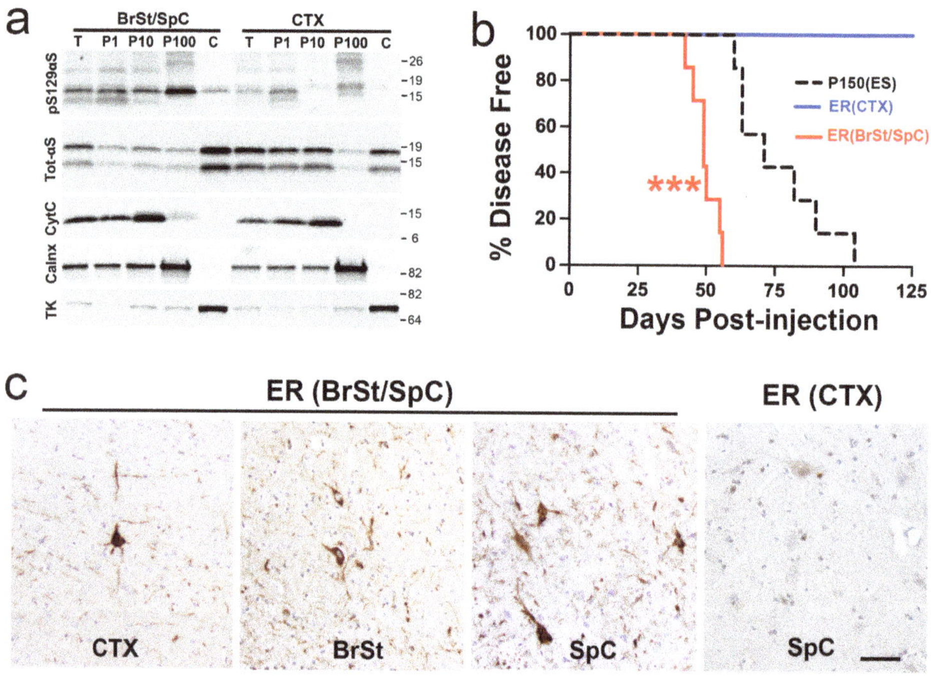

Figure 6. ER-enriched fraction from ES *TgA53T* mice, containing ER-associated αS, is highly pathogenic. (**a**) Subcellular fractions from BrSt/SpC and CTX were analyzed for pS129 αS, total αS, and organelle markers (Cytochrome C, CytC-mitochondria; Calnexin, Calnx-ER; Transketolase, KT-Cytosol). Calnexin-enriched ER fraction (P100) is relatively free of CytC and TK. pS129 αS is enriched in P100 from BrSt/SpC. (**b**) Kaplan–Meier survival curve of mice injected with P100 (ER) fractions from BrSt/SpC (BS/SC) or CTX. For reference, curve for P150-injected animals from Figure 4c is also shown. *** $p < 0.001$, ER (BrSt) vs. P150 (ES), Mantel–Cox log-rank test. P150 (ES), $n = 7$; ER (BrSt/SpC), $n = 7$; ER (CTX), $n = 6$. (**c**) Neuropathological analysis shows that ER (BS/SC) fraction induces pS129 αS pathology in *TgA53T* mice, particularly in BrSt and SpC. In contrast, ER (CTX) fraction does not cause pS129 pathology. Abbreviations: endoplasmic reticulum, ER; end-stage, ES; α-synuclein, αS; brainstem, BrSt/BS; spinal cord, SpC/SC; cortex, CTX. Scale bar: 50 μm.

3. Discussion

In this report, we provide a comparative analysis of αS pathology induced by the inoculation of lysates from *TgA53T* mice. Specifically, we examined the induction of αS pathology as a function of different lysate solubilities and locations of inoculation. Like that reported with αS PFF models [25], we show that regardless of inoculation site, pathogenic lysates from brains containing α-synucleinopathy lead to a remarkably consistent pattern of ES pathology in the *TgA53T* model. This indicates that the regional differences in the vulnerability to develop αS pathology with aging is maintained even when α-synucleinopathy is initiated by injections of pathogenic αS species. We also show that, in vivo, multiple αS species with different solubilities induce very similar αS pathologies.

Following the staging of PD pathology by Braak and colleagues [4], a series of in vitro and in vivo studies have confirmed that αS pathology can spread from an initial site of αS aggregate induction to other neurons and brain regions [5,9,10,12,13]. It is also established that lysates from murine or human tissues with αS pathology can induce early αS pathology in αS Tg mouse models [9,12]. Further, numerous studies have shown that injections of αS PFF into the CNS or peripheral tissue can induce αS pathology in both αS Tg mice and wild type mice [10–13]. While studies have generally assumed that the spreading of α-synucleinopathy occurs along the interconnected networks [5,11], it is notable that while human α-synucleinopathy may progress rostrally from the DMV, the interconnected aspect has not been demonstrated. Further, some studies using rodents show deviations from the simple cell-to-cell transmission model [25]. Thus, while the transmission of αS pathology via the interconnected network may be valid, not all neurons within the connected network are equally vulnerable to developing αS pathology. We show that progressive α-synucleinopathy is induced by injections with pathogenic αS species at different injection sites. However, the pattern of α-synucleinopathy following injections at different sites is remarkably consistent with BrSt and SpC neurons being most vulnerable to developing α-synucleinopathy. Our time course study also shows that at intermediate stages, independent of where pathogenic αS is injected, αS pathology is more prominent in BrSt and SpC neurons. These results indicate that the spread of αS pathology following pathogenic αS injections is determined, in part, by differences in neuronal vulnerability to the development of αS pathology. One potential factor could be the expression levels of endogenous αS and βS as βS is known to inhibit αS aggregation [20–22] and mouse αS is not efficiently templated by human αS aggregates [19].

Analysis of intermediate-stage mice for coincidence of αS pathology and glial activation indicates that αS pathology occurs before the microglial or astrocyte activation. This result agrees with a prior study demonstrating a temporal relationship between the onset of αS pathology and microglial activation following IM injections of αS PFF [27]. We extended the prior studies by showing that no obvious glial activation occurs near the αS pathology at the initial stages of pathology (Figures 4 and S5). Thus, if microglia promote αS propagation [24,33], such processes are likely to occur at more advanced stages of αS pathology.

Most in vivo studies on the transmission of α-synucleinopathies have relied on αS PFF or oligomers derived from recombinant αS as the initiating factor for pathology. Although PFF can efficiently induce pathology, the biological relevance of an in vitro aggregated recombinant αS is unclear as the structure of αS PFF could be divergent from in vivo pathology. Because toxic species of αS in human PD and in vivo models, such as the *TgA53T* model, are likely to be a complex mixture of various αS species [34–36], characterizing the pathology induced by tissue-derived αS species will have relevance for the in vivo pathogenesis of α-synucleinopathy. This view is supported by the differential induction of pathology in huαS Tg models by various human brain-derived lysates [37]. In *TgA53T* models, αS forms soluble oligomers, particularly in the ER, which increasingly aggregate to become insoluble as the disease progresses [30,31]. This, taken together with the fact that pathogenic αS PFFs are insoluble, we initially expected that the insoluble fraction from ESL will contain most of the disease-initiating activity. Thus, we expected P150, enriched in mature pS129αS-containing aggregates, to aggressively induce αS pathology following inoculation. Surprisingly, the S150 from ES mice, despite being biochemically similar to the S150 from AS mice, is capable of inducing αS pathology in vivo. Although the exact nature of the soluble "toxic" species is unknown, our results indicate that S150 contains potentially pathogenic FILA1+ αS oligomers. Pathogenic soluble aggregates have been identified with tau pathology from the human brain [38,39], suggesting that seeding overt pathology by soluble aggregates/oligomers may be a general feature of α-synucleinopathy and tauopathy. Regardless, our results show that different species of αS with varying degrees of solubility can induce identical patterns of pathology. We also show that, consistent with the pathogenic importance of ER-associated αS pathology [30,32], microsomes from ES *TgA53T* mice are highly pathogenic in vivo. This result is consistent with prior studies showing that ER fractions from TgA53T mice are highly toxic to cultured neurons [32]. While the basis for the enhanced toxicity of ER-associated αS is under active study, the greater abundance of both FILA1+ and A11+ oligomers in the ER fraction [30] may be a contributing factor. Moreover, the interaction of αS aggregates with specific ER-associated components [40] and/or lipids [41] may also contribute to the differential pathogenicity of the ER fraction. Overall, these data indicate that further separating soluble and insoluble αS species by size and/or subcellular localization could be a fruitful approach for identifying the elusive transmissible αS species. For example, it would be significant to fractionate S150 lysate using Size Exclusion Chromatography to determine the oligomeric nature of the toxic S150 species.

In summary, our study indicates that differential neuronal vulnerability contributes to patterns of α-synucleinopathy. Thus, simple spreading of αS pathology via cell-to-cell connection cannot fully account for the progression of α-synucleinopathy. Further, our evidence supports the view that multiple αS species are capable of initiating αS pathology in vivo and it will be important to fully define the disease relevance of these different pathogenic αS species.

4. Materials and Methods

4.1. Transgenic Mouse Lines

The transgenic mouse line over-expressing A53T mutant human αS under control of the mouse prion protein promoter has been described previously (*TgA53T*, line G2-3 and line H5) [15]. *TgA53T(G2-3)* mice develop a progressive, fatal neurological disease at ~10–14 months of age and show marked αS pathology by end-stage disease. *TgA53T(H5)* lines express less αS (~50%) and do not spontaneously develop pathology. Young, asymptomatic mice aged 3–6 months were used for all experiments. Animals showing complete hindlimb paralysis were considered end-stage. All study methods were approved in full by the University of Minnesota Institutional Animal Care and Use Committee and were consistent with the National Institutes of Health Office of Laboratory Animal Welfare Policy.

4.2. Inoculation Material

Lysates used for all injections were generated from tissues of *TgA53T* mice at either 4 months of age (asymptomatic lysate, ASL) or end-stage (end-stage lysate, ESL) (For protocol, see 10.17504/protocols.io.4r3l22d2ql1y/v1; made available: 20 March 2024; last accessed: 27 June 2025). BrSt and spinal cord (SpC) were combined and processed together, as both regions demonstrate robust αS pathology at end-stage [15]. Preparation of the

3000× *g* lysate (S3000) has been detailed previously [9]. Briefly, affected tissue previously harvested and stored at −80 °C was sonicated in sterile saline (0.9% NaCl; 1:10 w/vol). The homogenate was centrifuged for five minutes (3000× *g*, 4 °C) and the recovered supernatant was the 3000× *g* lysate (S3000). To obtain highly soluble (S150) and insoluble (P150) fractions, S3000 fraction was centrifuged for 45 min (150,000× *g*, 4 °C). The resulting supernatant was termed the S150 fraction. The pellet was washed and resuspended in sterile saline by sonication (3x10s pulses) in half of the original volume ultracentrifuged and used as the P150 fraction.

Endoplasmic Reticulum (ER)-enriched fractions were obtained from *TgA53T* mice as previously detailed [32]. Briefly, fresh tissues were homogenized in a 1:10 (wt/vol) volume of lysis buffer (250 mM sucrose, 20mM HEPES, 10mM KCl, 1.5mM $MgCl_2$, 2 mM EDTA, protease-inhibitor cocktail) and centrifuged at 1000× *g*. The supernatant was centrifuged at 10,000× *g* to pellet mitochondria and the supernatant was centrifuged at 100,000× *g* to obtain ER (pellet) and cytosol (supernatant).

Human αS preformed fibrils (PFFs) were generated from in vitro aggregation of recombinant wild type human αS monomers as previously described [16] (For protocol see: dx.doi.org/10.17504/protocols.io.81wgbxjx3lpk/v1; made available: 20 March 2024; last accessed: 27 June 2025). Human αS PFF were stored at 5 mg/mL and were sonicated and diluted to desired concentration prior to use.

4.3. Inoculation of *TgA53T* Mice

All injections into the brain were performed unilaterally into the right hemisphere/side in *TgA53T* mice at 3–6 months of age (For protocol, see dx.doi.org/10.17504/protocols.io.dm6gp3738vzp/v1; made available: 20 March 2024; last accessed: 27 June 2025). Animals were stereotaxically injected with 2.5 µg of total protein in 2.5 µL. The injections occurred at a rate of 0.1 µL per minute using a 28g needle attached via tubing to a Hamilton syringe controlled by a constant pressure syringe pump (Harvard Apparatus, Holliston, MA, USA). The needle was kept in place for five minutes post-injection. The stereotaxic coordinates used for the injection sites were as follows: intracortical/intrastriatal (IC/IS): 2.0 mm lateral from the midline, +0.2 mm relative to bregma, and 0.8 and 2.6 mm deep from the dura; BrSt: 0.2 mm lateral from the midline, −7.34 mm from bregma, 3.75 mm deep from the dura. Intramuscular (IM) injections of αS PFF were performed as described [16]. Briefly, isoflurane anesthetized mice (~3.5-months-old) were bilaterally injected into the bicep femoralis muscle bilaterally with 5 µg of PFF.

Starting at 60 days following PFF injections, the mice were evaluated for disease onset and progression three days per week (Monday, Wednesday, Friday). Disease onset was identified by an imbalance in gait leading to a wobbling phenotype and the end-stage was characterized as complete hindlimb paralysis. Animals were euthanized at either pre-determined time points or upon complete hindlimb paralysis. Half of the brain was used for immunohistochemistry and was immediately drop-fixed in 4% paraformaldehyde and remained in fixative for at least 48 h. The other half was dissected into gross brain regions, including the cortex and brainstem, and either snap-frozen on dry ice and stored at −80 °C or processed immediately. Half of the spinal cords collected remained in the spinal column and were drop-fixed in 4% paraformaldehyde and half were removed and frozen or processed in the same manner as the brains.

4.4. Immunohistochemistry and Mapping of αS Pathology

All fixed brains were embedded in paraffin and serially sectioned (6 µm) in the sagittal plane. Four to six sections taken every 12th section, starting at first section from midline containing hippocampus, were processed using standard immunohistochemical (IHC) techniques for DAB (Covance) detection as previously described [15–17] (For protocol, see dx.doi.org/10.17504/protocols.io.kxygxzqyzv8j/v1; made available: 14 December.2023; last accessed: 27 June 2025). Sections were counterstained with hematoxylin and imaged with a Leica DM2500 microscope using Leica Application Suite v4.1 imaging software (RRID: SCR_016555; https://www.leica-microsystems.com/products/microscope-software/p/leica-application-suite/, accessed on 20 March 2025). PhosphoSerine-129 αS (pS129αS)-positive inclusions/cell bodies and neurites were mapped at three lateral positions: 0.225, 1.10, and 1.525 mm from the midline using Image J v1.54e/Fiji (NIH; RRID: SCR_003070; https://imagej.net, accessed on 20 March 2025).

4.5. Immunofluorescence Staining

Paraffin sections of sagittal brain (BrSt-injected) and SpC (IM-injected) were processed for immunofluorescent staining as previously described [16] (For protocol, see dx.doi.org/10.17504/protocols.io.36wgq3x33lk5/v1; made available: 20 March 2024; last accessed: 27 June 2025). Briefly, the tissues were first deparaffinized followed by 30 min of antigen retrieval. Then, 3% hydrogen peroxide was used to quench endogenous peroxidases followed by blocking using 100% background sniper solution for 13 min. Sections were then incubated in primary antibodies with 5% sniper diluted in TRIS overnight at 4 °C (pS129αS, 015-25191, 1:500; Iba1, 019-19741, 1:100; Fujifilm-Wako-USA, Richmond, VA USA; GFAP, Z0334, 1:200 Dako Cytomation, Glostrup, DK) followed by incubation in secondary fluorophore-conjugated antibody (Donkey anti-mouse 488, Donkey anti-rabbit 594). Sections were imaged using a Leica Stellaris 8 confocal microscope.

4.6. Western Blot (Immunoblot) Analysis

Lysate preparation and immunoblot analysis were performed as previously described [16,17] (For protocol, see dx.doi.org/10.17504/protocols.io.3byl4qpqjvo5/v1; made available: 20 March 2024; last accessed: 27 June 2025). Briefly, brain tissues from mice were dissected and stored at −80 °C and frozen tissues were homogenized in TNE buffer [Tris-HCl 50 mM, NaCl 150 mM and EDTA 5 mM, and protease and phosphatase inhibitors (HALT; Thermo-Fisher; Waltham, MA, USA)]. To obtain total lysate, homogenates were solubilized with 0.5% NP40, 0.5% DOC, and 1% SDS. For TX-100 non-ionic detergent fractionation, TNE homogenates were added to 1% TX-100 and sonicated prior to centrifugation at 100,000× g for 20 min. Supernatant was collected as the soluble fraction while the pellet was washed and centrifuged again in 1% TX-100. The resulting pellet was then reconstituted in TNE with detergents as the TX-100 insoluble fraction [16,34].

The protein content of the lysates was determined using the BCA assay (Pierce, Thermo; Rockford, IL, USA) and samples were prepared to equal protein concentrations in reducing, SDS-sample, Laemmli buffer (Boston BioProducts; Ashland, MA, USA). For Western blot analysis, protein lysates were run on Criterion™ TGX™ gels (BioRad, Hercules, CA, USA) and transferred onto nitrocellulose membranes. Proteins on membranes were detected using appropriate primary antibodies followed by horseradish peroxidase (HRP)-conjugated secondary antibodies (Invitrogen, Carlsbad, CA, USA). Membranes were then developed using chemiluminescent substrates (Bio-Rad, Hercules, CA, USA and Thermo) and imaged using the ImageQuant LAS 4000 detection system (GE Life Sciences, Piscataway, NJ, USA). Densitometry on Western blot images was analyzed using the ImageQuant TL 8.1 software (GE Life Sciences; RRID: SCR_018374; https://www.cytivalifesciences.com/en/us/shop/protein-analysis/molecular-imaging-for-proteins/imaging-software/imagequant-tl-10-2-analysis-software-p-28619, accessed on 20 March 2025).

4.7. Antibodies

Primary antibodies used for immunohistochemistry and immunoblot analysis are listed in Supplemental Table S1.

4.8. Statistical Analysis

All analyses were completed using GraphPad Prism 10 (RRID: SCR_002798; https://www.graphpad.com, accessed on 20 March 2025). Values are expressed as mean ± SD/SEM as indicated. Differences between means were analyzed by Student's *t*-test, one-way ANOVA, or two-way ANOVA. ANOVA analyses were combined with the Tukey post hoc test for multiple comparisons (Prism 10; GraphPad Software, Boston, MA, USA). Log-rank (Mantel–Cox) test was used to assess significant differences in survival curves. An α value of 0.05 and adjusted p values for multiple comparisons were used to determine significance.

Supplementary Materials: The following supporting information can be downloaded at https://www.mdpi.com/article/10.3390/ijms26136254/s1.

Author Contributions: J.B. and M.K.L. conceived and designed the study. J.B., J.M., H.M.-M., and M.K.L. were responsible for the acquisition of the data for Figures 1–3, 5, 6, S1–S4 and S6–S9. J.B. and M.K.L. were responsible for assembling Figures 1–3, 5, 6, S1–S4 and S6–S9. S.C.V. was responsible for the data acquisition and assembling Figures 4 and S5. J.B., S.C.V., and M.K.L. were responsible for data analysis and writing/editing of the manuscript. All authors have read and agreed to the published version of the manuscript.

Funding: This work was funded by Grants R01 NS086074, R01 NS092093, and ASAP-024407. J.B., J.M., H.H.M., and M.K.L. were supported by R01 NS086074, R01 NS092093. S.C.V. and M.K.L. were supported by ASAP-024407. ASAP-02447 is administered by the Michael J. Fox Foundation (MJFF) on behalf of the Aligning Science Across Parkinson's (ASAP) initiative.

Institutional Review Board Statement: All study methods involving animals were approved in full by the University of Minnesota Institutional Animal Care and Use Committee (2112-3900A, approval date: 21 February 2025) and were consistent with the National Institutes of Health Office of Laboratory Animal Welfare Policy.

Informed Consent Statement: Not applicable.

Data Availability Statement: The datasets and images used and/or analyzed, as well as a table of key resources (KRT), for the current study are available on Zenodo (https://zenodo.org/doi/10.5281/zenodo.10819209; made available: 14 March 2024; last accessed: 27 June 2025).

Acknowledgments: We thank Cindy Croy for help in generating the transgenic mouse cohorts and Poul Henning Jensen, Aarhus University for sharing the FILA-1 antibody.

Conflicts of Interest: The authors declare no conflicts of interest.

References

1. Emamzadeh, F.N.; Surguchov, A. Parkinson's Disease: Biomarkers, Treatment, and Risk Factors. *Front. Neurosci.* **2018**, *12*, 612. [CrossRef]
2. Vila, M.; Przedborski, S. Genetic clues to the pathogenesis of Parkinson's disease. *Nat. Med.* **2004**, *10* (Suppl. S7), S58–S62. [CrossRef] [PubMed]
3. Obeso, J.A.; Rodriguez-Oroz, M.C.; Goetz, C.G.; Marin, C.; Kordower, J.H.; Rodriguez, M.; Hirsch, E.C.; Farrer, M.; Schapira, A.H.; Halliday, G. Missing pieces in the Parkinson's disease puzzle. *Nat. Med.* **2010**, *16*, 653–661. [CrossRef] [PubMed]
4. Braak, H.; Del Tredici, K.; Rub, U.; de Vos, R.A.; Jansen Steur, E.N.; Braak, E. Staging of brain pathology related to sporadic Parkinson's disease. *Neurobiol. Aging* **2003**, *24*, 197–211. [CrossRef]
5. Killinger, B.A.; Kordower, J.H. Spreading of alpha-synuclein—Relevant or epiphenomenon? *J. Neurochem.* **2019**, *150*, 605–611. [CrossRef]
6. Del Tredici, K.; Rub, U.; De Vos, R.A.; Bohl, J.R.; Braak, H. Where does parkinson disease pathology begin in the brain? *J. Neuropathol. Exp. Neurol.* **2002**, *61*, 413–426. [CrossRef]

7. Li, J.Y.; Englund, E.; Holton, J.L.; Soulet, D.; Hagell, P.; Lees, A.J.; Lashley, T.; Quinn, N.P.; Rehncrona, S.; Bjorklund, A.; et al. Lewy bodies in grafted neurons in subjects with Parkinson's disease suggest host-to-graft disease propagation. *Nat. Med.* **2008**, *14*, 501–503. [CrossRef] [PubMed]
8. Kordower, J.H.; Chu, Y.; Hauser, R.A.; Freeman, T.B.; Olanow, C.W. Lewy body-like pathology in long-term embryonic nigral transplants in Parkinson's disease. *Nat. Med.* **2008**, *14*, 504–506. [CrossRef]
9. Luk, K.C.; Kehm, V.M.; Zhang, B.; O'Brien, P.; Trojanowski, J.Q.; Lee, V.M. Intracerebral inoculation of pathological alpha-synuclein initiates a rapidly progressive neurodegenerative alpha-synucleinopathy in mice. *J. Exp. Med.* **2012**, *209*, 975–986. [CrossRef]
10. Luk, K.C.; Kehm, V.; Carroll, J.; Zhang, B.; O'Brien, P.; Trojanowski, J.Q.; Lee, V.M. Pathological alpha-synuclein transmission initiates Parkinson-like neurodegeneration in nontransgenic mice. *Science* **2012**, *338*, 949–953. [CrossRef]
11. Henderson, M.X.; Cornblath, E.J.; Darwich, A.; Zhang, B.; Brown, H.; Gathagan, R.J.; Sandler, R.M.; Bassett, D.S.; Trojanowski, J.Q.; Lee, V.M.Y. Spread of alpha-synuclein pathology through the brain connectome is modulated by selective vulnerability and predicted by network analysis. *Nat. Neurosci.* **2019**, *22*, 1248–1257. [CrossRef]
12. Chung, H.K.; Ho, H.A.; Perez-Acuna, D.; Lee, S.J. Modeling alpha-Synuclein Propagation with Preformed Fibril Injections. *J. Mov. Disord.* **2019**, *12*, 139–151. [CrossRef]
13. Sacino, A.N.; Brooks, M.; Thomas, M.A.; McKinney, A.B.; Lee, S.; Regenhardt, R.W.; McGarvey, N.H.; Ayers, J.I.; Notterpek, L.; Borchelt, D.R.; et al. Intramuscular injection of alpha-synuclein induces CNS alpha-synuclein pathology and a rapid-onset motor phenotype in transgenic mice. *Proc. Natl. Acad. Sci. USA* **2014**, *111*, 10732–10737. [CrossRef] [PubMed]
14. Braak, H.; Del Tredici, K.; Bratzke, H.; Hamm-Clement, J.; Sandmann-Keil, D.; Rub, U. Staging of the intracerebral inclusion body pathology associated with idiopathic Parkinson's disease (preclinical and clinical stages). *J. Neurol.* **2002**, *249* (Suppl. 3), iii1–iii5. [CrossRef] [PubMed]
15. Lee, M.K.; Stirling, W.; Xu, Y.; Xu, X.; Qui, D.; Mandir, A.S.; Dawson, T.M.; Copeland, N.G.; Jenkins, N.A.; Price, D.L. Human alpha-synuclein-harboring familial Parkinson's disease-linked Ala-53 --> Thr mutation causes neurodegenerative disease with alpha-synuclein aggregation in transgenic mice. *Proc. Natl. Acad. Sci. USA* **2002**, *99*, 8968–8973. [CrossRef] [PubMed]
16. Vermilyea, S.C.; Christensen, A.; Meints, J.; Singh, B.; Martell-Martinez, H.; Karim, M.R.; Lee, M.K. Loss of tau expression attenuates neurodegeneration associated with alpha-synucleinopathy. *Transl. Neurodegener.* **2022**, *11*, 34. [CrossRef]
17. Karim, M.R.; Liao, E.E.; Kim, J.; Meints, J.; Martinez, H.M.; Pletnikova, O.; Troncoso, J.C.; Lee, M.K. alpha-Synucleinopathy associated c-Abl activation causes p53-dependent autophagy impairment. *Mol. Neurodegener.* **2020**, *15*, 27. [CrossRef]
18. Teravskis, P.J.; Covelo, A.; Miller, E.C.; Singh, B.; Martell-Martinez, H.A.; Benneyworth, M.A.; Gallardo, C.; Oxnard, B.R.; Araque, A.; Lee, M.K.; et al. A53T mutant alpha-synuclein induces tau dependent postsynaptic impairment independent of neurodegenerative changes. *J. Neurosci.* **2018**, *38*, 9754–9767. [CrossRef]
19. Luk, K.C.; Covell, D.J.; Kehm, V.M.; Zhang, B.; Song, I.Y.; Byrne, M.D.; Pitkin, R.M.; Decker, S.C.; Trojanowski, J.Q.; Lee, V.M. Molecular and Biological Compatibility with Host Alpha-Synuclein Influences Fibril Pathogenicity. *Cell Rep.* **2016**, *16*, 3373–3387. [CrossRef]
20. Hashimoto, M.; Rockenstein, E.; Mante, M.; Mallory, M.; Masliah, E. beta-Synuclein inhibits alpha-synuclein aggregation: A possible role as an anti-parkinsonian factor. *Neuron* **2001**, *32*, 213–223. [CrossRef]
21. Brown, J.W.; Buell, A.K.; Michaels, T.C.; Meisl, G.; Carozza, J.; Flagmeier, P.; Vendruscolo, M.; Knowles, T.P.; Dobson, C.M.; Galvagnion, C. beta-Synuclein suppresses both the initiation and amplification steps of alpha-synuclein aggregation via competitive binding to surfaces. *Sci. Rep.* **2016**, *6*, 36010. [CrossRef] [PubMed]
22. Li, X.; Yu, L.; Liu, X.; Shi, T.; Zhang, Y.; Xiao, Y.; Wang, C.; Song, L.; Li, N.; Liu, X.; et al. beta-synuclein regulates the phase transitions and amyloid conversion of alpha-synuclein. *Nat. Commun.* **2024**, *15*, 8748. [CrossRef] [PubMed]
23. Xu, B.; He, W.; Fan, F.; Chen, S.; Zhu, M.; Hou, Y.; Zheng, L.; Yu, H.; Liu, Y. beta-synuclein blocks alpha-synuclein condensate fusion to disrupt the maturation of phase separation. *Cell Rep.* **2025**, *44*, 115761. [CrossRef]
24. George, S.; Rey, N.L.; Tyson, T.; Esquibel, C.; Meyerdirk, L.; Schulz, E.; Pierce, S.; Burmeister, A.R.; Madaj, Z.; Steiner, J.A.; et al. Microglia affect alpha-synuclein cell-to-cell transfer in a mouse model of Parkinson's disease. *Mol. Neurodegener.* **2019**, *14*, 34. [CrossRef] [PubMed]
25. Sorrentino, Z.A.; Brooks, M.M.T.; Hudson, V., 3rd; Rutherford, N.J.; Golde, T.E.; Giasson, B.I.; Chakrabarty, P. Intrastriatal injection of alpha-synuclein can lead to widespread synucleinopathy independent of neuroanatomic connectivity. *Mol. Neurodegener.* **2017**, *12*, 40. [CrossRef]
26. Sorrentino, Z.A.; Giasson, B.I.; Chakrabarty, P. alpha-Synuclein and astrocytes: Tracing the pathways from homeostasis to neurodegeneration in Lewy body disease. *Acta Neuropathol.* **2019**, *138*, 1–21. [CrossRef] [PubMed]
27. Sorrentino, Z.A.; Xia, Y.; Funk, C.; Riffe, C.J.; Rutherford, N.J.; Ceballos Diaz, C.; Sacino, A.N.; Price, N.D.; Golde, T.E.; Giasson, B.I.; et al. Motor neuron loss and neuroinflammation in a model of alpha-synuclein-induced neurodegeneration. *Neurobiol. Dis.* **2018**, *120*, 98–106. [CrossRef]

28. Lindersson, E.; Beedholm, R.; Hojrup, P.; Moos, T.; Gai, W.; Hendil, K.B.; Jensen, P.H. Proteasomal inhibition by alpha-synuclein filaments and oligomers. *J. Biol. Chem.* **2004**, *279*, 12924–12934. [CrossRef]
29. Kayed, R.; Head, E.; Sarsoza, F.; Saing, T.; Cotman, C.W.; Necula, M.; Margol, L.; Wu, J.; Breydo, L.; Thompson, J.L.; et al. Fibril specific, conformation dependent antibodies recognize a generic epitope common to amyloid fibrils and fibrillar oligomers that is absent in prefibrillar oligomers. *Mol. Neurodegener.* **2007**, *2*, 18. [CrossRef]
30. Colla, E.; Jensen, P.H.; Pletnikova, O.; Troncoso, J.C.; Glabe, C.; Lee, M.K. Accumulation of Toxic alpha-Synuclein Oligomer within Endoplasmic Reticulum Occurs in alpha-Synucleinopathy In Vivo. *J. Neurosci.* **2012**, *32*, 3301–3305. [CrossRef]
31. Colla, E.; Coune, P.; Liu, Y.; Pletnikova, O.; Troncoso, J.C.; Iwatsubo, T.; Schneider, B.L.; Lee, M.K. Endoplasmic Reticulum Stress Is Important for the Manifestations of alpha-Synucleinopathy In Vivo. *J. Neurosci.* **2012**, *32*, 3306–3320. [CrossRef] [PubMed]
32. Colla, E.; Panattoni, G.; Ricci, A.; Rizzi, C.; Rota, L.; Carucci, N.; Valvano, V.; Gobbo, F.; Capsoni, S.; Lee, M.K.; et al. Toxic properties of microsome-associated alpha-synuclein species in mouse primary neurons. *Neurobiol. Dis.* **2018**, *111*, 36–47. [CrossRef] [PubMed]
33. Thi Lai, T.; Kim, Y.E.; Nguyen, L.T.N.; Thi Nguyen, T.; Kwak, I.H.; Richter, F.; Kim, Y.J.; Ma, H.I. Microglial inhibition alleviates alpha-synuclein propagation and neurodegeneration in Parkinson's disease mouse model. *NPJ Park. Dis.* **2024**, *10*, 32. [CrossRef] [PubMed]
34. Li, W.; West, N.; Colla, E.; Pletnikova, O.; Troncoso, J.C.; Marsh, L.; Dawson, T.M.; Jakala, P.; Hartmann, T.; Price, D.L.; et al. Aggregation promoting C-terminal truncation of alpha-synuclein is a normal cellular process and is enhanced by the familial Parkinson's disease-linked mutations. *Proc. Natl. Acad. Sci. USA* **2005**, *102*, 2162–2167. [CrossRef]
35. Sonustun, B.; Altay, M.F.; Strand, C.; Ebanks, K.; Hondhamuni, G.; Warner, T.T.; Lashuel, H.A.; Bandopadhyay, R. Pathological Relevance of Post-Translationally Modified Alpha-Synuclein (pSer87, pSer129, nTyr39) in Idiopathic Parkinson's Disease and Multiple System Atrophy. *Cells* **2022**, *11*, 906. [CrossRef] [PubMed]
36. Zhang, S.; Zhu, R.; Pan, B.; Xu, H.; Olufemi, M.F.; Gathagan, R.J.; Li, Y.; Zhang, L.; Zhang, J.; Xiang, W.; et al. Post-translational modifications of soluble alpha-synuclein regulate the amplification of pathological alpha-synuclein. *Nat. Neurosci.* **2023**, *26*, 213–225. [CrossRef]
37. Lloyd, G.M.; Sorrentino, Z.A.; Quintin, S.; Gorion, K.M.; Bell, B.M.; Paterno, G.; Long, B.; Prokop, S.; Giasson, B.I. Unique seeding profiles and prion-like propagation of synucleinopathies are highly dependent on the host in human alpha-synuclein transgenic mice. *Acta Neuropathol.* **2022**, *143*, 663–685. [CrossRef] [PubMed]
38. Kumar, M.; Quittot, N.; Dujardin, S.; Schlaffner, C.N.; Viode, A.; Wiedmer, A.; Beerepoot, P.; Chun, J.E.; Glynn, C.; Fernandes, A.R.; et al. Alzheimer proteopathic tau seeds are biochemically a forme fruste of mature paired helical filaments. *Brain* **2024**, *147*, 637–648. [CrossRef]
39. Mate De Gerando, A.; Welikovitch, L.A.; Khasnavis, A.; Commins, C.; Glynn, C.; Chun, J.E.; Perbet, R.; Hyman, B.T. Tau seeding and spreading in vivo is supported by both AD-derived fibrillar and oligomeric tau. *Acta Neuropathol.* **2023**, *146*, 191–210. [CrossRef]
40. Stojkovska, I.; Wani, W.Y.; Zunke, F.; Belur, N.R.; Pavlenko, E.A.; Mwenda, N.; Sharma, K.; Francelle, L.; Mazzulli, J.R. Rescue of alpha-synuclein aggregation in Parkinson's patient neurons by synergistic enhancement of ER proteostasis and protein trafficking. *Neuron* **2022**, *110*, 436–451.e11. [CrossRef]
41. Makasewicz, K.; Linse, S.; Sparr, E. Interplay of alpha-synuclein with Lipid Membranes: Cooperative Adsorption, Membrane Remodeling and Coaggregation. *JACS Au* **2024**, *4*, 1250–1262. [CrossRef] [PubMed]

International Journal of
Molecular Sciences

MDPI

Article

Targeted Gut Microbiota Modulation Enhances Levodopa Bioavailability and Motor Recovery in MPTP Parkinson's Disease Models

Penghui Ai [1,†], Shaoqing Xu [2,†], Yuan Yuan [1], Ziqi Xu [1], Xiaoqin He [1], Chengjun Mo [1], Yi Zhang [1], Xiaodong Yang [1,*] and Qin Xiao [1,*]

[1] Department of Neurology and Institute of Neurology, Ruijin Hospital, Shanghai Jiao Tong University School of Medicine, Shanghai 200025, China; aipenghui@sjtu.edu.cn (P.A.); yybzcmu@163.com (Y.Y.); una_xu@sjtu.edu.cn (Z.X.); 18176329672@163.com (X.H.); mochengjun97@163.com (C.M.); zhangyi9102@126.com (Y.Z.)

[2] Department of Geriatrics, Ruijin Hospital, Shanghai Jiao Tong University School of Medicine, Shanghai 200025, China; shaoqingfx@163.com

* Correspondence: yxd12586@rjh.com.cn (X.Y.); xq10537@rjh.com.cn (Q.X.)

† These authors contributed equally to this work.

Abstract: Emerging evidence highlights the gut microbiota as a pivotal determinant of pharmacological efficacy. While *Enterococcus faecalis* (*E. faecalis*)-derived tyrosine decarboxylases (*tyrDCs*) are known to decarboxylate levodopa (L-dopa), compromising systemic bioavailability, the causal mechanisms underlying microbiota-mediated pharmacodynamic variability remain unresolved. In our study, we employed antibiotic-induced microbiota depletion and fecal microbiota transplantation (FMT) to interrogate microbiota-L-dopa interactions in MPTP-induced Parkinson's disease (PD) mice. The study demonstrated that antibiotic-mediated microbiota depletion enhances L-dopa bioavailability and striatal dopamine (DA) level, correlating with improved motor function. To dissect clinical heterogeneity in the L-dopa response, PD patients were stratified into moderate responders and good responders following standardized L-dopa challenges. In vitro bioconversion assays revealed greater L-dopa-to-DA conversion in fecal samples from moderate responders versus good responders. FMT experiments confirmed mice receiving good-responder microbiota exhibited enhanced L-dopa bioavailability, higher striatal DA concentrations, and a heightened therapeutic effect of L-dopa relative to moderate-responder recipients. Collectively, our study provided evidence that the gut microbiota directly modulates L-dopa metabolism and microbial composition determines interindividual therapeutic heterogeneity. Targeted microbial modulation—through precision antibiotics or donor-matched FMT—is a viable strategy to optimize PD pharmacotherapy, supporting the potential for microbiota-targeted adjuvant therapies in PD management.

Keywords: Parkinson's disease; gut microbiota; levodopa; fecal microbiota transplantation; pharmacomicrobiomics

1. Introduction

Parkinson's disease (PD), recognized as the second most common neurodegenerative disorder worldwide, is neuropathologically characterized by the progressive loss of dopaminergic neurons in the substantia nigra pars compacta (SNpc) and the presence of Lewy bodies containing aggregated α-synuclein (α-syn) [1]. The resulting dopaminergic deficit manifests clinically as cardinal motor symptoms including resting tremors, bradykinesia, and muscular rigidity [2]. Levodopa (L-dopa), the metabolic precursor of dopamine (DA), has remained the gold-standard therapy for the treatment of PD since its introduction in the 1960s [3,4]. Following intestinal absorption, L-dopa undergoes enzymatic conversion to DA via aromatic L-amino acid decarboxylase (AADC) within striatal terminals. However, its clinical utility is severely constrained by extensive peripheral metabolism. Over 95% of orally administered L-dopa is prematurely decarboxylated in the gastrointestinal tract and vasculature, resulting in low bioavailability and dose-limiting complications. Although the co-administration of peripheral AADC inhibitors (e.g., carbidopa) reduces extracerebral metabolism, approximately 50% of L-dopa still undergoes premature catabolism, highlighting unmet needs in therapeutic optimization [5,6].

Emerging evidence implicates the gut microbiota as a critical modulator of PD pathogenesis. Studies suggest that PD patients exhibit distinct microbiota compositions compared to healthy individuals [7]. Beyond disease modulation, the gut microbiota has recently emerged as a key determinant of drug pharmacokinetics through the direct enzymatic modification of xenobiotics and indirect regulation of host metabolic enzymes [8]. Notable examples include the *Eggerthella lenta (E. lenta)*-mediated inactivation of digoxin via cardiac glycoside reductases and *Faecalibacterium prausnitzii (F. prausnitzii)*-associated tacrolimus resistance in transplant recipients [9–11]. Moreover, research has revealed that the gut microbiota influences drug metabolism by modulating the expression and activity of metabolic enzymes in the liver and intestines [12]. Gnotobiotic studies demonstrate that microbiota depletion significantly alters hepatic cytochrome P450 (CYP3A) expression, suggesting microbiota-endobiotic crosstalk [13].

Of particular relevance to PD therapeutics, multiple bacterial species exhibit L-dopa-metabolizing capacity independent of host AADC activity. Studies discovered that tyrosine decarboxylases (*tyrDCs*), predominately found in *Enterococcous faecalis (E. faecalis)*, decarboxylated L-dopa to DA irreversibly. Subsequent bacterial metabolism by *E. lenta* further transforms dopamine into m-tyramine via dehydroxylation [5]. Moreover, *Clostridium sporogenes (C. sporogenes)* facilitates L-dopa deamination via aromatic amino acid transaminases [14]. Additionally, *Helicobacter pylori (H. pylori)* infection and small intestinal bacterial overgrowth (SIBO) impair L-dopa absorption through altered gastric motility, leading to the limitation of L-dopa efficacy [15–17]. While many studies have identified correlations between microbiota composition and drug metabolism or efficacy, one of the most significant hurdles in pharmacomicrobiomics is determining the causal relationships between the microbiota and drug response. By mapping individual microbiota profiles, clinicians could optimize drug therapies for PD patients, minimizing side effects and improving treatment outcomes. However, the pharmacomicrobiomic interactions in PD remain unestablished and the therapeutic potential of microbiota modulation in PD management has been underexplored.

To address these questions, we employed antibiotic-induced microbiota depletion and microbial reconstitution via fecal microbiota transplantation (FMT) in MPTP-induced PD mice. Our results demonstrated that microbiota manipulation through antibiotic treatment or FMT regulate L-dopa bioavailability and its therapeutic effects for motor symptom amelioration. These findings establish causal microbiota–drug interactions and provide preclinical rationale for developing microbiota-targeted adjuvants in PD treatment.

2. Results

2.1. Antibiotic-Induced Gut Microbiota Depletion Improved L-Dopa Bioavailability

To evaluate the effect of gut microbiota on L-dopa bioavailability, the mice were divided into SPF + MPTP (n = 5) and PGF + MPTP (n = 5) groups. After a 4-week antibiotic treatment, quantitative PCR analysis demonstrated a >90% reduction in bacterial gene copies of pseudo-germ-free (PGF) mice compared to specific pathogen-free (SPF) controls (Figure S1). Following the intraperitoneal injection of MPTP, mice were orally administered a single dose of L-dopa/carbidopa (10/30 mg/kg) via gavage, with serial blood sampling conducted for pharmacokinetic profiling (Figure 1A). Notably, PGF + MPTP mice exhibited higher serum L-dopa concentrations at 15, 60, and 90 min timepoints versus SPF + MPTP controls. Pharmacokinetic modeling revealed significant increases in the maximum plasma concentration (C_{max}) and systemic exposure (area under the curve from 0 to 2 h, $AUC_{0–2h}$) while the elimination half-life (T1/2) remained comparable between groups (Figure 1C). Critically, striatal DA quantification showed higher striatal DA levels in PGF + MPTP mice than in SPF + MPTP counterparts (Figure 1D). These results demonstrated that the elimination of gut microbiota effectively increased the availability of L-dopa, thus increasing functional central nervous system (CNS) delivery.

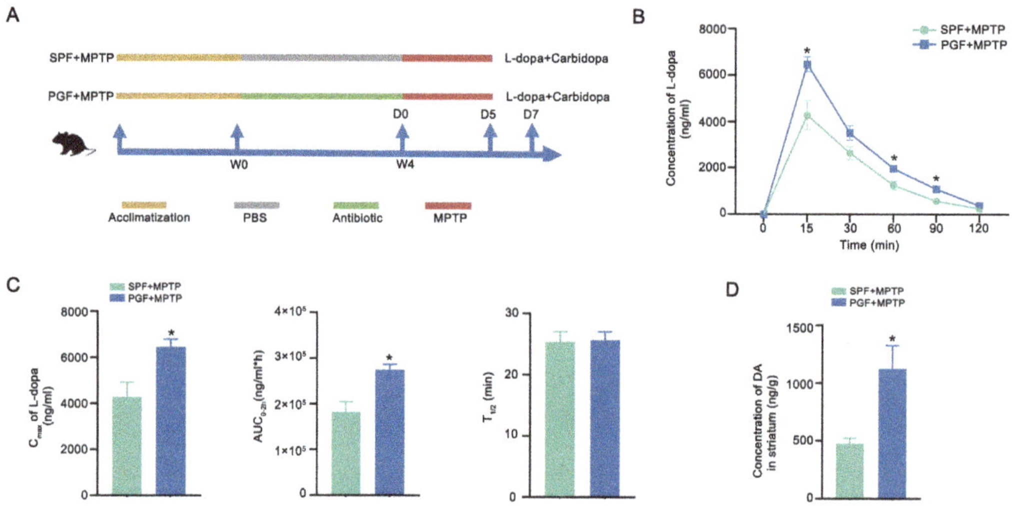

Figure 1. Antibiotic depletion of gut microbiota improved L-dopa bioavailability. (**A**) Schematic diagram of the experiment. (**B**) Serum concentration–time curves of serum L-dopa in each group. (**C**) C_{max}, $AUC_{0–2h}$, and $T_{1/2}$ of serum L-dopa in each group. (**D**) Striatal DA concentration in each group. In each group, n = 5. Data are represented as means ± SEMs. *t*-test was used to determine significance. * $p < 0.05$. C_{max}: maximum serum concentration. $AUC_{0–2h}$: area under the curve from 0 to 2 h. $T_{1/2}$: half-life. L-dopa: levodopa. DA: dopamine.

2.2. Antibiotic-Induced Gut Microbiota Depletion Enhanced Therapeutic Effect of L-Dopa in PD Mice

To evaluate the therapeutic impact of microbiota-L-dopa interactions, we implemented a 7-day therapeutic regimen (L-dopa/carbidopa) in MPTP-induced PD mice (Figure 2A). Behavioral tests revealed that SPF + MPTP mice showed a prolonged pole descent latency and reduced rotarod retention time compared with the control group. L-dopa and carbidopa treatment ameliorated the MPTP-induced impairment of motor function while the depletion of gut microbiota further enhanced the motor function recovery in the PGF + MPTP + L-dopa group compared with the SPF + MPTP + L-dopa group (Figure 2B). The level of striatal DA was significantly elevated in the SPF + MPTP + L-dopa group compared with untreated PD mice (SPF + MPTP group) whereas PGF + MPTP + L-dopa cohorts showed a greater increase compared to the SPF + MPTP + L-dopa group (Figure 2C). Further analysis of dopaminergic neurons in the SNpc showed a significant reduction in the MPTP-treated group compared to the SPF group. However, neither L-dopa therapy nor L-dopa therapy following microbiota depletion resulted in the recovery of dopaminergic neuron levels, indicating that motor improvements were driven by enhanced L-dopa pharmacodynamics rather than neuronal rescue (Figure 2D). In conclusion, removing gut microbiota can enhance the therapeutic effects of L-dopa by increasing its bioavailability.

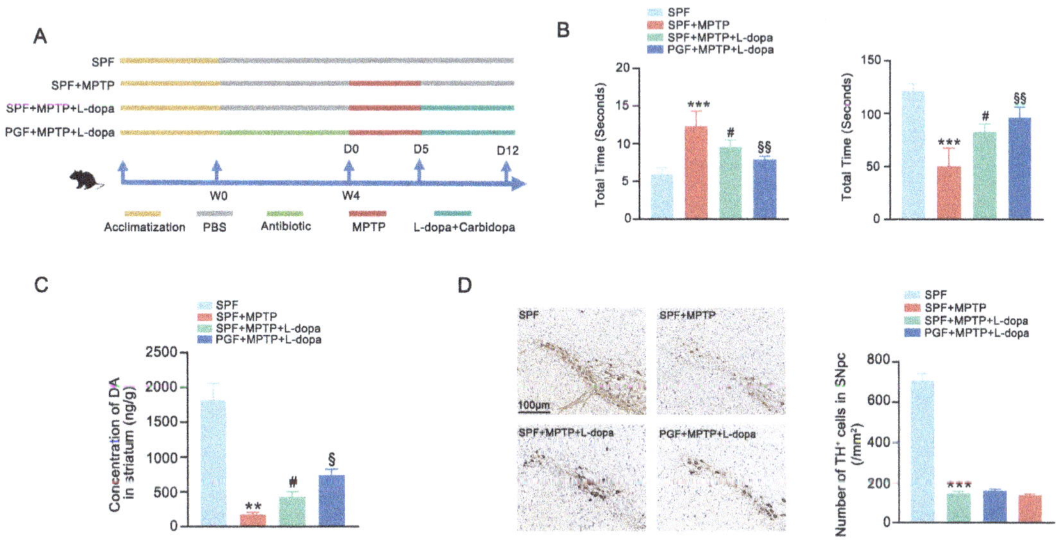

Figure 2. Antibiotic depletion of gut microbiota improved therapeutic effect of L-dopa in PD mice. (**A**) Schematic diagram of the experimental procedure. (**B**) Behavioral test of each group. The left figure shows the pole test and the right figure shows the rotarod test. (**C**) Striatal DA concentration in each group. (**D**) The left figure shows representative images of immunohistochemistry of TH^+ neurons in the SNpc of mice in each group. The scale bar is 100 μm. The right figure shows the comparison of the number of TH^+ neurons between groups. In each group, n = 5. Data are represented as means ± SEMs. The comparison between the two groups was conducted through *t*-test. The comparisons among multiple groups were performed through one-way analysis of variance followed by Tukey's post hoc tests. * means versus SPF group, # means versus SPF + MPTP group, and § means versus SPF + MPTP + L-dopa group. ** $p < 0.01$, *** $p < 0.001$, # $p < 0.05$, § $p < 0.05$, and §§ $p < 0.01$. DA: dopamine. TH^+: tyrosine-hydroxylase-positive. SNpc: substantia nigra pars compacta.

2.3. FMT from Good Responders Enhanced L-Dopa Bioavailability and Therapeutic Effect

Clinical observations revealed significant interindividual variability in L-dopa responsiveness among PD patients in our previous study [18]. To investigate microbiota-driven pharmacodynamic heterogeneity in PD, five moderate responders and five good responders to L-dopa were enrolled from our previously published L-dopa responsiveness cohort. There were no differences in clinical characteristics, medication dosages, and lifestyle variables (Table S1). The metagenomic sequencing of gut microbiota showed that the Chao1, Shannon, and Simpson indexes did not differ between the two groups, indicating similar fecal microbiota richness and diversity (Figure 3A). However, the principal coordinate analysis (PCoA) score plots revealed distinct clustering patterns, suggesting differences in the composition of gut microbiota between groups (Figure 3B). At the bacterial species level, the abundance of *E. faecalis* and *Christensenella minuta (C. minuta)* was significantly higher in moderate responders compared to good responders. Meanwhile, the abundance of *Ruminococcus bicirculans (R. bicirculans)* and *Clostridium citroniae (C. citroniae)* was greater in good responders (Figure 3C). Similarly, linear discriminant analysis effect size (LEfSe) revealed that *C. minuta*, *Parabacteroides merdae (P. merdae)*, and *E. faecalis* were enriched in moderate responders whereas *C. citroniae* was enriched in good responders (Figure 3D). Furthermore, quantitative real-time PCR analysis confirmed the increased abundance of *E. faecalis* in moderate responders (Figure 3E). To confirm the differential capacity metabolization of L-dopa by the feces of two groups, ex vivo fecal fermentation assays demonstrated that the microbiota of moderate responders metabolized L-dopa greatly, associated with excessive DA generation compared to good responders (Figure 3F).

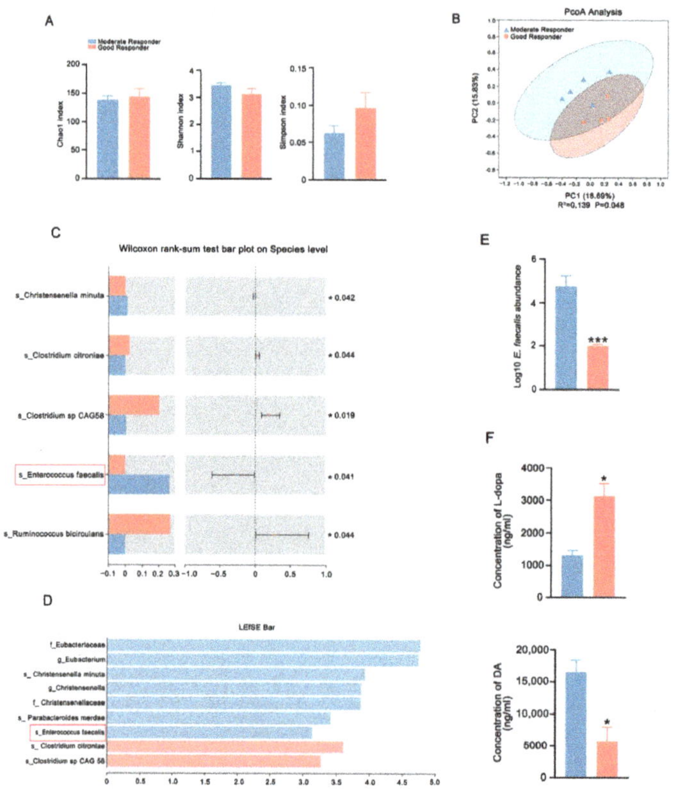

Figure 3. Comparison of gut microbial characteristics between the moderate responders and good responders. (**A**) Analysis of alpha-diversity in fecal microbiota of good responders and moderate responders by Chao1, Shannon, and Simpson estimators. (**B**) PCoA in fecal microbiota of good responders and moderate responders using weighted UniFrac metric based on relative abundance of OTUs. (**C**) The differential microbiota abundance of moderate responders and good responders at the species level. (**D**) LEfSe analysis computed for differentially abundant species in the feces of moderate responders and good responders. LDA scores > 2.0 are shown. (**E**) The abundances of *E. faecalis* in the feces of moderate responders and in good responders detected by qPCR analysis. (**F**) The L-dopa-metabolizing capacity of gut microbiota from good responders and moderate responders detected by fecal culture in vitro. In each group, n = 5. Data are presented as means ± SEMs. *t*-test was used to determine significance between two groups. * $p < 0.05$, and *** $p < 0.001$. PCoA: principal coordinate analysis. LEfSe: linear discriminant analysis effect size. LDA: linear discriminant analysis. *E. faecalis*: *Enterococcus faecalis*. L-dopa: levodopa. DA: dopamine.

Next, mice were colonized with donor microbiota via daily FMT for 7 days to observe the effects on the metabolism of L-dopa in vivo (Figure 4A). Notably, qPCR analysis indicated that the abundance of *E. faecalis* in the feces of mice receiving moderate-responder microbiota (MPTP + MR) was higher than in mice recipients from good-responder microbiota (MPTP + GR) (Figure 4B). Additionally, the feces of MPTP + MR mice recapitulated the metabolic signature of human donors as evidence showing enhanced L-dopa metabolism and DA production in the periphery compared to that from MPTP + good responders (Figure 4C). Conversely, mice recipients from good-responder microbiota showed revealed heightened L-dopa bioavailability with higher C_{max} and $AUC_{0\text{–}2h}$ in the pharmacokinetic profiling in vivo (Figure 4D,E). The levels of striatal DA in recipients from good responders were higher versus moderate-responder recipients (Figure 4F). These results demonstrate that gut microbiota from good responders exhibit reduced L-dopa metabolic capacity, and consequently, FMT from these individuals enhance L-dopa bioavailability in PD mice.

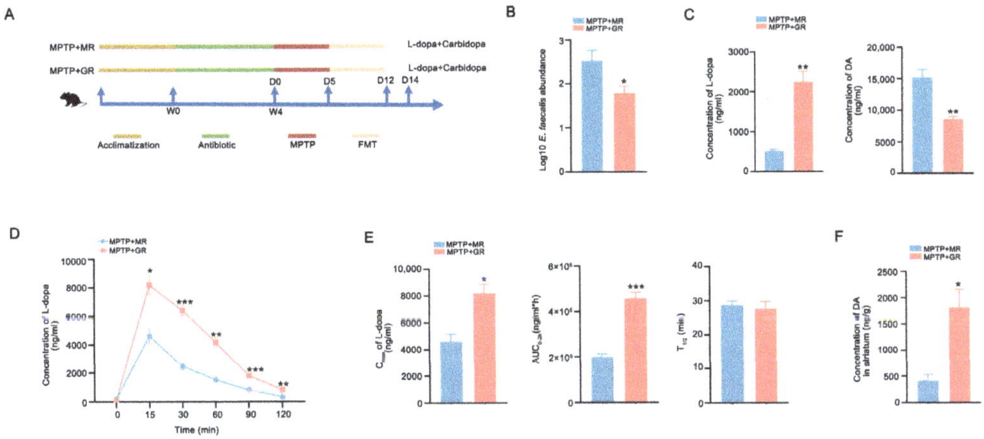

Figure 4. FMT from good responders improved L-dopa bioavailability in PD mice. (**A**) Schematic diagram of the experiment. (**B**) The abundances of *E. faecalis* in the feces of MPTP + MR and MPTP + GR mice detected by qPCR analysis. (**C**) The L-dopa-metabolizing capacity of gut microbiota from the MPTP + MR and MPTP + GR mice detected by fecal culture in vitro. (**D**) Serum concentration–time curves of L-dopa in two groups of mice after FMT. (**E**) The C_{max}, $AUC_{0\text{–}2h}$, and $T_{1/2}$ of L-dopa concentration in the sera of two groups of mice after FMT. (**F**) Striatal DA concentration in two groups of mice after FMT. In each group, n = 5. Data are presented as means ± SEMs. *t*-test was used to determine significance between two groups. * $p < 0.05$, ** $p < 0.01$, and *** $p < 0.001$. L-dopa: levodopa. FMT: fecal microbiota transplantation. C_{max}: maximum serum concentration. $AUC_{0\text{–}2h}$: area under the curve from 0 to 2 h. $T_{1/2}$: half-life. DA: dopamine.

The causal links between microbial composition and L-dopa therapeutic outcomes remain to be further elucidated (Figure 5A). Behavioral assessments revealed that good responder FMT recipients (MPTP + GR + L-dopa) exhibited shorter pole descent latency and longer rotarod endurance compared to recipients of moderate responders (MPTP + MR + L-dopa) (Figure 5B). Paralleling these behavioral improvements, DA quantification demonstrated higher striatal DA concentrations in good-responder recipients (Figure 5C). The quantification of TH^+ neuron showed no differences, confirming that the therapeutic benefits might be mediated through enhanced L-dopa bioavailability (Figure 5D). These data demonstrated that microbiota composition directly modulates L-dopa bioavailability and clinical efficacy. The transplantation of feces with a good response to L-dopa can improve the utilization and enhances the therapeutic effect of L-dopa in the PD model.

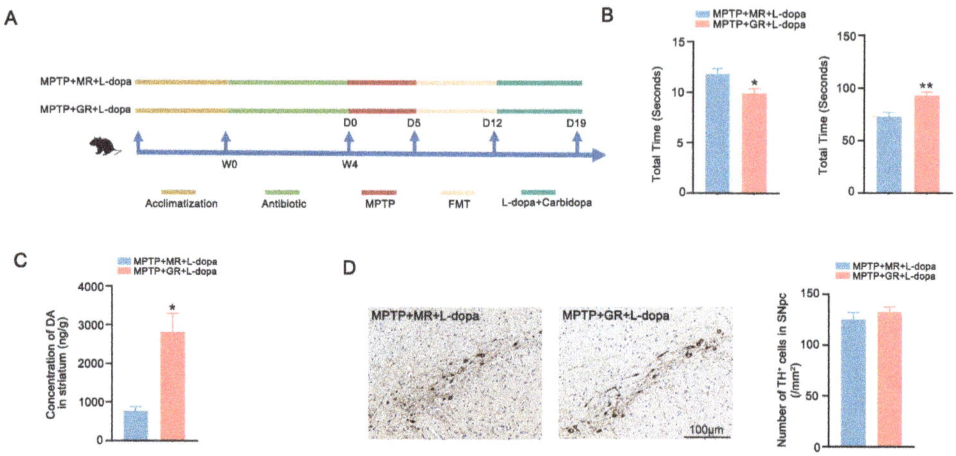

Figure 5. FMT from good responder enhances the therapeutic effect of L-dopa. (**A**) Schematic diagram of the experiment. (**B**) Behavioral test results of the two groups of mice after microbiota transplantation. The left figure shows the pole test and the right figure shows the rotarod test. (**C**) Striatal DA concentration in the two groups of mice after microbiota transplantation. (**D**) Representative images of immunohistochemistry of TH^+ neurons in the SNpc of the two groups of mice after microbiota transplantation. The scale bar is 100 μm. The right figure shows the comparison of the number of TH^+ neurons between the two groups. In each group, n = 5. Data are presented as means ± SEMs. *t*-test was used to determine significance between the two group. * $p < 0.05$; ** $p < 0.01$. DA: dopamine. TH: tyrosine hydroxylase. SNpc: substantia nigra pars compacta.

3. Discussion

The intricate interplay between gut microbiota and L-dopa metabolism has emerged as a pivotal factor influencing therapeutic efficacy in Parkinson's disease [19,20]. Our findings provide compelling experimental evidence that gut microbiota directly modulates L-dopa bioavailability and the targeted manipulation of gut microbiota via antibiotic-induced depletion or that FMT significantly enhances L-dopa bioavailability and amplifies its therapeutic benefits on motor dysfunction in PD models. These results not only align with prior observations of microbial contributions to L-dopa metabolism but also introduce novel insights into microbiota-targeted interventions as a strategy to optimize PD treatment.

L-dopa absorption is restricted to the proximal small intestine (duodenum and jejunum), where it is transported into the bloodstream via a competitive transport system shared with other large neutral amino acids (LNAA) [21]. Dietary proteins, when metabolized to amino acids, competitively inhibit L-dopa absorption, highlighting the critical impact of meal composition and timing on therapeutic efficacy [22]. Additionally, peripheral enzymatic degradation by AADC and catechol-O-methyltransferase (COMT) converts L-dopa into DA and 3-O-methyldopa (3-OMD), further limiting systemic bioavailability [23]. Emerging evidence implicates specific gut microbial communities in regulating L-dopa pharmacokinetics. SIBO may influence L-dopa metabolism by increasing small intestine permeability and inflammation, potentially exacerbating motor fluctuations and unpredictable variations in PD patients [24]. These symptoms can be improved following the successful eradication of SIBO [25,26]. Similarly, *H. pylori* infection might delay gastric emptying, resulting in delayed L-dopa entry and impaired L-dopa transport in the duodenum, therefore interfering with the efficiency of L-dopa—a deficit ameliorated by pathogen clearance [15,27,28]. Our findings extend these observations by demonstrating that commensal microbiota directly metabolize L-dopa. Antibiotic-induced gut microbiota depletion significantly enhances L-dopa bioavailability with increased C_{max} and

$AUC_{0\text{–}2h}$ values of L-dopa and increases striatal DA levels, confirming microbial activity as an intestinal "first-pass" metabolic barrier. We suppose the improvement in L-dopa absorption in our study was likely driven by the suppressed activity of bacterial *tyrDCs*, notably those encoded by *E. faecalis* [5,29]. In fact, prior research has identified a compound, (S)-α-fluoromethyltyrosine (AFMT), which is recognized as a mechanism-based inhibitor of pyridoxal phosphate-dependent decarboxylases. In vitro studies have demonstrated that AFMT inhibits L-dopa decarboxylation by *TyrDC* and *E. faecalis*, as well as in complex gut microbiota samples from PD patients. In vivo studies have shown that AFMT effectively reduces the microbial metabolism of L-dopa, thereby enhancing its bioavailability in mice. These findings underscore the potential for developing L-dopa-based combination therapies targeting enzymes produced by gut microbiota to prevent microbial drug metabolism [5]. Notably, the unchanged $T_{1/2}$ of L-dopa suggests that microbiota-mediated metabolism occurs pre-systemically in the gut rather than altering hepatic or renal clearance pathways. Critically, microbiota-depleted PD mice exhibited improved motor performance post L-dopa treatment. This benefit was directly related to elevated striatal DA levels rather than dopaminergic neuron preservation. This was because we found no differences in TH^+ neurons in the substantia nigra between the SPF + MPTP + L-dopa and PGF + MPTP + L-dopa groups. Collectively, various bacterial pathways are involved in L-dopa metabolism, expanding the range of potential pharmacological targets to enhance L-dopa absorption. Furthermore, researchers have identified a strain of the gut Actinobacterium, *Eggerthella lentu (E. lentu)*, that is capable of selectively removing the para hydroxyl group of dopamine to yield m-tyramine, which may influence the side effects associated with peripheral L-dopa decarboxylation such as rapid hypertensive crisis. Thus, targeted gut microbiota modulation for the metabolism of L-dopa not only increases drug availability but also could reduce adverse drug responses [5,30]. Strategic microbiota modulation—via transient depletion might be a complementary strategy for suppressing microbial decarboxylation to optimize L-dopa treatment. The growing field of pharmacomicrobiomics underscores the need to characterize host–microbe–drug interactions at strain-specific resolution [8,31]. Further studies are needed to better elucidate the interaction between gut microbiota and L-dopa for individualized treatment.

Clinical experience reveals significant interindividual heterogeneity in the treatment effectiveness and medication responses of L-dopa [32–34]. While the disease duration, motor subtypes, and pharmacogenomic variations (e.g., COMT, dopamine receptor (DRD2), and solute carrier family 6 member 3 (SLC6A3) polymorphisms) contribute to therapeutic variability, results on the link between investigated genes and drug response phenotypes remain inconclusive, suggesting that other factors might be involved [35–38]. Multiple studies have revealed that gut microbiota interferes with pharmaceutical treatment [39,40]. Our FMT experiments have established gut microbiota as a causal association of the heterogeneity of L-dopa response for the first time. In vitro fecal cultures revealed that microbiota of moderate responders exhibited heightened L-dopa-to-DA conversion, mirroring excessive peripheral decarboxylation linked to reduced central drug bioavailability. Conversely, transplanting good-responder microbiota into PGF mice recapitulated the donors' metabolic profile, as evidenced by the reduced degradation of L-dopa in vitro culture by microbiota from the recipient mice. Mice receiving the gut microbiota from good responders achieved enhanced L-dopa bioavailability, higher striatal DA levels, and superior motor outcomes compared to moderate-responder FMT recipients. The FMT experiment revealed that the composition of the donor's gut microbiota influences the metabolic capacity and therapeutic efficacy of L-dopa in the recipient, underscoring the causal role of microbial composition in pharmacodynamic variability.

The beneficial effects of FMT have been clearly shown in several PD animal models, showing improvements in motor behavior, pathology, and gut microbiota features [41,42]. Currently, two preliminary studies observed improved both motor and non-motor symptoms, especially constipation, following FMT being administered [43,44]. The effectiveness of FMT on PD symptoms may be attributed to the alteration of gut microbiota and functional pathways [45,46]. In our study, we found that FMT treatment enhances the efficacy of PD medication, leading to improvements in PD symptoms. Donors with good L-dopa responsiveness exhibited gut microbiota with the reduced metabolism of L-dopa, which facilitates its absorption. Mechanically, FMT may restore gut microbiota balance and maintain intestinal homeostasis, thereby potentially enhancing drug transport efficiency [47].

Our findings position gut microbiota modulation as a tractable strategy to enhance L-dopa pharmacokinetics and clinical outcomes in Parkinson's disease. The development of microbiota-guided interventions—such as targeted microbiota depletion or donor-selected FMT—could address the persistent challenge of therapeutic heterogeneity. However, limitations remain in our study. First, broad-spectrum microbiota depletion risks long-term dysbiosis and immune dysregulation, necessitating precision and targeted approaches. Additionally, while FMT from good responders enhanced drug bioavailability in our models, interindividual variability in microbial engraftment and functional durability requires rigorous characterization. Thirdly, establishing whether specific microbial species are directly influencing drug outcomes or whether these associations are incidental remains a key research challenge. Lastly, our results were obtained from a relatively limited sample size, and future studies will require larger samples to further validate and confirm these findings.

4. Materials and Methods

4.1. Study Approvals

This study was approved by the Research Ethics Committee of Ruijin Hospital, Shanghai Jiao Tong University School of Medicine. All participants included in this study were informed of the study's purpose. All participants were provided informed consent forms and signed them.

4.2. Mouse Treatment

All animal procedures were approved by the Animal Ethics Committee of Ruijin Hospital, Shanghai Jiao Tong University School of Medicine and conducted in accordance with NIH guidelines. Mice were maintained under specific pathogen-free (SPF) conditions with suitable temperature and 12h light/dark cycles.

Mouse Experiment 1: Ten male C57BL/6 mice were randomly divided into two groups: SPF + MPTP (n = 5) group was housed in an SPF-level environment and was treated with MPTP by intraperitoneal for 5 days to induce Parkinsonism. PGF + MPTP (n = 5) group received a quadruple-antibiotic cock-tail (1 g/L ampicillin, 1 g/L neomycin, 1 g/L metronidazole, 0.5 g/L vancomycin) in drinking water for 14 days, which was confirmed as >90% microbiota depletion by weekly fecal qPCR and then followed by MPTP treatment for 5 days. On day 7, L-dopa/carbidopa (10/30 mg/kg) was orally gavaged, followed by serial blood sampling (0–120 min) from the posterior orbital venous plexus. Serum separation involved 1 h clotting at 25 °C, centrifugation (3000× *g*, 15 min, 4 °C), and storage at −80 °C for pharmacokinetic analysis. Experimental timeline and design are detailed in Figure 1A.

Mouse Experiment 2: Twenty male C57BL/6 mice were randomly assigned to four groups following acclimatization: SPF (n = 5), SPF + MPTP (n = 5), SPF + MPTP + L-dopa (n = 5), and PGF + MPTP + L-dopa (n = 5). Mice in the SPF group were housed in an SPF-level environment and orally administered sterile PBS throughout the experimental period. The SPF + MPTP group received oral administration of sterile PBS from week 1 to week 4, followed by MPTP treatment for 5 consecutive days and subsequent oral administration of sterile PBS for 1 week. The SPF + MPTP + L-dopa group received sterile PBS from week 1 to week 4, followed by MPTP treatment for 5 days and subsequent daily gavage administration of L-dopa/carbidopa (10/30 mg/kg) for 1 week. The PGF + MPTP + L-dopa group underwent a 4-week antibiotic regimen (composition as described in Experiment 1), with weekly fecal qPCR monitoring, followed by MPTP treatment for 5 days. After PD model validation, this group was administered L-dopa/carbidopa (10/30 mg/kg) daily via gavage for 7 days. Following gavage treatment, all groups underwent behavioral tests, including pole climbing and rotarod tests, to evaluate motor function. The experimental design is schematically illustrated in Figure 2A.

Mouse Experiment 3: A total of 10 matched patients were stratified into moderate responders (<40% UPDRS-III improvement post-dose) and good responders (≥40% UPDRS-III improvement post-dose) according to clinical L-dopa responsiveness detected by L-dopa challenge experiment from our previously published L-dopa responsiveness cohort. All participants were strictly matched for potential confounding factors that could potentially influence L-dopa response, including clinical characteristics, medication dosages, and lifestyle variables. Ten male C57BL/6 mice were randomly divided into MPTP + MR (n = 5) and MPTP + GR (n = 5) groups and established using fecal donors from moderate responders and good responders. Following 28-day antibiotic-induced PGF status, all mice received MPTP neurotoxin (30 mg/kg/day ×5, i.p.). PD model validation preceded daily FMT administration (200 μL/dose) for 7 days using donor stools from moderate- or good-L-dopa responders. Post FMT, mice received oral L-dopa/carbidopa (10/30 mg/kg) with serial retro-orbital blood sampling (50 μL/timepoint) at 0–120 min intervals. Serum processing involved 30 min clotting at 25 °C, centrifugation (3000× *g*, 15 min, 4 °C), and storage at −80 °C for pharmacokinetic profiling. Experimental workflow is detailed in Figure 4A.

Mouse Experiment 4: Ten male C57BL/6 mice were randomly divided into MPTP + MR + L-dopa (n = 5) and MPTP + GR + L-dopa (n = 5) groups. Following 28-day antibiotic-induced PGF status and MPTP neurotoxin administration (30 mg/kg/day ×5, i.p.), mice received donor-matched FMT via oral gavage (200 μL/day ×7) using fecal samples from PD patients stratified by L-dopa responsiveness. All the group were administered L-dopa/carbidopa (10/30 mg/kg/day ×7) after FMT with motor function assessed 24 h post-final dose via pole test and rotarod test. The experimental schematic diagram is presented in Figure 5A.

4.3. *Quantitative Analysis of L-Dopa and DA*

(I) Standard Preparation

1. Stock Solutions: L-dopa and DA single-component stock solutions (10 ppm) prepared in 10 mM HCl and acetonitrile, respectively.
2. Standard Mixed External Standard Solution 1: We diluted L-dopa (1 ppm) and DA (1 ppm) with acetonitrile to prepare a mixed working solution.
3. Standard Mixed External Standard Solution 2: We diluted L-dopa (100 ppb) and DA (100 ppb) with acetonitrile to prepare a mixed working solution.
4. Standard Mixed Internal Standard Working Solution: We diluted L-dopa-d3 (1 ppm, MCE, America) and dopamine-d4 (1 ppm, MCE, America) with acetonitrile to prepare a mixed internal-standard working solution.
5. Calibration Standards: We took 200 μL, 100 μL, and 50 μL of Standard Mixed External Standard Solution 1 and 200 μL and 100 μL of Standard Mixed External Standard Solution 2. We added 100 μL of the mixed internal-standard working

solution to each tube, vortexed thoroughly, and analyzed using the instrument to obtain five points for the standard curve.

(II) Sample Pre-treatment

1. Culture Supernatant Sample Treatment: We centrifuged the culture supernatant at 13,000 rpm for 15 min at 4 °C. We collected 450 μL of the supernatant, added 50 μL of the mixed internal-standard working solution, vortexed thoroughly, filtered through a 0.22 μm membrane, and analyzed using the instrument.
2. Mouse Serum Sample Treatment: We took 10 μL of serum sample, added 20 μL of the mixed internal-standard working solution, and added 70 μL of 40% acetonitrile trifluoroacetic acid solution. We vortexed for 1 min, centrifuged at 15,000 rpm for 20 min at 4 °C, collected the supernatant, and analyzed using the instrument.
3. Mouse Brain Tissue Sample Treatment: Weigh 0.1 g of brain tissue, add 100 μL of 10 mM HCl solution, followed by 80 μL of 40% acetonitrile trifluoroacetic acid solution, and finally add 20 μL of the mixed internal-standard working solution. We homogenized using magnetic beads for 10 min, froze at −80 °C for 30 min, then centrifuged at 15,000 rpm for 20 min at 4 °C. We collected the supernatant, filtered through a 0.22 μm membrane, and analyzed using the instrument.

(III) Ultra-High-Performance Liquid Chromatography–Tandem Mass Spectrometry (HPLC-MS/MS, Shimadzu LCMS-8050, Kyoto, Japan) Conditions

1. Liquid Phase Conditions: Mobile phase A: 2.55 mM ammonium acetate, 0.25% formic acid in water; mobile phase B: acetonitrile. Chromatographic column: YMC TICLIC-C18 (4.6 × 100 mm, 1.9 μm). Injector temperature: 10 °C; injection volume: 10 μL; flow rate: 0.2 mL/min; gradient: started with 15% B phase.
2. Mass Spectrometry Conditions: Ionization mode: electrospray ionization (ESI) in positive and negative ion modes; spray voltage: ±3.5 kV; nebulizer temperature: 560 °C; sheath gas pressure: 30 psi; auxiliary gas pressure: 20 psi; DL tube temperature: 350 °C; ion detection mode: multiple-reaction monitoring (MRM).

4.4. Behavioral Tests

4.4.1. Pole Test

A custom-designed wooden pole (50 cm height × 1 cm diameter) with a spherical wooden platform (diameter: 3 cm) affixed to the apex was vertically positioned within the home cage. Mice were gently placed in an inverted position on the top ball. The descent latency taken from the top of the pole into the home cage was recorded.

4.4.2. Rotarod Test

Mice were placed on the rotarod apparatus and tested at a speed of 30 rpm for a maximum duration of 120 s. The latency to fall, defined as the time when the mice first fell off the rod, was recorded.

All mice underwent a 3-day pretraining period prior to the behavioral test. Each mouse performed three trials in the testing, with a 1 h interval between trials.

4.5. Immunohistochemical (IHC) Staining

For immunohistochemistry analysis, deparaffinized tissue sections were subjected to antigen retrieval by boiling in 10 mM citric acid buffer (pH 6.0) for 20 min. Subsequently, the sections were treated with 3% hydrogen peroxide for 10 min at room temperature to block endogenous peroxidase activity, followed by three washes with PBS. Sections were then blocked with 5% goat serum for 1 h at room temperature and incubated overnight at 4 °C with primary antibodies: tyrosine hydroxylase (CST, 58844S, 1:200). After three additional washes with PBS, the sections were incubated with horseradish peroxidase (HRP)-conjugated secondary antibody for 1 h at room temperature. Finally, the slides were stained with 3,3′-diaminobenzidine (DAB) and cover-slipped for microscopic examination.

4.6. Participant Recruitment

Total of 10 PD patients were enrolled in this study in the clinic of the Department of Neurology and Institute of Neurology, Ruijin Hospital, Shanghai, China. PD was diagnosed by a movement disorder specialist on the basis of the United Kingdom Parkinson's Disease Society Brain Bank criteria [48]. The exclusion criteria were as follows: (1) atypical or secondary Parkinsonism, Alzheimer's disease, or other central nervous system diseases; (2) levodopa intolerance; (3) history of deep brain stimulation, ablative, or lesioning procedures; (4) history of gastrointestinal tract absorption disorders including celiac disease and lactose intolerance; (5) chronic illnesses (e.g., abnormal liver enzymes, renal dysfunction, inflammatory bowel disease, malignancy, endocrine diseases, hematological diseases, or heart failure); (6) the use of antibiotics, probiotics, or proton pump inhibitors within the past 3 months; or (7) investigator judgement that the candidate was not suited to participate in the study. All participants signed the written informed consent forms.

4.7. Data and Sample Collection

Age, sex, and body mass index (BMI) were recorded for participants. Clinical features were collected, including disease duration, Hoehn and Yahr stage (H-Y stage), and Movement Disorder Society—Unified Parkinson's Disease Rating Scale (MDS-UPDRS) [49]. Levodopa-equivalent daily dosages were calculated using classical methods. Fecal samples were collected with sterile fecal collection containers, and each sample was divided into aliquots as soon as possible followed by immediate storage at −80 °C prior to the following processes.

4.8. L-Dopa Challenge Experiment

For the specific method of the L-dopa challenge experiment, we referred to our group's previous study [18]. Briefly, PD patients discontinued DA receptor agonists for 36 h and all other anti-Parkinsonian medications, including L-dopa, for 12 h prior to the experiment. Motor symptoms during the off-period were assessed using Part III of the Movement Disorder Society—Unified Parkinson's Disease Rating Scale (MDS-UPDRS). At 8:00 a.m., patients received a single dose of L-dopa equivalent to 1.5 times their usual morning dose of anti-Parkinson's medication. In this study, patients were administered 200/50 mg Madopar (L-dopa/benserazide), and motor symptoms were evaluated every 30 min based on MDS-UPDRS Part III scores for a total duration of 5 h.

The formula for calculating L-dopa efficacy is as follows:

$$\text{L-dopa efficacy (\%)} = [(\text{Off-period MDS-UPDRS III score} - \text{Peak MDS-UPDRS III score})/\text{Off-period MDS-UPDRS III score}] \times 100\%.$$

4.9. Metagenomic Sequencing

DNA extract was fragmented to an average size of about 400 bp using Covaris M220 (Gene Company Limited, Shanghai, China) for paired-end library construction. Paired-end library was constructed using NEXTFLEX Rapid DNA-Seq (Bioo Scientific, Austin, TX, USA). Paired-end sequencing was performed on Illumina NovaSeq™ (Illumina Inc., San Diego, CA, USA) at Majorbio Bio-Pharm Technology Co., Ltd. (Shanghai, China) according to the manufacturer's instructions (www.illumina.com).

The data were analyzed on the online platform of Majorbio Cloud Platform (www.majorbio.com). Briefly, the raw sequencing reads were trimmed of adapters, and low-quality reads (length < 50 bp, with a quality value < 20 or having N bases) were removed by fastp (https://github.com/OpenGene/fastp, accessed on 30 October 2023, version 0.20.0) [50]. Reads were aligned to the human genome by BWA (http://bio-bwa.sourceforge.net, accessed on 30 October 2023, version 0.7.17) and any hits associated with the reads and their mated reads were removed [51].

The quality-filtered data were assembled with MEGAHIT (https://github.com/voutcn/megahit, accessed on 30 October 2023, version 1.1.2) [52]. Contigs with a length $\geq$ 300 bp were selected as the final assembling results. Open reading frames (ORFs) from each assembled contigs were predicted using Prodigal (https://github.com/hyattpd/Prodigal, version 2.6.3) and a length $\geq$ 100 bp of ORFs was retrieved [53]. Sequence analysis was performed using the UPARSE software (version 11) package with the UPARSE-OTU and UPARSE-OTUref algorithms. Sequences with $\geq$97% similarity were assigned to the same operational taxonomic units (OTUs).

4.10. LEfSe

Linear discriminant analysis effect size (LEfSe) was examined to determine the significant differences in microbiota at species level. The parameters were set as default: p values $\leq$ 0.05 for significance and linear discriminant analysis (LDA) score > 3.

4.11. Fecal Sample DNA Extraction and Quantitative Real-Time PCR

The TIANamp Stool DNA Kit (TIANGEN Biotech Co., Ltd., Beijing, China) was employed for the extraction of fecal bacterial DNA, following the manufacturer's protocol. Microbial-species-specific primers targeting the 16S rRNA gene of *E. faecalis* (forward: CGCTTCTTTCCTCCCGAGT; reverse: GCCATGCGGCATAAACTG) were designed using Primer Premier 5.0 software. All real-time PCR amplifications were performed in a total reaction volume of 20 μL, comprising 1 μL of template genomic DNA, 10 μL of 2× T5 Fast qPCR Mix (SYBR Green I), and 10 μM of each primer. The amplification program consisted of an initial denaturation at 95 °C for 1 min, followed by 40 cycles of denaturation at 95 °C for 15 s, annealing at 60 °C for 15 s, and extension at 72 °C for 30 s. Serial dilutions of the standard plasmid DNA were used as templates to generate the real-time PCR standard curve. All quantifications were performed in triplicate, and the mean Ct value was utilized to calculate the copy numbers of *E. faecalis*. The abundances of the *E. faecalis* were expressed as log10 copies per gram of dry weight feces.

4.12. Metabolic Assay of L-Dopa by Fecal Incubation In Vitro

Fecal samples from PD patients and mice were processed for metabolic analysis. Specifically, approximately 1 g of feces from each PD patient group was resuspended in 10 mL of sterile phosphate-buffered saline (PBS) while 500 mg of mouse feces was resuspended in 5 mL of sterile PBS. The fecal suspensions were vigorously vortexed for 5 min and then allowed to settle at room temperature for 10 min to separate large particles and insoluble materials. Subsequently, the suspensions were diluted 1:100 into 5 mL of MEGA medium supplemented with 1 mmol of L-dopa. The mixtures were cultured under strict anaerobic conditions at 37 °C for 48 h. After incubation, the supernatant was collected, and the reaction was quenched by adding methanol. The supernatant was then centrifuged at 13,000 rpm for 10 min at 4 °C, and the resulting supernatant was subjected to HPLC-MS/MS for the quantification of L-dopa and DA [5].

4.13. Fecal Microbiota Transplantation

FMT was conducted following established protocols [54]. Briefly, fresh fecal samples were collected from two groups of PD patients. The feces were resuspended in sterile PBS to achieve a concentration of 0.1 g/mL and thoroughly mixed. The suspension was centrifuged at 500× *g* for 5 min at 4 °C to remove large insoluble particles, and the supernatant was collected. An equal volume of 50% sterile glycerol was added to the supernatant, which was then thoroughly mixed and aliquoted. The aliquots were stored at −80 °C until use. For the transplantation procedure, each mouse received 200 μL of the fecal suspension via daily gavage for 7 consecutive days. Control group mice were administered an equivalent volume of PBS using the same protocol.

4.14. Statistical Analysis

All statistical analyses were performed using GraphPad Prism 8.0 (GraphPad, Boston, MA, USA). Continuous data are presented as means ± standard errors of the mean (SEM). Comparisons between two groups were analyzed using two-tailed unpaired *t*-tests. For comparisons involving multiple groups, one-way analysis of variance (ANOVA) was conducted, followed by Tukey's post hoc test. Categorical variables are expressed here in frequency (percentage) form and were analyzed using chi-square tests. All statistical tests were two-tailed, and a p-value < 0.05 was considered statistically significant.

5. Conclusions

Our study established the gut microbiota as a critical modifier of L-dopa pharmacokinetics and clinical efficacy in Parkinson's disease. Targeted microbial modulation through antibiotic depletion or precision FMT enhances L-dopa bioavailability, correlating with improved striatal DA levels and motor outcomes. These insights advocate for integrating microbiota into PD treatment to maximize therapy benefits, urging a paradigm shift toward integrated host–microbe therapeutics.

Supplementary Materials: The following supporting information can be downloaded at: https://www.mdpi.com/article/10.3390/ijms26115282/s1.

Author Contributions: Conceptualization, Q.X. and X.Y.; methodology, X.Y. and P.A.; validation, P.A. and S.X.; investigation, P.A. and S.X.; resources, P.A., Y.Y., Z.X. and X.H.; data curation, P.A., C.M. and Y.Z.; writing—original draft preparation, P.A.; writing—review and editing, P.A., S.X. and X.Y.; supervision, Q.X. and X.Y.; project administration, Q.X.; funding acquisition, Q.X. and X.Y. All authors have read and agreed to the published version of the manuscript.

Funding: This work was funded by the National Natural Science Foundation of China (Grant Nos. 82371251 and 82171246), the National Key R&D Program of China (Grant No. 2022YFE0210100), the Shanghai Rising-Star Program (Grant No. 22QA1405700), and the Shanghai Sailing Program (22YF1440200).

Institutional Review Board Statement: The study was conducted in accordance with the Declaration of Helsinki and approved by the Research Ethics Committee, Ruijin Hospital, Shanghai Jiao Tong University School of Medicine (27 December 2019, 2019-226) for studies involving humans. The animal study protocol was approved by the Animal Ethics Committee of Ruijin Hospital, Shanghai Jiao Tong University School of Medicine (25 May 2023, RJ2023018).

Informed Consent Statement: Informed consent was obtained from all subjects involved in the study. Written informed consent has been obtained from the patients to publish this paper.

Data Availability Statement: The original contributions presented in the study are included in the article/Supplementary Materials; further inquiries can be directed to the corresponding authors.

Conflicts of Interest: The authors declare no conflicts of interest.

References

1. Dorsey, E.R.; Sherer, T.; Okun, M.S.; Bloem, B.R. The Emerging Evidence of the Parkinson Pandemic. *J. Park. Dis.* **2018**, *8*, S3–S8. [CrossRef] [PubMed]
2. Armstrong, M.J.; Okun, M.S. Diagnosis and Treatment of Parkinson Disease: A Review. *JAMA* **2020**, *323*, 548–560. [CrossRef] [PubMed]
3. Morgan, J.P.; Bianchine, J.R.; Spiegel, H.E.; Rivera-Calimlim, L.; Hersey, R.M. Metabolism of Levodopa in Patients with Parkinson's Disease: Radioactive and Fluorometric Assays. *Arch. Neurol.* **1971**, *25*, 39–44. [CrossRef]
4. Cacabelos, R. Parkinson's Disease: From Pathogenesis to Pharmacogenomics. *Int. J. Mol. Sci.* **2017**, *18*, 551. [CrossRef]
5. Maini Rekdal, V.; Bess, E.N.; Bisanz, J.E.; Turnbaugh, P.J.; Balskus, E.P. Discovery and inhibition of an interspecies gut bacterial pathway for Levodopa metabolism. *Science* **2019**, *364*, eaau6323. [CrossRef]
6. Muleiro Alvarez, M.; Cano-Herrera, G.; Osorio Martínez, M.F.; Vega Gonzales-Portillo, J.; Monroy, G.R.; Murguiondo Pérez, R.; Torres-Ríos, J.A.; van Tienhoven, X.A.; Garibaldi Bernot, E.M.; Esparza Salazar, F.; et al. A Comprehensive Approach to Parkinson's Disease: Addressing Its Molecular, Clinical, and Therapeutic Aspects. *Int. J. Mol. Sci.* **2024**, *25*, 7183. [CrossRef]
7. Caputi, V.; Giron, M.C. Microbiome-Gut-Brain Axis and Toll-like Receptors in Parkinson's Disease. *Int. J. Mol. Sci.* **2018**, *19*, 1689. [CrossRef]
8. Tsunoda, S.M.; Gonzales, C.; Jarmusch, A.K.; Momper, J.D.; Ma, J.D. Contribution of the Gut Microbiome to Drug Disposition, Pharmacokinetic and Pharmacodynamic Variability. *Clin. Pharmacokinet.* **2021**, *60*, 971–984. [CrossRef]
9. Haiser, H.J.; Gootenberg, D.B.; Chatman, K.; Sirasani, G.; Balskus, E.P.; Turnbaugh, P.J. Predicting and Manipulating Cardiac Drug Inactivation by the Human Gut Bacterium *Eggerthella lenta*. *Science* **2013**, *341*, 295–298. [CrossRef]
10. Guo, Y.; Crnkovic, C.M.; Won, K.J.; Yang, X.; Lee, J.R.; Orjala, J.; Lee, H.; Jeong, H. Commensal Gut Bacteria Convert the Immunosuppressant Tacrolimus to Less Potent Metabolites. *Drug Metab. Dispos. Biol. Fate Chem.* **2019**, *47*, 194–202. [CrossRef]
11. Lee, J.R.; Muthukumar, T.; Dadhania, D.; Taur, Y.; Jenq, R.R.; Toussaint, N.C.; Ling, L.; Pamer, E.; Suthanthiran, M. Gut microbiota and tacrolimus dosing in kidney transplantation. *PLoS ONE* **2015**, *10*, e0122399. [CrossRef] [PubMed]
12. Claus Sandrine, P.; Ellero Sandrine, L.; Berger, B.; Krause, L.; Bruttin, A.; Molina, J.; Paris, A.; Want Elizabeth, J.; de Waziers, I.; Cloarec, O.; et al. Colonization-Induced Host-Gut Microbial Metabolic Interaction. *mBio* **2011**, *2*, 1–8. [CrossRef]
13. Björkholm, B.; Bok, C.M.; Lundin, A.; Rafter, J.; Hibberd, M.L.; Pettersson, S. Intestinal Microbiota Regulate Xenobiotic Metabolism in the Liver. *PLoS ONE* **2009**, *4*, e6958. [CrossRef]
14. van Kessel, S.P.; de Jong, H.R.; Winkel, S.L.; van Leeuwen, S.S.; Nelemans, S.A.; Permentier, H.; Keshavarzian, A.; El Aidy, S. Gut bacterial deamination of residual levodopa medication for Parkinson's disease. *BMC Biol.* **2020**, *18*, 137. [CrossRef]
15. Pierantozzi, M.; Pietroiusti, A.; Galante, A.; Sancesario, G.; Lunardi, G.; Fedele, E.; Giacomini, P.; Stanzione, P. Helicobacter pylori-induced reduction of acute levodopa absorption in Parkinson's disease patients. *Ann. Neurol.* **2001**, *50*, 686–687. [CrossRef]
16. Lee, W.Y.; Yoon, W.T.; Shin, H.Y.; Jeon, S.H.; Rhee, P.-L. Helicobacter pylori infection and motor fluctuations in patients with Parkinson's disease. *Mov. Disord.* **2008**, *23*, 1696–1700. [CrossRef]
17. Spencer, R.P. Intestinal absorption of amino acids. Current concepts. *Am. J. Clin. Nutr.* **1969**, *22*, 292–299. [CrossRef]
18. Zhang, Y.; He, X.; Mo, C.; Liu, X.; Li, J.; Yan, Z.; Qian, Y.; Lai, Y.; Xu, S.; Yang, X.; et al. Association Between Microbial Tyrosine Decarboxylase Gene and Levodopa Responsiveness in Patients with Parkinson Disease. *Neurology* **2022**, *99*, e2443–e2453. [CrossRef]
19. Menozzi, E.; Schapira, A.H.V. The Gut Microbiota in Parkinson Disease: Interactions with Drugs and Potential for Therapeutic Applications. *CNS Drugs* **2024**, *38*, 315–331. [CrossRef] [PubMed]
20. Zhang, Y.; Chengjun, M.; Penghui, A.; Xiaoqin, H.; Qin, X.; and Yang, X. Pharmacomicrobiomics: A new field contributing to optimizing drug therapy in Parkinson's disease. *Gut Microbes* **2025**, *17*, 2454937. [CrossRef]
21. Contin, M.; Martinelli, P. Pharmacokinetics of levodopa. *J. Neurol.* **2010**, *257*, 253–261. [CrossRef] [PubMed]
22. Wang, L.; Xiong, N.; Huang, J.; Guo, S.; Liu, L.; Han, C.; Zhang, G.; Jiang, H.; Ma, K.; Xia, Y.; et al. Protein-Restricted Diets for Ameliorating Motor Fluctuations in Parkinson's Disease. *Front. Aging Neurosci.* **2017**, *9*, 206. [CrossRef]
23. Warren Olanow, C.; Torti, M.; Kieburtz, K.; Leinonen, M.; Vacca, L.; Grassini, P.; Heller, A.; Heller, E.; Stocchi, F. Continuous versus intermittent oral administration of levodopa in Parkinson's disease patients with motor fluctuations: A pharmacokinetics, safety, and efficacy study. *Mov. Disord.* **2019**, *34*, 425–429. [CrossRef]
24. Leta, V.; Klingelhoefer, L.; Longardner, K.; Campagnolo, M.; Levent, H.Ç.; Aureli, F.; Metta, V.; Bhidayasiri, R.; Chung-Faye, G.; Falup-Pecurariu, C.; et al. Gastrointestinal barriers to levodopa transport and absorption in Parkinson's disease. *Eur. J. Neurol.* **2023**, *30*, 1465–1480. [CrossRef]
25. Fasano, A.; Bove, F.; Gabrielli, M.; Petracca, M.; Zocco, M.A.; Ragazzoni, E.; Barbaro, F.; Piano, C.; Fortuna, S.; Tortora, A.; et al. The role of small intestinal bacterial overgrowth in Parkinson's disease. *Mov. Disord.* **2013**, *28*, 1241–1249. [CrossRef]
26. Tan, A.H.; Mahadeva, S.; Thalha, A.M.; Gibson, P.R.; Kiew, C.K.; Yeat, C.M.; Ng, S.W.; Ang, S.P.; Chow, S.K.; Tan, C.T.; et al. Small intestinal bacterial overgrowth in Parkinson's disease. *Park. Relat. Disord.* **2014**, *20*, 535–540. [CrossRef]

27. Pierantozzi, M.; Pietroiusti, A.; Sancesario, G.; Lunardi, G.; Fedele, E.; Giacomini, P.; Frasca, S.; Galante, A.; Marciani, M.G.; Stanzione, P. Reduced L-dopa absorption and increased clinical fluctuations in Helicobacter pylori-infected Parkinson's disease patients. *Neurol. Sci.* **2001**, *22*, 89–91. [CrossRef]
28. Pierantozzi, M.; Pietroiusti, A.; Brusa, L.; Galati, S.; Stefani, A.; Lunardi, G.; Fedele, E.; Sancesario, G.; Bernardi, G.; Bergamaschi, A.; et al. Helicobacter pylori eradication and l-dopa absorption in patients with PD and motor fluctuations. *Neurology* **2006**, *66*, 1824–1829. [CrossRef]
29. van Kessel, S.P.; Frye, A.K.; El-Gendy, A.O.; Castejon, M.; Keshavarzian, A.; van Dijk, G.; El Aidy, S. Gut bacterial tyrosine decarboxylases restrict levels of levodopa in the treatment of Parkinson's disease. *Nat. Commun.* **2019**, *10*, 310. [CrossRef]
30. Whitfield, A.C.; Moore, B.T.; Daniels, R.N. Classics in Chemical Neuroscience: Levodopa. *ACS Chem. Neurosci.* **2014**, *5*, 1192–1197. [CrossRef] [PubMed]
31. Flowers, S.A.; Bhat, S.; Lee, J.C. Potential Implications of Gut Microbiota in Drug Pharmacokinetics and Bioavailability. *Pharmacother. J. Hum. Pharmacol. Drug Ther.* **2020**, *40*, 704–712. [CrossRef] [PubMed]
32. Nutt, J.G.; Carter, J.H.; Lea, E.S.; Sexton, G.J. Evolution of the response to levodopa during the first 4 years of therapy. *Ann. Neurol.* **2002**, *51*, 686–693. [CrossRef] [PubMed]
33. Sethi, K. Levodopa unresponsive symptoms in Parkinson disease. *Mov. Disord.* **2008**, *23*, S521–S533. [CrossRef]
34. He, X.; Lai, Y.; Mo, C.; Zhang, Y.; Ai, P.; Xu, S.; Qian, Y.; Xiao, Q.; Yang, X. Association between Fecal Bile Acids and Levodopa Response in Patients with Parkinson's Disease. *Microorganisms* **2024**, *12*, 1432. [CrossRef]
35. Sampaio, T.F.; dos Santos, E.U.D.; de Lima, G.D.C.; dos Anjos, R.S.G.; da Silva, R.C.; Asano, A.G.C.; Asano, N.M.J.; Crovella, S.; de Souza, P.R.E. MAO-B and COMT Genetic Variations Associated with Levodopa Treatment Response in Patients with Parkinson's Disease. *J. Clin. Pharmacol.* **2018**, *58*, 920–926. [CrossRef]
36. Devos, D.; Lejeune, S.; Cormier-Dequaire, F.; Tahiri, K.; Charbonnier-Beaupel, F.; Rouaix, N.; Duhamel, A.; Sablonnière, B.; Bonnet, A.-M.; Bonnet, C.; et al. Dopa-decarboxylase gene polymorphisms affect the motor response to l-dopa in Parkinson's disease. *Park. Relat. Disord.* **2014**, *20*, 170–175. [CrossRef]
37. Moreau, C.; Meguig, S.; Corvol, J.-C.; Labreuche, J.; Vasseur, F.; Duhamel, A.; Delval, A.; Bardyn, T.; Devedjian, J.-C.; Rouaix, N.; et al. Polymorphism of the dopamine transporter type 1 gene modifies the treatment response in Parkinson's disease. *Brain* **2015**, *138*, 1271–1283. [CrossRef]
38. He, X.; Shi, K.; Mo, C.; Zhang, Y.; Xiao, Q.; Yang, X. Contribution of dopaminergic polymorphisms to levodopa treatment response and drug concentration in Chinese patients with Parkinson's disease. *Clin. Park. Relat. Disord.* **2025**, *12*, 100333. [CrossRef]
39. Koppel, N.; Maini Rekdal, V.; Balskus, E.P. Chemical transformation of xenobiotics by the human gut microbiota. *Science* **2017**, *356*, eaag2770. [CrossRef]
40. Spanogiannopoulos, P.; Bess, E.N.; Carmody, R.N.; Turnbaugh, P.J. The microbial pharmacists within us: A metagenomic view of xenobiotic metabolism. *Nat. Rev. Microbiol.* **2016**, *14*, 273–287. [CrossRef]
41. Zhao, Z.; Ning, J.; Bao, X.-Q.; Shang, M.; Ma, J.; Li, G.; Zhang, D. Fecal microbiota transplantation protects rotenone-induced Parkinson's disease mice via suppressing inflammation mediated by the lipopolysaccharide-TLR4 signaling pathway through the microbiota-gut-brain axis. *Microbiome* **2021**, *9*, 226. [CrossRef] [PubMed]
42. Sun, M.-F.; Zhu, Y.-L.; Zhou, Z.-L.; Jia, X.-B.; Xu, Y.-D.; Yang, Q.; Cui, C.; Shen, Y.-Q. Neuroprotective effects of fecal microbiota transplantation on MPTP-induced Parkinson's disease mice: Gut microbiota, glial reaction and TLR4/TNF-α signaling pathway. *Brain Behav. Immun.* **2018**, *70*, 48–60. [CrossRef]
43. Kuai, X.-Y.; Yao, X.-H.; Xu, L.-J.; Zhou, Y.-Q.; Zhang, L.-P.; Liu, Y.; Pei, S.-F.; Zhou, C.-L. Evaluation of fecal microbiota transplantation in Parkinson's disease patients with constipation. *Microb. Cell Factories* **2021**, *20*, 98. [CrossRef]
44. Xue, L.-J.; Yang, X.-Z.; Tong, Q.; Shen, P.; Ma, S.-J.; Wu, S.-N.; Zheng, J.-L.; Wang, H.-G. Fecal microbiota transplantation therapy for Parkinson's disease: A preliminary study. *Medicine* **2020**, *99*, e22035. [CrossRef]
45. Sadowski, K.; Zając, W.; Milanowski, Ł.; Koziorowski, D.; Figura, M. Exploring Fecal Microbiota Transplantation for Modulating Inflammation in Parkinson's Disease: A Review of Inflammatory Markers and Potential Effects. *Int. J. Mol. Sci.* **2024**, *25*, 7741. [CrossRef]
46. Domínguez Rojo, N.; Blanco Benítez, M.; Cava, R.; Fuentes, J.M.; Canales Cortés, S.; González Polo, R.A. Convergence of Neuroinflammation, Microbiota, and Parkinson's Disease: Therapeutic Insights and Prospects. *Int. J. Mol. Sci.* **2024**, *25*, 11629. [CrossRef]
47. Zhong, Z.; Ye, M.; Yan, F. A review of studies on gut microbiota and levodopa metabolism. *Front. Neurol.* **2023**, *14*, 1046910. [CrossRef]
48. Daniel, S.E.; Lees, A.J. Parkinson's Disease Society Brain Bank, London: Overview and research. *J. Neural Transm. Suppl.* **1993**, *39*, 165–172.
49. Tomlinson, C.L.; Stowe, R.; Patel, S.; Rick, C.; Gray, R.; Clarke, C.E. Systematic review of levodopa dose equivalency reporting in Parkinson's disease. *Mov. Disord.* **2010**, *25*, 2649–2653. [CrossRef]
50. Chen, S.; Zhou, Y.; Chen, Y.; Gu, J. fastp: An ultra-fast all-in-one FASTQ preprocessor. *Bioinformatics* **2018**, *34*, i884–i890. [CrossRef]

51. Li, H.; Durbin, R. Fast and accurate short read alignment with Burrows-Wheeler transform. *Bioinformatics* **2009**, *25*, 1754–1760. [CrossRef] [PubMed]
52. Beghini, F.; McIver, L.J.; Blanco-Miguez, A.; Dubois, L.; Asnicar, F.; Maharjan, S.; Mailyan, A.; Manghi, P.; Scholz, M.; Thomas, A.M.; et al. Integrating taxonomic, functional, and strain-levelprofiling of diverse microbial communities with bioBakery 3. *eLife* **2021**, *10*, e65088. [CrossRef] [PubMed]
53. Hyatt, D.; Chen, G.L.; Locascio, P.F.; Land, M.L.; Larimer, F.W.; Hauser, L.J. Prodigal: Prokaryotic gene recognition and translation initiation site identification. *BMC Bioinform.* **2010**, *11*, 119. [CrossRef] [PubMed]
54. Jia, L.; Jiang, Y.; Wu, L.; Fu, J.; Du, J.; Luo, Z.; Guo, L.; Xu, J.; Liu, Y. Porphyromonas gingivalis aggravates colitis via a gut microbiota-linoleic acid metabolism-Th17/Treg cell balance axis. *Nat. Commun.* **2024**, *15*, 1617. [CrossRef]

International Journal of
Molecular Sciences

MDPI

Article

Study of Intracellular Peptides of the Central Nervous System of Zebrafish (*Danio rerio*) in a Parkinson's Disease Model

Louise O. Fiametti [1,2], Camilla A. Franco [1], Leticia O. C. Nunes [1,2], Leandro M. de Castro [3] and Norival A. Santos-Filho [1,2,*]

1 Institute of Chemistry, São Paulo State University, Araraquara 14800-060, Brazil; louise.fiametti@unesp.br (L.O.F.); camilla.franco@unesp.br (C.A.F.); leticia.catarin@unesp.br (L.O.C.N.)
2 School of Pharmaceutical Sciences, São Paulo State University, Araraquara 14800-903, Brazil
3 Institute of Biosciences, São Paulo State University, São Vicente 11380-972, Brazil; leandro.mantovani@unesp.br
* Correspondence: norival.santos-filho@unesp.br

Abstract: Although peptides have been shown to have biological functions in neurodegenerative diseases, their role in Parkinson's disease has been understudied. A previous study by our group, which used a 6-hydroxydopamine zebrafish model, suggested that nine intracellular peptides may play a part in this condition. In this context, our aim is to better understand the role of five of these nine peptides. The selection of peptides was made based on their precursor proteins, which are fatty acid binding protein 7, mitochondrial ribosomal protein S36, MARCKS-related protein 1-B, excitatory amino acid transporter 2 and thymosin beta-4. The peptides were chemically synthesized in solid phase and characterized by high-performance liquid chromatography and mass spectrometry. Circular dichroism was performed to determine the secondary structure of each peptide, which showed that all five peptides maintain a random structure in the aqueous solutions that were studied. Two molecules show a helical profile in trifluoroethanol, a known structuring agent. Cell viability by the MTT assay indicates that all five peptides are not cytotoxic in all concentrations tested in both mouse and human cell lines. Behavioral assay using a 6-OHDA zebrafish larvae model suggest that all peptides help in the recovery of motor function with 24 h treatment at two concentrations. Three peptides showed a complete recovery from the 6-OHDA-induced motor impairment. Further studies are needed to better understand the mechanism of action of these peptides and whether they are truly a potential ally against Parkinson's disease.

Keywords: Parkinson's disease; intracellular peptides; zebrafish larvae

1. Introduction

Parkinson's disease (PD) is the second most common neurodegenerative disease in the world [1]. Its cause remains unknown and there is no cure [2]. Treatment options include a variety of pharmacological interventions, but patient adherence to medication is usually low due to a variety of side effects including nausea, orthostatic hypotension and narcolepsy, among others [3]. Levodopa, a drug introduced into the market in 1960, still provides the most significant improvement in motor symptoms, but its long-term use can worsen involuntary and uncontrolled muscle movements (dyskinesia) and motor fluctuations [4]. Thus, it is important to better understand PD and continue the research for new treatments.

A model that has been widely used in PD research is 6-hydroxydopamine (6-OHDA), a neurotoxin that is structurally similar to dopamine but is toxic to dopaminergic neurons, causing oxidative stress and neuroinflammation [5,6]. It is possible to treat animals with 6-OHDA, such as rodents [7–9] and fish, including zebrafish [10–12].

The use of zebrafish has become an alternative for the development of new drugs thanks to the improvement and standardization of the model in studies involving various behavioral repertoires in the field of neuropsychiatry [13]. For neuroscience in particular, the use of zebrafish as a model has allowed the correlation of physiological processes and behavioral responses with underlying molecular pathways or signaling cascades. This allows for a better understanding of the homeostasis mechanisms of the central nervous system, as well as pathological processes, such as in the case of neurodegenerative diseases [14].

In this context, a previous study by Fiametti et al. [10] used peptidomics to compare the peptide profile of 6-OHDA or saline-treated adult zebrafish brain. They showed that nine out of one hundred eighteen intracellular peptides identified in zebrafish brain were significantly altered in the 6-OHDA group. This suggests that these peptides may be involved in neurodegenerative processes, but further studies are needed to better understand if and how this involvement actually occurs. Five of the nine peptides mentioned were chosen for further studies. The peptides were selected based on their precursor proteins.

The first peptide chosen was GVGFATRQV (p-FABP7), which derived from the fatty acid binding protein 7 brain a (FABP7) [10]. According to Islam et al. [15], increased FABP7 elevates the expression of antioxidant enzymes and protects against apoptosis signaling. Ebrahimi et al. [16] showed that FABP7 deficiency causes abnormalities in dendritic neuronal morphology and affects excitatory synaptic functions. Islam et al. [15] also showed that another protein of the same family, fatty acid binding protein 3 (FABP3), is critical for the uptake of alpha-synuclein into dopaminergic neurons, a molecule found in Lewy bodies [2].

The second sequence that was selected was TLDNEMDYIQRGGPE (p-MRPS36), derived from mitochondrial ribosomal protein S36 (MRPS36) [10]. A study by Hevler et al. [17] suggests that MRPS36 evolved as an E3 adaptor protein, functionally replacing the peripheral subunit binding domain in the eukaryotic 2-oxoglutarate dehydrogenase complex. In other words, MRPS36 may have a role in the generation of NADH in aerobic respiration. Given the correlation between neurodegenerative diseases and reduced mitochondrial activity [18], MRPS36 can possibly have a role in PD.

The third peptide, TKQTEETNSTPAPSEQKE (p-MARCKS1B), is derived from the C-terminal region of the myristoylated alanine-rich C kinase substrate protein 1-B (MARCKS-related protein 1-B) [10]. The proteins of this family control cell movement by regulating the actin cytoskeleton, which is involved in the formation of filopodia and lamellipodia, and are highly conserved among vertebrates. MARCKS-related proteins play various roles related to brain growth, neuronal migration, neurite outgrowth, neurotransmitter release and synaptic plasticity [19].

The fourth sequence chosen, HESHLEPIE (p-EEAT2), is derived from the excitatory amino acid transporter 2 (EEAT2) [10], a protein with an essential role in the recycling of glutamate, the most abundant neurotransmitter in the mammalian brain [20]. A study by Trudler et al. [21] suggests that alpha-synuclein oligomers present in Lewy bodies induce glutamate release from astrocytes, leading to excessive extrasynaptic N-methyl-D-aspartate receptors, which are associated with neurodegeneration. Thus, glutamate recycling has an important role in controlling a neurodegenerative scenario.

Lastly, TIEQEKQAGSS (p-Tbeta4) is derived from thymosin beta-4 (Tbeta4) [10], a 43 amino acid peptide whose main function is to regulate actin polymerization, which is necessary for organogenesis and cell motility [22]. In addition, studies with Tbeta4 suggest that it is a peptide that supports myocardial recovery in post-infarction cases through cell migration [23,24], as well as in corneal [25] and liver recovery [26]. There is also evidence that Tbeta4 helps in the treatment of multiple sclerosis by restoring neuronal functionality [27], as well as in the improvement of injuries induced by oxygen-glucose deprivation [28] and reversing cognitive impairment in rats [29]. In other words, Tbeta4 may be an ally in neuronal recovery, and this study looks towards understanding if this derived peptide is also an ally.

As with other therapeutic strategies being studied, peptides may present some concerns such as stability, toxicity and immunogenicity [30]. However, they have some advantages over other small molecules, such as high specificity, high biological activities and high membrane penetration ability [31]. Such advantages result in about 100 peptides approved by the Food and Drug Administration (FDA) for various applications [32]. Thus, the hypothesis that one or more peptides altered in the study by Fiametti et al. [10] may have a beneficial effect in PD treatment is worthy of further investigation.

In order to perform further studies, all five peptides were synthesized, purified and characterized as described below. Cell viability by the MTT assays and behavioral assays in a 6-OHDA zebrafish larvae model were carried out to better understand the potential action of these peptides.

2. Results and Discussion

2.1. Synthesis, Purification and Characterization of Peptides

Peptides used in this study are the result of degradation by protein metabolism [10], which results in an acid C-terminal. For this reason, Wang resins were chosen for synthesis since, after cleavage in acid medium, their final peptide results in a COOH C-terminal. All peptides were synthesized, purified to >95% purity, and successfully characterized, as shown in Figure 1.

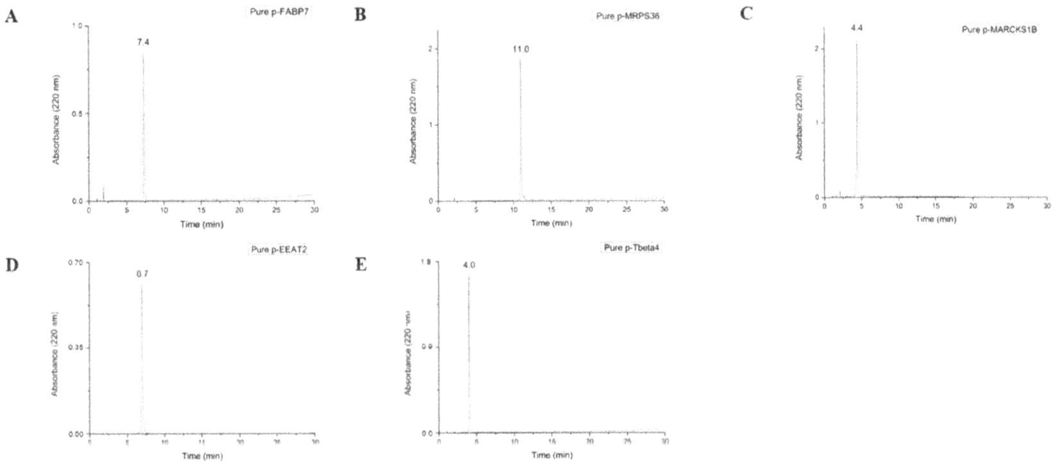

Figure 1. Chromatograms of pure peptides p-FABP7 (**A**), p-MRPS36 (**B**), p-MARCKS1B (**C**), p-EEAT2 (**D**) and p-Tbeta4 (**E**). The degree of purity of the fractions was determined on a Shimadzu chromatograph, using a 0.46 × 25 cm C18 reversed-phase analytical column (Kromasil). Solutions used for purification and verification of the purity degree of the peptides were 0.045% TFA in ultrapure water (solution A) and 0.036% TFA in acetonitrile (solution B) in a 30 min run at a flow rate of 1 mL/min in a 5–95% of solution B gradient.

2.2. Mass Spectrometry

Mass spectrometry was used to confirm that the synthesized and purified material was the material of interest (Figure 2). To perform this, the mass/charge ratio of each peptide was calculated and compared to the value of each peak shown in the spectrum obtained. If the numbers for the mass/charge ratio and at least one peak in the spectrum are the same, this indicates that the molecular weight of the material is the material of interest, or in other words, that the synthesis was performed correctly.

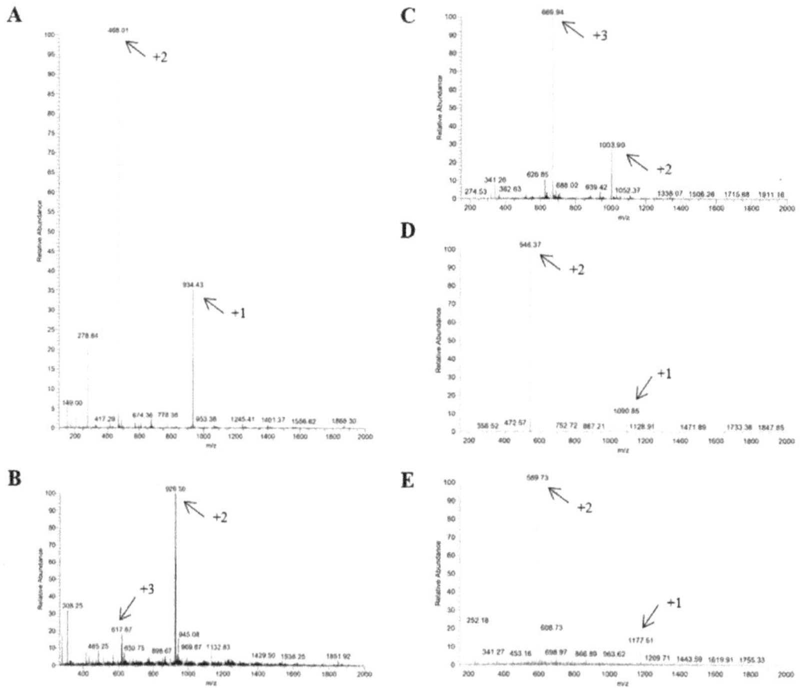

Figure 2. Mass spectra of p-FABP7 (**A**), p-MRPS36 (**B**), p-MARCKS1B (**C**), p-EEAT (**D**) and p-Tbeta4 (**E**). Arrows indicate peaks that confirm correct synthesis according to mass/charge ratio. Mass spectra were obtained by direct injection electrospray ionization in positive detection mode.

2.3. Circular Dichroism

A circular dichroism (CD) spectrum indicates an all-helix structure when there are two negative bands at ~222 and ~208 nm, plus a positive band at ~190 nm. An all-β-sheet molecule spectrum shows a negative band between 210 and 220 nm and a positive band between 195 and 200 nm. A random-structured molecule has a negative band at 200 nm [33]. In this study, CD was performed at three different pHs (5, 7.4 and 10) in order to observe possible structure modification and peptide stability and also in trifluoroethanol (TFE), a solvent that is known to be structuring and stabilizing [34,35] (Figure 3).

FABP7 has a β-sheet folding for most of its structure but presents two close regions in helix, one of which happens to be where p-FABP7 is located [36]. However, the CD spectrum of p-FABP7 (Figure 3A) indicates a random structure for this peptide in all pHs studied and also in TFE.

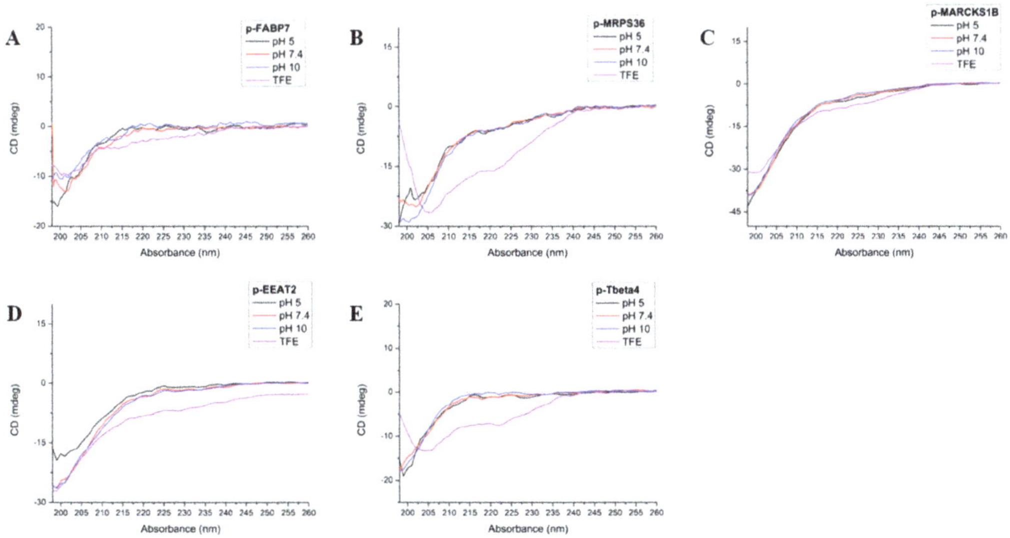

Figure 3. Circular dichroism spectra of p-FABP7 (**A**), p_MRSP36 (**B**), p-MARCKS1B (**C**), p-EEAT2 (**D**) and p-Tbeta4 (**E**) at pHs 5 (black), 7.4 (red), 10 (blue) and in TFE (violet).

MRPS36 has a random structure with four distant regions that evolve into an α-helix [17]. The CD spectrum of p-MRPS36 (Figure 3B) shows that this peptide remains with a random structure in all pHs studied, but in TFE the profile changes to a helix profile, showing that this configuration is possible.

The CD spectra for p-MARCKS1B (Figure 3C) and p-EEAT2 (Figure 3D) are similar: both show that in the three pHs studied and also in TFE both peptides remain in a random form. MARCKS-related protein 1B has a mostly random structure [37], while EEAT2 has a mostly α-helix structure [38].

Regarding Tbeta4, the precursor 43-amino acid peptide has a mostly random form but has two helixes throughout its structures [39]. The CD of p-Tbeta4 (Figure 3E) indicates that the derived peptide maintains a random structure at the pHs studied, but in TFE the profile changes to an α-helix.

2.4. Cell Viability by MTT Assay

Neuro 2A is a mouse neuroblast cell line [40] widely used in neuroscience research, while SHSY5Y is a cloned subline of human neuroblastoma [41] and is the most commonly used cell line in in vitro research about PD [42]. To investigate whether the peptides are cytotoxic, the MTT cytotoxicity assay was performed on both cell lines (Figures 4 and 5). Peptide concentrations started at 512 µM and followed a serial dilution down to 1 µM. DMSO 10% was used as positive control.

Treatment in Neuro 2A cells (Figure 4) indicates that all peptides do not present cytotoxicity. This includes all concentrations tested, even the highest at 512 µM. Figure 5 shows that the peptides did not show cytotoxicity in the human neuroblastoma cell line either.

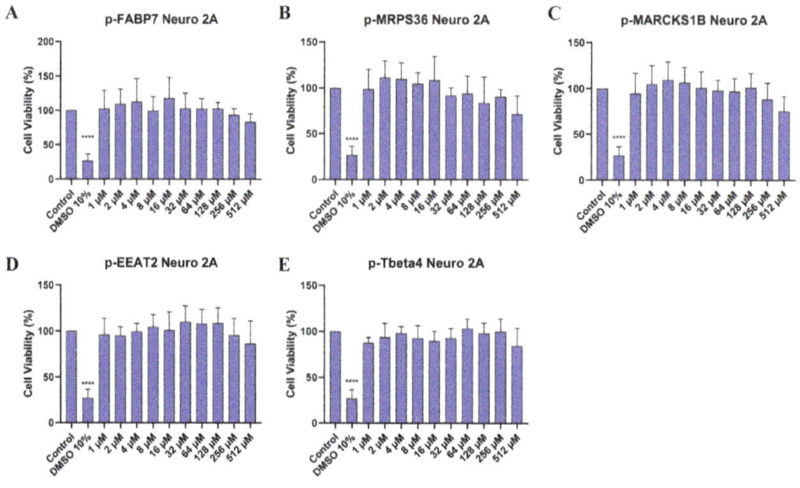

Figure 4. Cell viability (%) of Neuro 2A cells treated in different concentrations of p-FABP7 (**A**), p-MRPS36 (**B**), p-MARCKS1B (**C**), p-EAAT2 (**D**) and p-Tbeta4 (**E**). Experiments were performed in duplicate on three different occasions. ****, $p < 0.0001$.

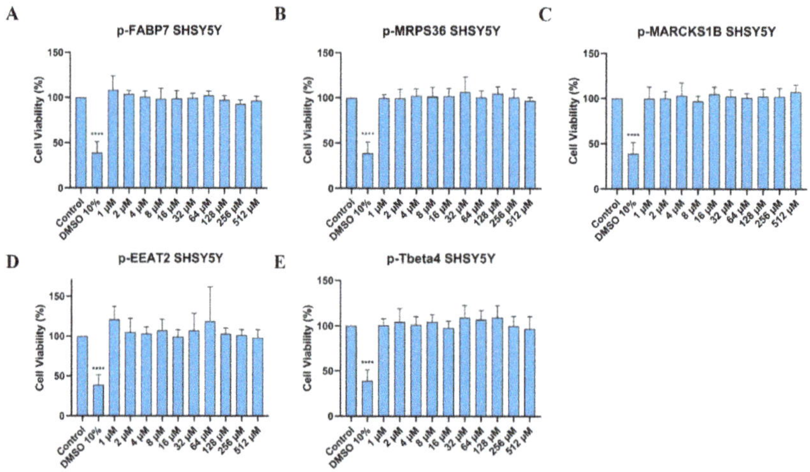

Figure 5. Cell viability (%) of SHSY5Y cells treated in different concentrations of p-FABP7 (**A**), p-MRPS36 (**B**), p-MARCKS1B (**C**), p-EAAT2 (**D**) and p-Tbeta4 (**E**). Experiments were performed in duplicate on three different occasions. ****, $p < 0.0001$.

2.5. Behavioral Assay

It is well described in the literature that neurotoxicity of 6-OHDA causes behavioral changes in animals such as rodents [7–9] and fish [11,12,43,44]. The main parameters observed in these studies are mean speed and distance traveled during a specific time period. This can be a strategy to either verify the efficacy of treatment with 6-OHDA or to observe the action of different treatments and if they have an effect on behavior.

As shown in Figures 6 and 7, treatment with all five peptides showed a significant improvement in mean speed (mm/s) and distance traveled (mm) compared to the 6-OHDA group at the two highest concentrations studied, 40 µM and 50 µM. Sequences p-FABP7, p-MPRS36 and p-Tbeta4 showed no significant difference compared to the control group at the highest concentration studied.

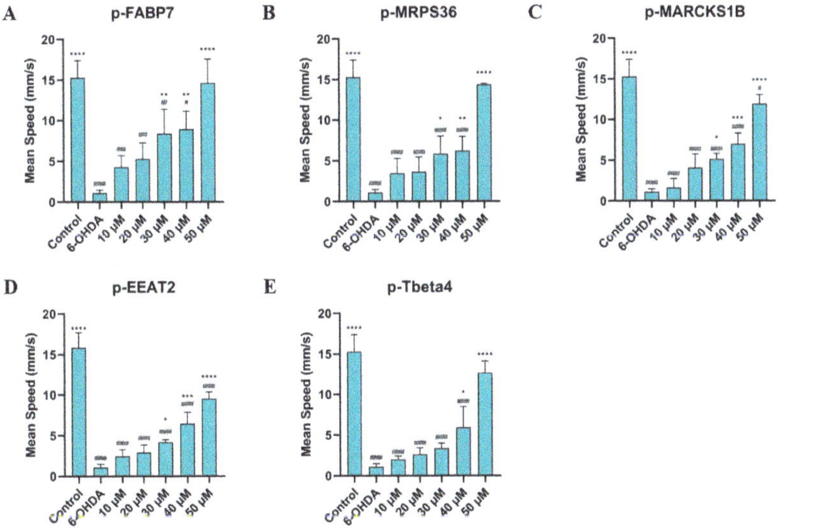

Figure 6. Mean speed of zebrafish treated with 750 μM for 24 h, then by p-FABP7 (**A**), p-MRPS36 (**B**), p-MARCKS1B (**C**), p-EEAT2 (**D**) and p-Tbeta4 (**E**) at concentrations of 10 μM, 20 μM, 30 μM, 40 μM and 50 μM for 24 h. Experiments were performed with 4 embryos per group on three different occasions, adding up to 12 embryos per group. #, $p < 0.01$, ##, $p < 0.001$, ###, $p < 0001$, ####, $p < 0.00001$ compared to the control group. *, $p < 0.01$; **, $p < 0.001$; ***, $p < 0.0001$; ****, $p < 0.00001$ compared to the 6-OHDA group.

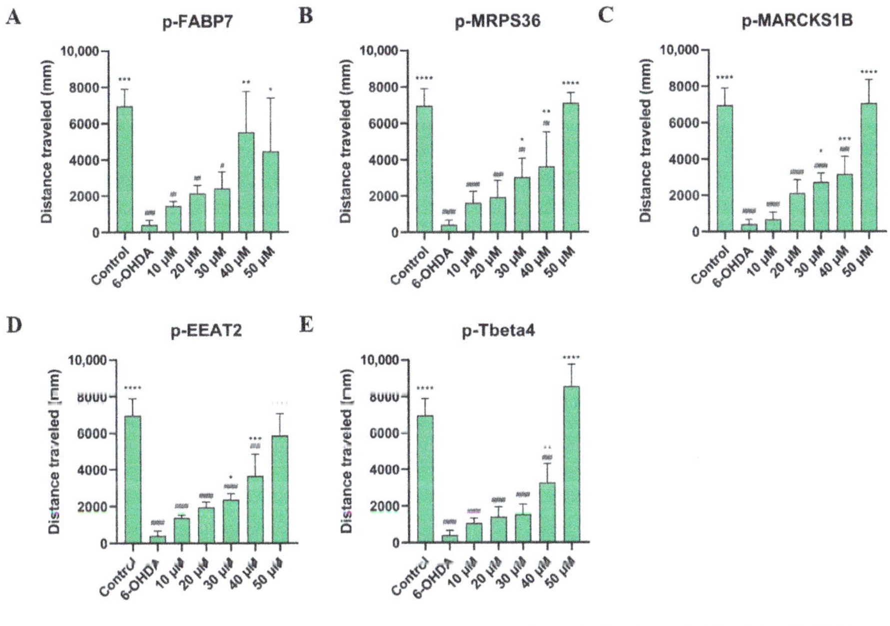

Figure 7. Distance traveled by zebrafish treated with 750 μM for 24 h, then by p-FABP7 (**A**), p-MRPS36 (**B**), p-MARCKS1B (**C**), p-EEAT2 (**D**) and p-Tbeta4 (**E**) at concentrations of 10 μM, 20 μM, 30 μM, 40 μM and 50 μM for 24 h. Experiments were performed with 4 embryos per group on three different occasions, adding up to 12 embryos per group. #, $p < 0.01$; ##, $p < 0.001$; ###, $p < 0001$; ####, $p < 0.00001$ compared to the control group. *, $p < 0.01$; **, $p < 0.001$; ***, $p < 0.0001$; ****, $p < 0.00001$ compared to the 6-OHDA group.

This suggests that the peptides help recover from the movement impairment caused by 6-OHDA. The results of the behavioral assay reinforce the lack of cytotoxicity of these peptides given that all groups observed had their physical capabilities improved, especially at higher concentrations. Three peptides, p-FABP7, p-MRPS36 and p-Tbeta4, showed statistically equal results to the control group in both parameters observed when treated with 50 μM, suggesting a complete recovery of mean speed and total distance swum. The precursor proteins of these three peptides have been shown to be involved in neurological functions in different scenarios, as mentioned above.

Further studies are needed to better understand the mechanism of action of these five peptides and whether or not these results are related to the role of each precursor protein. However, these initial data suggest that they are safe and effective molecules against 6-OHDA-Parkinson's-like motor symptoms.

3. Materials and Methods

3.1. Peptide Synthesis

Synthesis was performed by solid-phase peptide synthesis [45] following the fluorenylmethyloxycarboxyl (Fmoc) chemistry protocol. Wang resins previously coupled to the first amino acid were used for all five peptides (Table 1). Coupling was made with N,N-diisopropylcarbodiimilde (DIC) and N-hydroxybenzotriazole (HOBt).

Table 1. Resins used for synthesis of each peptide.

Peptide	Sequence	Precursor Protein	Resin
p-FABP7	GVGFATRQV	Fatty acid binding protein 7	Fmoc-Val-Wang
p-MRPS36	TLDLNEMDYIQRGGPE	Mitochondrial ribosomal protein S36	Fmoc-Glu(Otbu)-Wang
p-MARCKS1B	TKQTEETNSTPAPSEQKE	MARCKS-related protein 1B	Fmoc-Glu(Otbu)-Wang
p-EEAT2	HESHLEPIE	Excitatory amino acid transporter 2	Fmoc-Glu(Otbu)-Wang
p-Tbeta4	TIEQEKQAGSS	Thymosin beta-4	Fmoc-Ser(tBu)-Wang

3.2. Purification and Characterization

The synthesized peptides were purified by high-performance liquid chromatography (HPLC) on a Shimadzu chromatograph in semi-preparative mode on a C18 reversed-phase column (Phenomenex 1 × 25 cm) at a flow rate of 5 mL/min in different methods (Table 2). The purity degree of the fractions was determined on a Shimadzu chromatograph on an analytical column (0.4 × 25 cm) of C18 reversed phase (Kromassil) in a gradient of 5–95% of solution B in 30 min at a flow rate of 1 mL/min. Solutions used for purification and verification of the peptide purity were 0.045% trifluoroacetic acid (TFA) in ultrapure water (solution A) and 0.036% TFA in acetonitrile (solution B).

Table 2. Purification protocol of each peptide.

Peptide	Retention Time (min)	Solution B (%)	Purification Method
p-FABP7	9.9	29.7	5% to 45% of Solution B in 120 min
p-MRPS36	11.0	33.0	5% to 55% of Solution B in 120 min
p-MARCKS1B	4.4	13.2	1% to 35% of Solution B in 90 min
p-EEAT2	6.7	20.1	1% to 35% of Solution B in 90 min
p-Tbeta4	4.0	12	1% to 25% of Solution B in 120 min

3.3. Mass Spectrometry

To confirm the correct synthesis of peptides, mass spectra were obtained on an LCQ FLEET ThermoScientific mass spectrophotometer, with direct injection electrospray ionization, in positive detection mode.

3.4. Circular Dichroism

Circular dichroism (CD) spectroscopy was performed on a Jasco J-715 spectropolarimeter. CD spectra were obtained in aqueous solutions at pHs 5 (sodium acetate buffer 0.1 mol/L), 7.4 (PBS, phosphate buffer saline 0.027 mol/L) and 10 (borax buffer 0.1 mol/L) and in trifluoroethanol (TFE), a known structuring solvent [34,35].

3.5. Cell Viability by the MTT Assay

Neuro 2A and SHSY5Y cells were maintained in an incubator at 37 °C (atmosphere containing 95% air and 5% CO_2) in DMEM medium (Dulbecco's modified eagle medium, Gibco®, Grand Island, NY, USA) and DMEM/F12 medium (1:1 mixture of DMEM and Ham's F12 media, Gibco®, Grand Island, NY, USA), respectively, plus 100.000 units/L of penicillin, 0.1 g/L of streptomycin and 10% fetal bovine serum (FBS, Gibco®, Grand Island, NY, USA). After reaching ~80% confluency, cells were transferred to a 96-well plate, at 0.01×10^6 cell density. When ~80% confluency/well was reached, the medium was changed to DMEM or DMEM/F12 without antibiotics and FBS, plus each peptide in different concentrations. Dimethyl sulfoxide (DMSO) 10% was used as the positive control. After 24 h, the medium was removed, and each well was carefully washed with PBS. Then 5 mg/mL MTT (3-(4,5-dimethylthiazol-2-yl)-2-,5-diphenyltetrazolium bromide, Thermo Fisher Scientific, Waltham, MA, USA) in DMEM or DMEM/F12 was added to each well, and cells were incubated at 37 °C for 1 h. The medium was removed once more and substituted by DMSO. Plates were agitated on a shaker for approximately 15 min. Absorbance reading was performed at 570 nm on a VERSAMax spectrophotometer (Molecular Devices, LLC, San Jose, CA, USA). Statistical analysis was performed using a one-way analysis of variance (ANOVA) test followed by Dunnett's test for multiple comparisons. Data were statistically analyzed using GraphPad Prism Software Version 8.0.2 (GraphPad Software Inc., San Diego, CA, USA).

3.6. Animals

Wild-type adult *D. rerio* (zebrafish) were obtained from a licensed commercial distributor. They were maintained on a 14 h light–10 h dark photoperiod in fresh filtered water at 28 °C and fed twice a day with dry food and once a day with *Artemia* sp. Reproduction was performed in individual fish tanks in a 1 male–1 female proportion, with pebbles and artificial plants. Eggs were collected, selected and maintained in an incubator at 28 °C until 4 days post fertilization (dpf). This study followed the guidelines of the National Council for Animal Experimentation Control (CONCEA) and was approved by the Ethics Commission for Animal Use (CEUA) at the Bioscience Institute of São Paulo State University (São Vicente, Brazil: Protocol Number 06/2023, approved on 22 June 2023).

3.7. Behavioral Assay

Zebrafish (*Danio rerio*) larvae at 4 dpf were transferred to a 96-well plate, a single larva per well. Larvae were treated with 750 μM 6-OHDA in system water for 24 h [46] in an incubator at 28 °C. Then, solution was changed to system water with each peptide in different concentrations—10 μM, 20 μM, 30 μM, 40 μM and 50 μM [47]—and larvae were incubated at 28 °C for another 24 h. Finally, each larva was filmed for 5 min, and videos were processed by an automated tracking system (ToxTrac Version 2.96). Analysis parameters were mean speed (mm/s) and total distance swum (mm) (Figure 8) [45]. Statistical analysis was performed using a one-way analysis of variance (ANOVA) test followed by Dunnett's test for multiple comparisons. Data were statistically analyzed using GraphPad Prism Software (GraphPad Software Inc., San Diego, CA, USA).

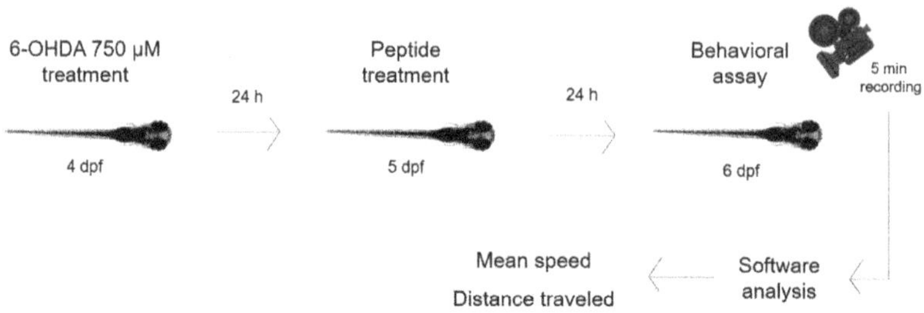

Figure 8. Zebrafish larva model. Zebrafish larvae were immersed in 6-OHDA 750 μM for 24 h, then in each peptide at 10 μM, 20 μM, 30 μM, 40 μM or 50 μM for 24 h. Videos were recorded for 5 min and then analyzed by software (ToxTrac Version 2.96). Parameters were mean speed and distance traveled.

4. Conclusions

All five peptides were successfully synthesized, purified and characterized. In aqueous solutions at pHs 5, 7.4 and 10, all peptides maintained a random secondary structure. In TFE, p-MRPS36 and p-Tbeta4 presented an α-helix profile. None of the peptides presented cytotoxicity in either cell line in all concentrations studied, which suggests that they can be used safely. Regarding zebrafish behavior, the performed assay indicates that all peptides help recover physical impairment caused by 6-OHDA in the two highest concentrations observed (40 μM and 50 μM). Three peptides showed no statistical difference when comparing treatment with 50 μM to the control group. This can possibly indicate that p-FABP7, p-MRPS36 and p-Tbeta4 led to a complete recovery of the physical impairment caused by 6-OHDA. Further studies are needed to better understand the mechanisms of these peptides and also to ensure the safety of these molecules for long-term treatment.

Author Contributions: Conceptualization, L.O.F. and L.M.d.C.; data curation, L.O.F., C.A.F. and L.O.C.N.; formal analysis, L.O.F. and C.A.F.; funding acquisition, N.A.S.-F.; investigation, L.O.F. and C.A.F.; methodology, L.O.F., C.A.F., L.O.C.N., L.M.d.C. and N.A.S.-F.; project administration, N.A.S.-F.; resources, N.A.S.-F.; supervision, L.M.d.C. and N.A.S.-F.; visualization, L.M.d.C.; writing—original draft, L.O.F.; writing—review and editing, L.O.F., C.A.F., L.O.C.N., L.M.d.C. and N.A.S.-F. All authors have read and agreed to the published version of the manuscript.

Funding: This work was supported by the Brazilian National Council for Scientific and Technological Development (grant 403232/2023-6) and the São Paulo Research Foundation (grant #2023/05219-2; grant #2023/03857-1). The funders had no role in the study design, data collection and analysis, decision to publish or preparation of the article.

Institutional Review Board Statement: This study followed the guidelines of the National Council for Animal Experimentation Control (CONCEA) and was approved by the Ethics Commission for Animal Use (CEUA) at the Bioscience Institute of São Paulo State University (São Vicente, Brazil: Protocol Number 06/2023, approved on 22 June 2023).

Informed Consent Statement: Not applicable.

Data Availability Statement: Data supporting the conclusions of this study can be made available upon request from the corresponding author.

Conflicts of Interest: The authors declare no conflicts of interest.

References

1. World Health Organization (WHO). *Parkinson Disease: A Public Health Approach: Technical Brief*; World Health Organization: Geneva, Switzerland, 2022.
2. Bloem, B.R.; Okun, M.S.; Klein, C. Parkinson's disease. *Lancet* **2021**, *397*, 2294–2303. [CrossRef] [PubMed]
3. Armstrong, M.J.; Okun, M.S. Diagnosis and treatment of Parkinson disease: A review. *JAMA* **2020**, *323*, 548–560. [CrossRef] [PubMed]
4. Bogetofte, H.; Alamyar, A.; Blaabjerg, M.; Meyer, M. Levodopa Therapy for Parkinson's Disease: History, Current Status and Perspectives. *CNS Neurol. Disord. Drug Targets* **2020**, *19*, 572–583. [CrossRef] [PubMed]
5. Ungerstedt, U. 6-Hydroxidopamine induced degeneration of central monoamine neurons. *Eur. J. Pharmacol.* **1968**, *5*, 107–110. [CrossRef] [PubMed]
6. Hernandez-Baltazar, D.; Zavala-Flores, L.M.; Villanueva-Olivo, A. The 6-hydroxidopamine model and parkinsonian pathophysiology: Novel findings in an older model. *Neurologia* **2017**, *32*, 533–539. [CrossRef]
7. Stefani, A.; Cerroni, R.; Pierantozzi, M.; D'Angelo, V.; Grani, L.; Spanetta, M.; Galati, S. Deep brain stimulation in Parkinson's disease patients and routine 6-OHDA rodent models: Synergies and pitfalls. *Eur. J. Neurosci.* **2020**, *53*, 2322–2343. [CrossRef]
8. Guimarães, R.P.; Ribeiro, D.L.; Santos, K.B.; Godoy, L.D.; Corrêa, M.R.; Padovan-Neto, E. The 6-hydroxydopamine Rat Model of Parkinson's Disease. *Vis. Exp.* **2021**, *176*, e62923. [CrossRef]
9. Slézia, A.; Hegedus, P.; Rusina, E.; Lengyel, K.; Solari, N.; Kaszas, A.; Balázsfi, D.; Botzanowski, B.; Acerbo, E.; Missey, F.; et al. Behavioral, neural and ultrastructural alterations in a graded-dose 6-OHDA mouse model of early-stage Parkinson's disease. *Sci. Rep.* **2023**, *13*, 19478. [CrossRef] [PubMed]
10. Fiametti, L.O.; Correa, C.N.; de Castro, L.M. Peptide Profile of Zebrafish Brain in a 6-OHDA-Induced Parkinson Model. *Zebrafish* **2021**, *18*, 55–65. [CrossRef] [PubMed]
11. Gomes, A.; Monteiro, S.M.; Venâncio, C.; Félix, L. Protective effects of 24-epibrassinolide against the 6-OHDA zebrafish model of Parkinson's disease. *Comp. Biochem. Physiol. Part C* **2023**, *269*, 109630. [CrossRef]
12. Romero, A.; Sanchez, A.; Jones, J.D.; Ledesma, K.; El-Halawany, M.S.; Hamouda, A.K.; Bill, B.R. Optimization of Zebrafish Larvae 6-OHDA Exposure for Neurotoxin Induced Dopaminergic Marker Reduction. *Zebrafish* **2024**, *21*, 287–293. [CrossRef] [PubMed]
13. Fontana, B.D.; Mezzomo, N.J.; Kalueff, A.V.; Rosenberg, D.B. The developing utility of zebrafish models of neurological and neuropsychiatric disorders: A critical review. *Exp. Neurol.* **2018**, *299*, 157–171. [CrossRef] [PubMed]
14. Chia, K.; Klimgsein, A.; Sieger, D.; Priller, J. Zebrafish as a model organism for neurodegenerative disease. *Front. Mol. Neurosci.* **2022**, *15*, 940484. [CrossRef] [PubMed]
15. Islam, A.; Kagawa, Y.; Miyazaki, H.; Shil, S.K.; Umaru, B.A.; Yasumoto, Y.; Yamamoto, Y.; Owada, Y. FABP7 protects astrocytes against ROS toxicity via Lipid Droplet Formation. *Mol. Neurobiol.* **2019**, *56*, 5763–5779. [CrossRef]
16. Ebrahimi, M.; Yamamoto, Y.; Sharifi, K.; Kida, H.; Kagawa, Y.; Yasumoto, Y.; Islam, A.; Miyazaki, H.; Shimamoto, C.; Maekawa, M.; et al. Astrocyte-Expressed FABP7 Regulates Dendritic Morphology and Excitatory Synaptic Function of Cortical Neurons. *Wiley Period.* **2015**, *64*, 48–62. [CrossRef] [PubMed]
17. Hevler, J.F.; Albanese, P.; Cabrera-Orefice, A.; Potter, A.; Jankevics, A.; Music, J.; Sheltema, R.A.; Brandt, U.; Arnold, S.; Heck, A.J.R. MRPS36 provides a structural link in the eukaryotic 2-oxoglutaate dehydrogenase complex. *Open Biol.* **2023**, *13*, 220363. [CrossRef]
18. Compagnoni, G.M.; Fonzo, A.D.; Corti, S.; Comi, G.P.; Bresolin, N.; Masliah, E. The Role of Mitochondria in Neurodegenerative Diseases: The Lesson from Alzheimer's Disease and Parkinson's Disease. *Mol. Neurobiol.* **2020**, *57*, 2959–2980. [CrossRef]
19. Amri, M.; Fitzgerald, U.; Schlosser, G. MARCKS and MARCKS-like proteins in development and regeneration. *J. Biomed. Sci.* **2018**, *25*, 43. [CrossRef]
20. Fontana, I.C.; Souza, D.G.; Souza, D.O.; Gee, A.; Zimmer, E.R.; Bongarzone, S. A Medicinal Chemistry Perspective on Excitatory Amino Acid Transporter 2, Dysfunction in Neurodegenerative Diseases. *J. Med. Chem.* **2023**, *66*, 2333–2347. [CrossRef]
21. Trudler, D.; Sanz-Blasco, S.; Eisele, Y.S.; Ghatak, S.; Bodhinathan, K.; Akhatar, M.W.; Lynch, W.P.; Piã-Crespo, J.C.; Talantova, M.; Kelly, J.W.; et al. α-synuclein Oligomers Induce Glutamate Release from Astrocytes and Excessive Extrasynaptic NMDAR Activity in Neurons, Thus Contributing to Synapse Loss. *Neurobiol. Dis.* **2021**, *41*, 2264–2273. [CrossRef] [PubMed]
22. Crockford, D. Development of Thymosin β4 for treatment of Patients with Ischemic Heart Disease. *Ann. N. Y. Acad. Sci.* **2007**, *1112*, 385–395. [CrossRef] [PubMed]
23. Cavasin, M.A. Theurapeutic Potential of Thymosin-β4 and its Derivative N-acetyl-seryl-aspartyl-lysyl-proline (Ac-SDKP) in Cardiac Healing After Infarction. *Am. J. Cardiovasc. Drugs* **2006**, *6*, 305–311. [CrossRef]
24. Hinkel, R.; El-Aouni, C.; Olson, T.; Horstkotte, J.; Mayer, S.; Muller, S.; Willhauck, M.; Spitzweg, C.; Glidehaus, F.-J.; Munzing, W.; et al. Thymosin β4 is an Essential Paracrine Factor of Embryonic Endothelial Progenitor Cell-Mediated Cardioprotection. *Circulation* **2008**, *117*, 2232–2240. [CrossRef] [PubMed]
25. Sosne, G.; Szlitzer, E.A.; Barret, R.; Kernacki, K.A.; Kleinman, H.; Hazlett, L.D. Thymosin Beta 4 Promotes Corneal Wound Healing and Decreases Inflammation in Vivo Following Alkali Injury. *Exp. Eye Res.* **2002**, *74*, 293–299. [CrossRef]

26. Yang, L.; Fu, W.-L.; Zhu, Y.; Wang, X.-G. Tβ4 suppresses linc-RNA-p21-mediated hepatic apoptosis and fibrosis by inhibiting PI3K-AKT-NF-κB pathway. *Gene* **2020**, *758*, 144946. [CrossRef] [PubMed]
27. Zhang, J.; Zhang, Z.G.; Morris, D.; Li, Y.; Roberts, C.; Elias, S.B.; Chopp, M. Neurological Functional Recovery After Thymosin Beta4 Treatment in Mice with Experimental Auto Encephalomyelitis. *Neuroscience* **2009**, *164*, 1887–1893. [CrossRef]
28. Zhang, Z.; Liu, S.; Huang, S. Thymosin β4 prevents oxygen-glucose deprivation/reperfusion-induced injury in rat cortical neurons. *Neuropsychiatr. Dis. Treat.* **2019**, *15*, 2385–2393. [CrossRef] [PubMed]
29. Wang, M.; Feng, L.-R.; Li, Z.-L.; Ma, K.G.; Chang, K.-W.; Chen, X.-L.; Yang, P.B.; Ji, S.-F.; Ma, Y.-B.; Han, H.; et al. Thymosin β reverses phenotyipic polarization of glial cells and cognitive impairment via negative regulation of NF-κB signaling axis in APP/PS1 mice. *J. Neuroinflammation* **2021**, *18*, 146. [CrossRef]
30. Baig, M.H.; Ahmad, K.; Saeed, M.; Alharbi, A.M.; Barreto, G.E.; Ashraf, G.M.; Choi, I. Peptide based therapeutics and their use for the treatment of neurodegenerative and other diseases. *Biomed. Pharmacother.* **2018**, *103*, 574–581. [CrossRef] [PubMed]
31. Vlieghe, P.; Lisowski, V.; Martinez, J.; Khrestchatisky, M. Synthetic theurapeutic peptides: Science and market. *Drug Discov.* **2010**, *15*, 40–56. [CrossRef]
32. Musaimi, O.A. Exploring FDA_Approved Frontiers: Insights into Natural and Engineered Peptide Analogues in the GLP-1, GIP, GHRH, CCK, ACTH and α-MSH realms. *Biomolecules* **2024**, *14*, 264. [CrossRef]
33. Ranjbar, B.; Gill, P. Circular dichroism techniques: Biomolecular and nanostructural analyses—A review. *Chem. Biol. Drug Des.* **2009**, *74*, 101–120. [CrossRef] [PubMed]
34. Rajan, R.; Balaram, P. A model for the interaction of trifluoroethanol with peptides and proteins. *Int. J. Pep. Protein Res.* **2009**, *48*, 328–336. [CrossRef] [PubMed]
35. Vincenzi, M.; Mercurio, F.A.; Leone, M. About TFE: Old and New Findings. *Curr. Protein Pept. Sci.* **2019**, *20*, 425–451. [CrossRef]
36. Lenz, S.; Bodnariuc, I.; Renaud-Young, M.; Butler, T.M.; MacCallum, J.L. Understanding FABP7 binding to fatty acid micelles and membranes. *Biophys. J.* **2023**, *122*, 603–615. [CrossRef] [PubMed]
37. Brudvig, J.J.; Weimer, J.M. X MARCKS the spot: Myristoylated alanine-rich C kinase substrate in neuronal function and disease. *Front. Cell. Neurosci.* **2015**, *9*, 407. [CrossRef] [PubMed]
38. Zhang, Z.; Chen, H.; Geng, Z.; Yu, Z.; Li, H.; Dong, Y.; Zhang, H.; Huang, Z.; Jiang, J.; Zhao, Y. Structural basis of ligand binding modes of human EEAT2. *Nat. Commun.* **2022**, *13*, 3329. [CrossRef] [PubMed]
39. Hoch, K.; Volk, D.E. Structures of Thymosin Proteins. In *Vitamins and Hormones*; Elsevier: Amsterdam, The Netherlands, 2016; Volume 102, ISSN 0083-6729.
40. ATTC CCL-131. Neuro-2a. Available online: https://www.atcc.org/products/ccl-131 (accessed on 23 November 2024).
41. ATTCC CRL-2266. SH-SY5Y. Available online: https://www.atcc.org/products/crl-2266 (accessed on 23 November 2024).
42. Ioghen, O.C.; Ceafalan, L.C.; Popescu, B.O. SHSY5Y Cell Line In Vitro Models for Parkinson Disease Research—Old Practice for New Trends. *J. Integr. Neurosci.* **2023**, *22*, 20. [CrossRef]
43. Matsui, H.; Ito, H.; Taniguchi, Y.; Inoue, H.; Takeda, S.; Takahashi, R. Proteasome inhibition in medaka brain induces the features of Parkinson's disease. *J. Neurochem.* **2010**, *115*, 178–187. [CrossRef]
44. Fiametti, L.O.; de Mello, F.R.; de Castro, L.M. Peptidomics od Zebrafish Brain in a 6-OHDA-induced Neurodegeneration Model. *Pept. Methods Strateg. Methods Mol. Biol.* **2024**, *2758*, 307–317. [CrossRef]
45. Merrifield, R.B. Solid phase peptide synthesis. I. The synthesis of a tetrapeptide. *J. Am. Chem. Soc.* **1963**, *85*, 2149–2154. [CrossRef]
46. Vaz, R.L.; Sousa, S.; Chapela, D.; van der Linde, H.C.; Willemsen, R.; Correia, A.; Outeiro, T.F.; Afonso, N.D. Identification of antiparkinsonian drugs in the 6-hydroxydopamine zebrafish model. *Pharmacol. Biochem. Behav.* **2017**, *189*, 172828. [CrossRef] [PubMed]
47. Prabha, N.; Guru, A.; Harikrisghnan, R.; Gatasheh, M.K.; Hatamleh, A.A.; Juliet, A.; Arockiaraj, J. Neuroprotective and antioxidant capability of RW20 peptide from histone acetyltransferases cused by oxidative stress-induced neurotoxicity in in vivo zebrafish larcal model. *J. King Saud Univ. Sci.* **2022**, *34*, 101861. [CrossRef]

International Journal of
Molecular Sciences

MDPI

Article

Association between Decreased SGK1 and Increased Intestinal α-Synuclein in an MPTP Mouse Model of Parkinson's Disease

Min Hyung Seo [1,†], Dasom Kwon [1,†], Soo-Hwan Kim [2,*] and Sujung Yeo [3,*]

1 Department of Meridian and Acupoint, College of Korean Medicine, Sang Ji University, Wonju 26339, Republic of Korea; cstcl@naver.com (M.H.S.); kdsom618@naver.com (D.K.)
2 Division of Biological Science and Technology, Yonsei University, 1 Yonseidae-gil, Wonju 26493, Republic of Korea
3 Research Institute of Korean Medicine, Sangji University, Wonju 26339, Republic of Korea
* Correspondence: soohwan@yonsei.ac.kr (S.-H.K.); pinkteeth@naver.com or yeosu@sangji.ac.kr (S.Y.); Tel.: +82-33-760-2787 (S.-H.K.); +82-33-738-7506 (S.Y.)
† These authors contributed equally to this work.

Abstract: Parkinson's disease (PD) is a globally common progressive neurodegenerative disease resulting from the loss of dopaminergic neurons in the brain. Increased α-synuclein (α-syn) is associated with the degeneration of dopaminergic neurons and non-motor symptoms like gastrointestinal disorders. In this study, we investigated the association between serum/glucocorticoid-related kinase 1 (SGK1) and α-syn in the colon of a PD mouse model. SGK1 and α-syn expression patterns were opposite in the surrounding colon tissue, with decreased SGK1 expression and increased α-syn expression in the PD group. Immunofluorescence analyses revealed the colocation of SGK1 and α-syn; the PD group demonstrated weaker SGK1 expression and stronger α-syn expression than the control group. Immunoblotting analysis showed that Na^+/K^+ pump ATPase α1 expression levels were significantly increased in the PD group. In SW480 cells with SGK1 knockdown using SGK1 siRNA, decreasing SGK1 levels corresponded with significant increases in the expression levels of α-syn and ATPase α1. These results suggest that SGK1 significantly regulates Na^+/K^+ pump ATPase, influencing the relationship between electrolyte balance and fecal formation in the PD mouse model. Gastrointestinal disorders are some of the major prodromal symptoms of PD. Therefore, modulating SGK1 expression could be an important strategy for controlling PD.

Keywords: Parkinson's disease; SGK1; α-synuclein; Na^+/K^+ pump ATPase; SW480

1. Introduction

Parkinson's disease (PD), first reported by James Parkinson in 1871 [1,2], is the second most common progressive neurodegenerative disease globally. Although it used to be a rare condition, its prevalence doubled from 1990 to 2015, affecting over six million people. It is projected that over 12 million people will have PD by 2040. PD is characterized by motor symptoms, such as bradykinesia, tremor, and rigidity, resulting from reduced dopaminergic neurons in the substantia nigra (SN) and non-motor symptoms, such as gastrointestinal disorders (constipation, etc.) [3–7]. Although the exact causes and mechanisms of PD are unclear, factors such as misfolding and accumulation of the synaptic protein α-synuclein (α-syn) are reportedly associated with the degeneration of dopaminergic neurons [8,9]. Non motor symptoms, such as gastrointestinal disorders, have also been associated with increased intestinal α-syn [10–12]. The elevated levels of α-syn in the intestines are likely to play a pivotal role in the underlying pathology of PD [13,14]. Increased intestinal α-syn expression has not been documented in Alzheimer's disease, and PD reportedly occurs selectively in a subset of the intestinal α-syn-positive population or may induce this expression [15]. Moreover, increased α-syn migrates to the brain and influences the pathology of PD [9,13,14,16]. This hypothesis of a gut–brain axis in PD was put forward by Braak et al. in 2003; they presented a staging system for PD based on α-syn deposition patterns identified in cases of autopsy [12,17].

Serum/glucocorticoid-related kinase 1 (SGK1), known for its activation by insulin and growth factors, is an enzyme that participates in various physiological and pathological processes [18]. Its functions include regulating cell survival, growth, differentiation, inflammation, electrolyte balance, and metabolism [18–21]. It is also involved in the pathophysiology of several diseases, including PD. Under oxidative stress, SGK1 exhibits a protective role [19,20,22]. Oxidative stress is a possible cause of neurodegenerative diseases, such as PD. The overexpression of SGK1 prevents cell death; therefore, SGK1 can overcome this oxidative stress [23–26].

Our previous study indicated that reduced SGK1 expression may significantly contribute to the upregulation of α-syn and may be associated with the death of dopaminergic cells in the substantia nigra (SN) and the muscles of mice with chronic Parkinsonism [27,28]. Gastrointestinal disorders, especially constipation, are major prodromal symptoms of PD [3,5]. Therefore, investigating the association between α-syn and elements of gastrointestinal disorders is important. SGK1 is related to Na^+/K^+ pump ATPase, and intestinal Na^+/K^+ pump ATPase α1 and ß1 have been reported to be present [29,30].

In this study, based on the report that 1-methyl 4-phenyl 1,2,3,6-tetrahydropyridine (MPTP) is a representative dopamine neuron toxin of the enteric nervous system inducing gastrointestinal dysfunction [6], the association between SGK1, a powerful stimulator of Na^+/K^+ pump ATPase, and α-syn was examined in the colon of an MPTP-induced PD model. This study hypothesized that non-motor symptoms of PD, such as constipation [7], could be caused by changes in SGK1 expression related to Na^+/K^+ pump and α-syn expression changes. A close association was reported between the basolateral Na^+-K^+-$2Cl^-$ cotransporter and intestinal water and mucus secretion, as well as its impact on the immune system [31]. As Alcian blue selectively binds to acidic mucopolysaccharides and mucins, Alcian blue staining was utilized in this study to confirm changes in mucus secretion in the colon of mice with MPTP-induced PD in response to alterations in fluid and electrolyte balance. In light of the changes in SGK1, alterations in α-syn and Na^+/K^+ pump-related factors were also studied in SW480 cells in this study.

2. Results

2.1. MPTP-Induced PD Model

The PD model induced by MPTP showed significantly decreased expression levels of tyrosine hydroxylase (TH) in the SN and striatum (ST, Figure 1A,B). The number of dopaminergic cells in the SN and the TH concentration in the ST were significantly reduced (Figure 1C). Consistently, immunoblot analysis showed markedly reduced TH expression levels in the SN and ST regions (Figure 1B,C). These results demonstrate the characteristics of PD.

2.2. Morphological Changes in the Colon in the MPTP-Induced PD Model

Alcian blue staining was performed to investigate the morphological changes in the colon in the MPTP-induced PD mouse model using colon tissues from the control (CTL) and MPTP-treated (MPTP) groups (Figure 2). The Alcian blue staining analyses showed that the expression of mucins was uneven in villi and microvilli of the MPTP group. The tissue matrix in the MPTP group was less sturdy than that in the CTL group. The results indicate that the density of the colon tissue significantly reduced in the MPTP group (Figure 2h).

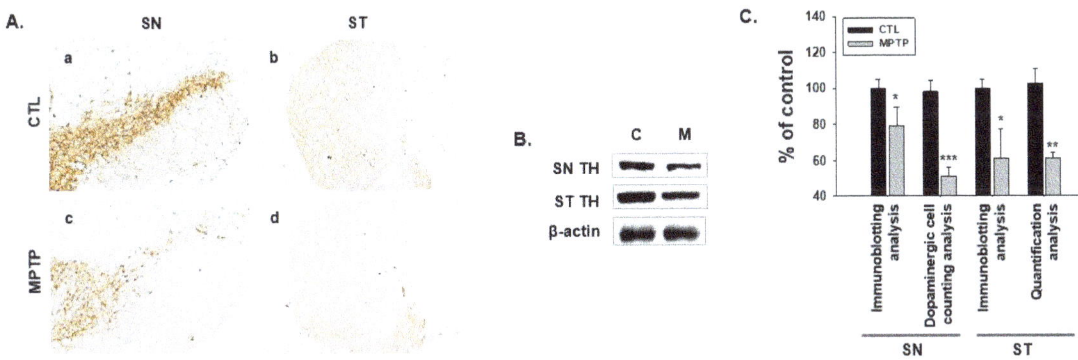

Figure 1. Changes in the expression patterns of tyrosine hydroxylase (TH) in the substantia nigra (SN) and striatum (ST) regions of the brain in 1-methyl-4-phenyl-1,2,3,6-tetrahydropyridine (MPTP)-induced Parkinson's disease model. (**A**) Immunohistochemistry analyses showing that TH expression decreased in SN and ST regions in the MPTP group (**c**,**d**) compared with that in the control (CTL) group (**a**,**b**). a and c panels; 100×, b and d panels; 40×. (**B**) Immunoblotting analyses of TH in SN and ST regions in the brain. TH expression significantly decreased in SN and ST regions in the MPTP group. (**C**) Graphs indicating the results of the immunoblotting analysis, dopaminergic cell counting analysis in SN, and TH expression quantification per area in ST. CTL, intraperitoneally injected with 100 μL 0.9% saline daily for 4 weeks; MPTP, intraperitoneally injected with MPTP-HCI (20 mg/kg) daily for 4 weeks ($n = 3$, * $p < 0.05$, ** $p < 0.005$, *** $p < 0.0005$).

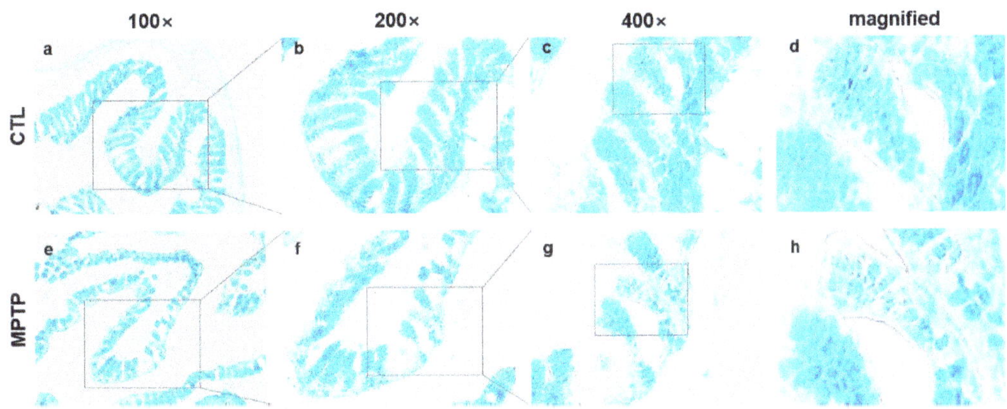

Figure 2. Alcian blue staining analysis in the colon of the 1-methyl-4-phenyl-1,2,3,6-tetrahydropyridine (MPTP)-induced Parkinson's disease model. ((**a**,**e**); 100×) Staining expression patterns were changed in the MPTP group compared with those in the control (CTL) group. ((**b**,**f**); 200×) Enlarged images of staining expression pattern changes in the villi in the CTL and MPTP groups are shown. ((**c**,**g**); 400×) Enlarged images of staining expression pattern changes in the microvilli in the CTL and MPTP groups are shown. (**d**,**h**) Magnified images of panels c and g. CTL, intraperitoneally injected with 100 μL 0.9% saline daily for 4 weeks; MPTP, intraperitoneally injected with MPTP-HCI (20 mg/kg) daily for 4 weeks.

2.3. SGK1 and α-Syn Expression in the Colon of MPTP-Induced PD Model

Immunohistochemical analysis of SGK1 in the colon of CTL and MPTP groups was performed. To verify the report that α-syn depositions are discovered in the gastrointestinal (GI) tract, α-syn in the colon was also analyzed immunohistochemically, and the expression patterns of SGK1 and α-syn were compared in the CTL and MPTP groups (Figure 3, Supplementary Figures S1 and S2). SGK1 and α-syn expression patterns were found to be opposite in the surrounding colon tissue, with the expression level of SGK1 decreasing (Figure 3c,f) and that of α-syn increasing in the MPTP group (Figure 3i,l).

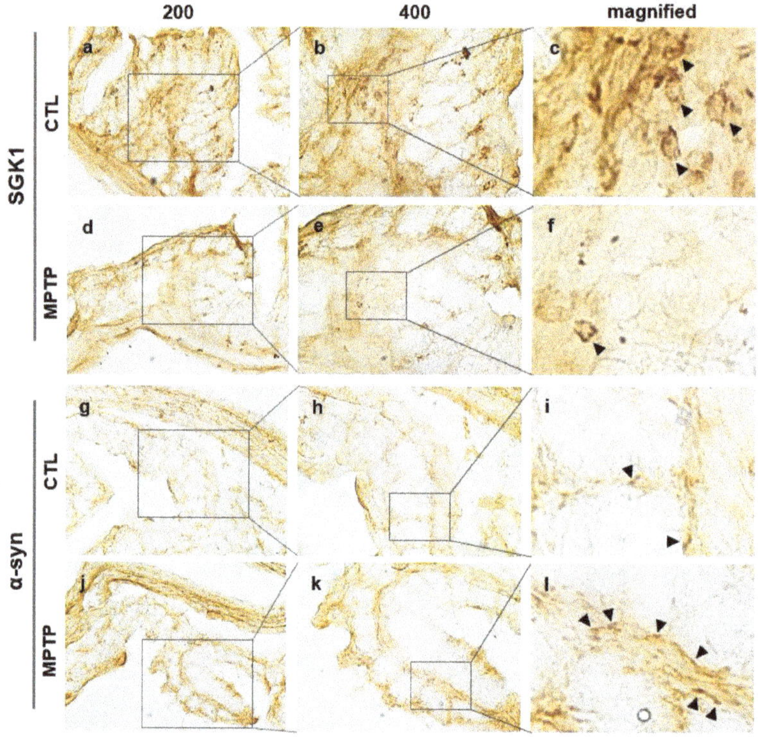

Figure 3. Changes in the expression pattern of serum/glucocorticoid-related kinase 1 (SGK1) and α-synuclein (α-syn) in the colon in 1-methyl-4-phenyl-1,2,3,6-tetrahydropyridine (MPTP)-induced Parkinson's disease model. Immunohistochemistry analyses of SGK1 (**a–f**) showing decreased SGK1 expression in the MPTP group (**d–f**) compared with that in the control (CTL) group (**a–c**). Immunohistochemistry analyses of α-syn (**g–i**) showing increased α-syn expression in the MPTP group (**j–l**) compared with that in the control (CTL) group (**g–i**). SGK1 and α-syn are indicated by black arrows in panels showing magnified images (**c**,**f**,**i**,**l**). CTL, intraperitoneally injected with 100 μL 0.9% saline daily for 4 weeks; MPTP, intraperitoneally injected with MPTP-HCl (20 mg/kg) daily for 4 weeks.

The expression patterns of SGK1 and α-syn were compared in the CTL and MPTP groups (Figure 4). The expression level of SGK1 significantly decreased, and that of α-syn significantly increased in the MPTP group compared with that in the CTL group (Figure 4B).

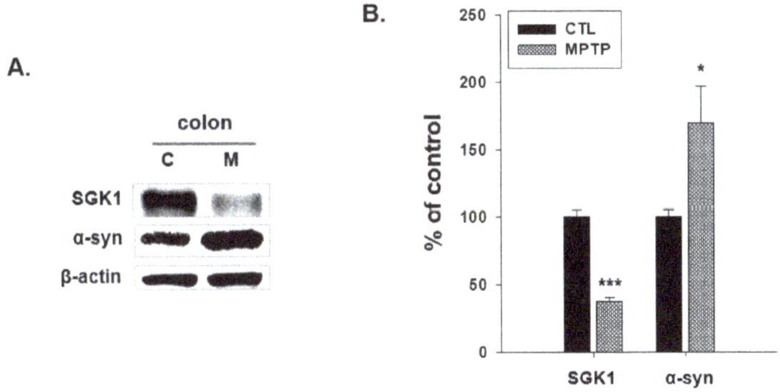

Figure 4. Changes in the expression pattern of serum/glucocorticoid-related kinase 1 (SGK1) and α-synuclein (α-syn) expression in the colon in 1-methyl-4-phenyl-1,2,3,6-tetrahydropyridine (MPTP)-induced Parkinson's disease model. (**A**) Immunoblot analyses showing decreased SGK1 expression. In contrast, α-syn expression increased in the colon in the MPTP group compared with that in the control (CTL) group. (**B**) Graphs of the immunoblot analysis results. CTL, intraperitoneally injected with 100 μL 0.9% saline daily for 4 weeks; MPTP, intraperitoneally injected with MPTP-HCI (20 mg/kg) daily for 4 weeks ($n = 3$, * $p < 0.05$, *** $p < 0.0005$).

Because of the opposite expression patterns of SGK1 and α-syn (Figure 4), immunofluorescence analyses of SGK1 and α-syn were also performed in the CTL and MPTP groups to identify the correlation (Figure 5). In the MPTP group, weaker SGK1 expression (Figure 5a,f) and stronger α-syn expression (Figure 5b,g) were observed than in the CTL group. The colocation of SGK1 and α-syn was confirmed in the merged panels (Figure 5c,h), and more brightly colocated SGK1 and α-syn in the MPTP group were also identified (see white arrows in Figure 5e,j).

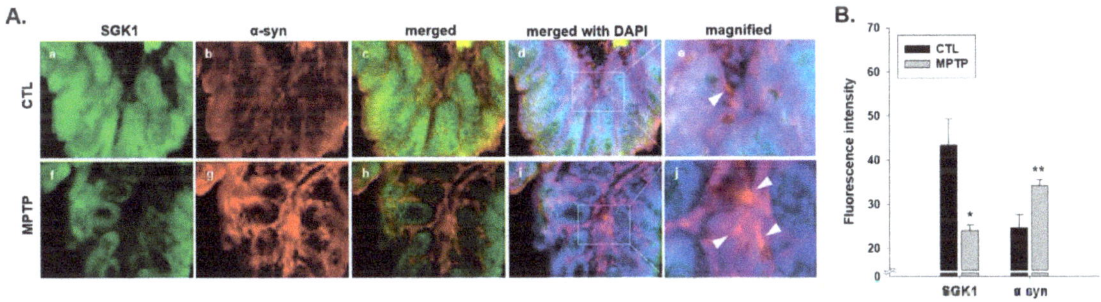

Figure 5. Immunofluorescence analyses of serum/glucocorticoid-related kinase 1 (SGK1) and α-synuclein (α-syn) in the colon in the 1-methyl-4-phenyl-1,2,3,6-tetrahydropyridine (MPTP)-induced Parkinson's disease model. (**A**) Immunofluorescence analyses of SGK1 and α-syn were performed in the control (CTL) (**a**–**e**) and MPTP groups (**f**–**j**). The expression of SGK1 decreased ((**a**,**f**); 400×), whereas that of α-syn increased ((**b**,**g**); 400×) in the MPTP group. SGK1 and α-syn were merged in each group ((**c**,**h**); 400×). Nuclei were stained with DAPI ((**d**,**i**); 400×). α-Syn is indicated by white arrows in panels showing magnified images (**e**,**j**). (**B**) The quantification of fluorescence intensity showing SGK1 and α-syn is indicated in a graph. CTL, intraperitoneally injected with 100 μL 0.9% saline daily for 4 weeks, MPTP; intraperitoneally injected with MPTP-HCI (20 mg/kg) daily for 4 weeks ($n = 3$, * $p < 0.05$, ** $p < 0.005$).

It can be inferred that these morphological changes may affect the function of the colon, and water absorption may be closely related to SGK1, which is involved in the Na^+/K^+ pump. Thus, the factors associated with SGK1 in the Na^+/K^+ pump and Na^+/K^+ pump ATPase α1 and β1 were examined in the colons of the CTL and MPTP groups. Immunoblotting analysis indicated that ATPase α1 expression levels were significantly increased, and ATPase β1 expression levels were significantly decreased in the MPTP group (Figure 6). Immunoblot results (Figure 6A) demonstrated a significantly increased expression of ATPase α1 and a decreased expression of ATPase β1 in the MPTP group (Figure 6B).

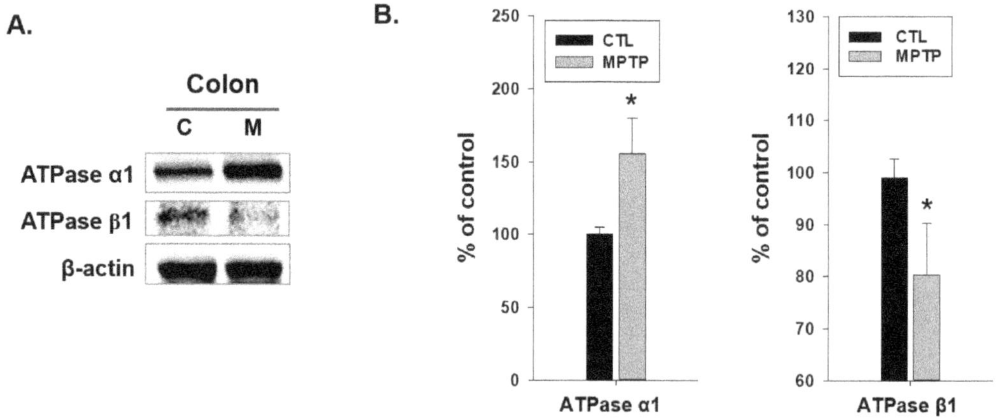

Figure 6. Changes in the expression pattern of Na^+/K^+ pump ATPase α1 and β1 in the colon in 1-methyl-4-phenyl-1,2,3,6-tetrahydropyridine (MPTP)-induced Parkinson's disease model. (**A**) Immunoblotting analyses of Na^+/K^+ pump ATPase α1 and β1 expression in the colon. ATPase α1 expression increased significantly, whereas ATPase β1 expression decreased in the colon in the MPTP group compared with that in the control (CTL) group. (**B**) Graphs indicating the immunoblotting analysis results ($n = 3$, * $p < 0.05$).

Changed expression levels of ATPase α1 and β1 were confirmed in the MPTP-induced PD model. It was also necessary to investigate SGK1-induced changes in ATPase α1 and β1 and the correlation between SGK1 and α-syn.

2.4. Expressional Changes in Factors Related to SGK1 Following SW480 Knockdown by SGK1 siRNA

Immunoblotting analysis was performed in SW480 cells with SGK1 knockdown to examine the changes in Na^+/K^+ pump ATPase α1, β1, and α-syn levels. Decreasing SGK1 levels corresponded with significant increases in the expression levels of α-syn and ATPase α1 (Figure 7). ATPase β1 decrease was significant in the SGK1 siRNA 10 group but not in the SGK1 siRNA 100 group (Figure 7B).

The results indicate that SGK1 affected α-syn and Na^+/K^+ pump ATPase α1 and β1 expression. Therefore, it can be inferred that SGK1 is an important factor influencing the increase in α-syn in the PD colon and affects the Na^+/K^+ pump system, leading to GI dysfunction associated with water absorption.

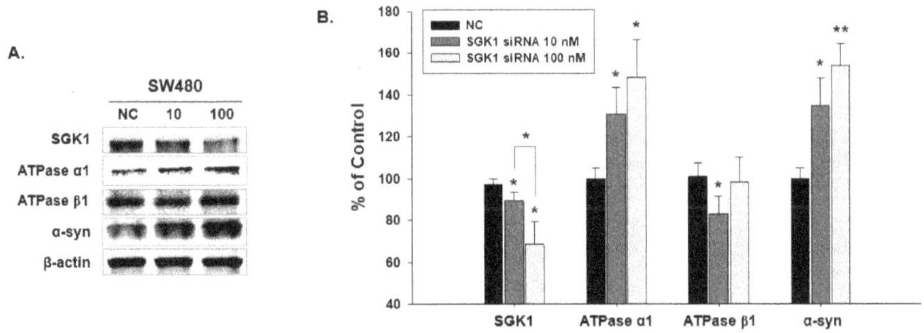

Figure 7. Changes in the expression pattern of serum/glucocorticoid-related kinase 1 (SGK1), α-synuclein (α-syn), and factors related to SGK1 in SW480 knocked down by SGK1 short interfering RNA (siRNA). (**A**) Immunoblotting analyses of SGK1, α-syn, and SGK1-related factors in SGK1-knockdown SW480 cells. Na^+/K^+ pump ATPase α1 and α-syn expression increased with decreasing SGK1 expression. The Na^+/K^+ pump ATPase β1 expression decreased in the 10 nM SGK1 siRNA group. (**B**) Bar graphs of the immunoblotting analysis results. NC, control siRNA treatment (100 nM for 2 days); siSGK1 10, SGK1 siRNA treatment (10 nM for 2 days); siSGK1 100, SGK1 siRNA treatment (100 nM for 2 days) ($n = 3$, * $p < 0.05$, ** $p < 0.005$).

3. Discussion

The exact cause and mechanism of PD remain unclear. Nonetheless, the elevated levels of α-syn in the intestines are likely to play a pivotal role in the underlying pathology of PD. α-Syn in the intestine is reported to migrate to the brain and influence the pathology of PD. Therefore, increased levels of α-syn in the intestine may also increase its transport to the brain, which may affect dopaminergic neurons. Additionally, intestinal α-syn is likely to play a role in gastrointestinal diseases, which are non-motor symptoms of PD.

In the current study, α-syn expression levels were increased and SGK1 expression levels were decreased in the colon of the MPTP-induced semi-chronic PD model. As reports have highlighted, a strong association of pervasive α-syn aggregations in the gastrointestinal tract with PD [8–11] was detected, and the SGK1 expression level was decreased in the colon. SGK1 and α-syn expression patterns were found to be opposite in the surrounding colon tissue, with the expression level of SGK1 decreasing and that of α-syn increasing in the MPTP group. Immunofluorescence analyses of SGK1 and α-syn showed the colocation of SGK1 and α-syn. In the MPTP group, weaker SGK1 and stronger α-syn expression were observed compared with that in the CTL group. Furthermore, the Alcian blue staining revealed an uneven expression of mucins in the villi and microvilli of the MPTP group. These morphological changes can affect the function of the colon, and it can be inferred that water absorption could be strongly associated with SGK1, which in turn is associated with the Na^+/K^+ pump. SGK1 increases Na^+/K^+ pump ATPase cell-surface expression [32]. Thus, the factors associated with SGK1 in the Na^+/K^+ pump, Na^+/K^+ pump ATPase α1 and β1, expressed in the colon, were examined in the colons of the CTL and MPTP groups. Immunoblotting analysis indicated that Na^+/K^+ pump ATPase α1 expression levels were significantly increased, and those of Na^+/K^+ pump ATPase β1 were significantly decreased in the MPTP group compared with those in the CTL group. In SW480 cells with SGK1 knockdown, decreasing SGK1 levels corresponded with significant increases in the expression levels of α-syn and Na^+/K^+ pump ATPase α1. Na^+/K^+ pump ATPase β1 decrease was also significant in the SGK1 siRNA 10 group. These results indicate that SGK1 affected the α-syn and Na^+/K^+ pump ATPase α1 and β1. Na^+/K^+ pump ATPase is reportedly involved in water balance in the colon [33,34]. Na^+-K^+-ATPase α1 plays a key role in sodium water transport [35]. Na^+/K^+ pump ATPase β1 is required for the structural and functional maturation of Na^+/K^+ pump ATPase α1 [36,37]. The increase in Na^+-K^+-ATPase α1 induced by SGK1 reduction may reduce the need for Na^+/K^+ pump ATPase β1, leading to a decrease in Na^+/K^+ pump ATPase β1. The abnormality of these intestinal functions is one of the major non-motor symptoms of PD. Moreover, a mutation affecting Na^+/K^+ pump ATPase is associated with neurodegenerative disorders like early-onset PD [38,39]. Therefore, it can be deduced that SGK1, affecting intestinal α-syn and the Na^+/K^+ pump system, is an important factor related to the pathology of PD.

Constipation is one of the important prodromal and intestinal non-motor symptoms of PD. In addition to bowel movement, fluid and electrolyte balance is an important factor in constipation. SGK1 is a powerful stimulator of Na^+/K^+ pump ATPase [40,41]. SGK1-sensitive functions do not require the presence of SGK1 but are significantly regulated by SGK1 [40]. Fluid and electrolyte balance changes due to reduced SGK1 may affect the colon. Furthermore, SGK1 influences potassium exchange and balance within the colon, which can further affect fecal characteristics. These results indicate that SGK1 may affect the colon by regulating Na^+/K^+ pump ATPase, which is involved in maintaining electrolyte balance and fecal formation in the PD mouse model.

In a previous study, SGK1 inhibition in astrocytes and microglia ameliorated the pathology and symptoms in an animal model of PD [42]. Brain-resident glia can contribute to the progression of neuronal loss in neurodegenerative disorders by establishing a disease-harmful inflammatory environment [43]. Therefore, SGK1 inhibition in astrocytes and microglia can induce the apoptosis of these cells, thereby exerting neuroprotective effects. The injection of SGK1-expressing adenovirus into the ST of MPTP-treated mice exerted a protective effect on dopaminergic neurons [44], and the overexpression of SGK1 prevented dopaminergic cell death [23–26]. The underlying mechanism may involve the inhibition of dopaminergic cell death via the SGK1-based regulation of apoptosis [21,43]. The upregulation of SKG1, which coincides with the onset of brain dopaminergic neuron death in PD mouse models [45], may regulate apoptosis. Acute treatment with MPTP leads to a strong upregulation of SKG1 in the cytoplasmic fraction [22]. The upregulation of SGK1 strongly correlates with the occurrence of cell death [22]. This upregulation of SKG1 has been reported to play a protective role during oxidative stress conditions [22]. Downregulation studies of SKG1 showed increased cell death after oxidative stress. SKG1 forms part of the molecular pathway of cell death and seems to exert a protective role [22]. Chronic treatment with MPTP showed decreased expression of SKG1-mRNA and protein. In chronic PD, the neuroprotective mechanism involving the increased expression of SKG1 no longer operates optimally, leading to worsened neurodegeneration [28]. Moreover, SGK induces a wide range of neurotrophic effects, including the induction of neuronal hypertrophy, protection against neuronal apoptosis, and neurotoxin-induced axonal regeneration [44,46]. Therefore, reduced SGK1 levels may imply that this neuroprotective effect is no longer exhibited and that the disease may show accelerated deterioration.

In our study, a decrease in SGK1 was observed in the colon of the PD mouse model. SGK1 decrease seems to be related to the increase in intestinal α-syn, which could migrate to the brain and influence the degeneration of dopaminergic neurons. Additionally, intestinal α-syn increase is related to intestinal non-motor symptoms. Particularly, SGK1-sensitive Na^+/K^+ pump ATPase α1 and β1 changes are inferred to be one of the elements of intestinal dysfunction because electrolyte balance and fecal formation, which are related to constipation, could be affected by SGK1 decrease. Therefore, as currently there are no treatment or prevention methods for PD, controlling the expression level of SGK1 can be a potential approach for treating/preventing PD. To confirm this, future studies must investigate the direct relationship between SGK1 reduction and constipation in PD.

4. Materials and Methods

4.1. MPTP Model of PD

Twelve 6-week-old male C57BL/6 mice (weighting 20–22 g) were obtained from Dae Han Bio-Link Co. (Daejeon, Republic of Korea). They were divided into two groups of six mice each: CTL and MPTP group. The CTL group was intraperitoneally injected with 0.9% saline (100 μL), whereas the MPTP group was intraperitoneally injected with MPTP-HCI (20 mg/kg, Sigma, St. Louis, MO, USA) in 0.9% saline solution (100 μL) once a day for

4 weeks to develop a chronic PD model. The day after the last MPTP injection, both groups of mice were anesthetized using Alfaxan (JUROX, Braeside, Australia) and perfused with cold 0.05 M sodium phosphate buffer to perform Western blotting. The Sangji University Animal Experimentation Committee approved all the animal protocols used in this study (#2021-9).

4.2. Immunohistochemistry

The mice were dissected after four weeks of treatment, and the tissues were fixed using 4% paraformaldehyde for one day at 4 °C and then dehydrated in 30% sucrose buffer for three days at 4 °C. For immunohistochemical analysis, whole SN, ST, and colonic tissues were processed in a cryomicrotome and incubated in 3% H_2O_2 with PBS (pH 7.4). The sections were then blocked using an avidin/biotin blocking kit (Vector Laboratories, Newark, CA, USA) for 1 h and incubated with M.O.M mouse Ig-blocking reagent (Vector Laboratories, Newark, CA, USA) for 1 h at room temperature; this was followed by incubation with a primary antibody. Each section was stained with anti-TH (1:2000; Santa Cruz Biotechnology, Dallas, TX, USA), anti-α-syn (1:1000; Novus Biologicals, Littleton, CO, USA), or anti-SGK1 (1:1000; Santa Cruz Biotechnology, Dallas, TX, USA). The sections were then treated with biotinylated anti-mouse IgG and avidin–biotin–peroxidase complexes. Diaminobenzidine hydrogen peroxide solution was used to visualize the antigens. The cells in the SN corresponded to dopaminergic cells. The concentration of TH in the ST was measured using Image J (version 1.52a; NIH, Bethesda, MD, USA).

4.3. Alcian Blue Staining

The colon tissues were incubated in 3% acetic acid solution for 3 min and in Alcian blue solution (pH 2.5) for 30 min. The sections were dehydrated using graded ethanol. The Alcian Blue Stain Kit (H-3501, Vector Laboratories, Newark, CA, USA) was used, and recommended procedures were followed.

4.4. Western Blotting

For analysis, SN, ST, and colonic tissues were homogenized using 20 mM radio-immunoprecipitation assay buffer and incubated on ice for 30 min. SW480 cells were lysed using a lysis buffer. After centrifugation at 12,000× *g* rpm at 4 °C for 20 min, an equal amount of protein supernatant was separated using sodium dodecyl sulfate-polyacrylamide gel electrophoresis and then transferred onto polyvinylidene difluoride membranes (Pall Life Science, Port Washington, NY, USA). The membranes were then blocked using 3% bovine serum albumin (BSA) for 1 h and 30 min and incubated with anti-TH (1:2000), anti-SGK1 (1:2000), anti-β-actin (1:2000; Santa Cruz Biotechnology, Dallas, TX, USA), anti-Na^+/K^+ pump ATPase α1 (1:2000; Santa Cruz Biotechnology, USA), anti-Na^+/K^+ pump ATPase β1 (1:2000; Santa Cruz Biotechnology, USA), or anti-α-syn (1:500; BD Biosciences, Franklin Lakes, NJ, USA) primary antibody. The membranes were washed three times with 0.1% Tris-buffered saline (20 mM Tris-HCl (pH 7.5) and 150 mM NaCl) containing 0.1% Tween-20 and treated with goat anti-mouse IgG (1:5000; Abcam, Cambridge, UK) as a secondary antibody for 1 h. Afterward, the membranes were treated with Pierce ECL Western detection reagents (GE Healthcare, Chicago, IL, USA), and the antigen bands were visualized with an Alliance Q9 Micro chemiluminescence imaging device (UVITEC, Cambridge, UK)

4.5. Immunofluorescence Microscopy

Cryosectioned colon regions were washed with cold PBS and incubated in 0.3% Triton X-100 for 30 min. After incubation in blocking buffer (1% BSA, 5% goat serum in PBS) for 1 h, sections were incubated with mouse anti-SGK1 and rabbit anti-α-syn primary antibodies. After overnight incubation at 4 °C, the sections were treated with anti-mouse IgG-Alexa Fluor 488 or anti-rabbit IgG-tetramethylrhodamine secondary antibodies. Then, colonic sections were treated with 4′,6-diamidino-2-phenylindole (1 μg/mL). A Nikon X-cite series 120Q microscope (Nikon, Tokyo, Japan) was used to visualize the samples.

4.6. Cell Lines and Cultures

The SW480 cell line (Korean Cell Line Bank, Seoul, Republic of Korea) was cultured in Roswell Park Memorial Institute medium (WELGENE, Seoul, Republic of Korea) containing 10% fetal bovine serum (GenDEPOT, Baker, TX, USA), 100 U/mL of penicillin, and 100 mg/mL streptomycin and maintained under standard culture conditions (5% CO_2, 37 °C).

4.7. siRNA Transfection

For transient knockdown, SGK1 siRNA (5′-AGG AGA ACA UCG AGC ACA ATT-3′) and negative control siRNA (5′-UUC UCC GAA CGU GUC ACG UTT-3′) were used. The siRNAs were purchased from Bioneer, Inc., Daejeon, Republic of Korea.

4.8. Statistical Analysis

Statistical comparisons were conducted using Student's *t*-test for two-group comparisons and one-way analysis of variance for three-group comparisons with SPSS 25 software (SPSS Inc. Released 2017). Differences yielding *p*-values of <0.05 were considered statistically significant. All values are expressed as means ± the standard error.

5. Conclusions

In conclusion, the expression of α-syn increased and that of SGK1 decreased in the colon of the MPTP-induced PD mouse model. The decreased expression of SGK1 induced an increase in α-syn and changes in the expression of Na^+/K^+ pump ATPase α1 and β1. These results suggest that reduced SGK1 may affect the colon by regulating Na^+/K^+ pump ATPase α1 and β1, thereby affecting electrolyte balance and fecal formation in the PD mouse model. Gastrointestinal disorders, especially constipation, are some of the major prodromal symptoms of PD. Therefore, modulating SGK1 expression could be an important strategy to control PD. This is the first study to investigate the relationship between SGK1 and intestinal changes in a PD mouse model.

Supplementary Materials: The following supporting information can be downloaded at: https://www.mdpi.com/article/10.3390/ijms242216408/s1.

Author Contributions: Conceptualization, S.Y.; methodology, S.Y., D.K. and M.H.S.; software, M.H.S.; validation, S.Y.; formal analysis, S.Y., D.K. and M.H.S.; investigation, S.Y.; resources, S.Y. and M.H.S.; data curation, S.Y. and M.H.S.; writing—original draft preparation, S.Y., D.K. and M.H.S.; writing—review and editing, S.Y. and S.-H.K.; visualization, S.Y.; supervision, S.Y.; project administration, S.Y. and S.-H.K.; funding acquisition, S.Y. and S.-H.K. All authors have read and agreed to the published version of the manuscript.

Funding: This research was funded by a National Research Foundation of Korea (NRF) grant funded by the Korean Government's Ministry of Science, ICT and Future Planning (MSIP; grant number 2020R1I1A3A04036500), and a grant from the Korea Health Technology R&D Project through the Korea Health Industry Development Institute (KHIDI), funded by the Ministry of Health and Welfare, Korea (grant number HF20C0019).

Institutional Review Board Statement: This study was approved by the Institutional Animal Care and Use Committee of Sang Ji University (approval code: 2021-9; date of approval: 5 March 2021).

Informed Consent Statement: Not applicable.

Data Availability Statement: All necessary data are included in the paper.

Conflicts of Interest: The authors declare no conflict of interest.

References

1. Tolosa, E.; Wenning, G.; Poewe, W. The diagnosis of Parkinson's disease. *Lancet Neurol.* **2006**, *5*, 75–86. [CrossRef]
2. Kalia, L.V.; Lang, A.E. Parkinson's disease. *Lancet* **2015**, *386*, 896–912. [CrossRef]
3. Jost, W.H.; Schimrigk, K. Constipation in Parkinson's disease. *Klin. Wochenschr.* **1991**, *69*, 906–909. [CrossRef] [PubMed]
4. Jost, W.H. Gastrointestinal motility problems in patients with Parkinson's disease. Effects of antiparkinsonian treatment and guidelines for management. *Drugs Aging* **1997**, *10*, 249–258. [CrossRef] [PubMed]
5. Pfeiffer, R.F. Gastrointestinal dysfunction in Parkinson's disease. *Lancet Neurol.* **2003**, *2*, 107–116. [CrossRef] [PubMed]
6. Anderson, G.; Noorian, A.R.; Taylor, G.; Anitha, M.; Bernhard, D.; Srinivasan, S.; Greene, J.G. Loss of enteric dopaminergic neurons and associated changes in colon motility in an MPTP mouse model of Parkinson's disease. *Exp. Neurol.* **2007**, *207*, 4–12. [CrossRef]
7. Sakakibara, R.; Odaka, T.; Uchiyama, T.; Asahina, M.; Yamaguchi, K.; Yamaguchi, T.; Yamanishi, T.; Hattori, T. Colonic transit time and rectoanal videomanometry in Parkinson's disease. *J. Neurol. Neurosurg. Psychiatry* **2003**, *74*, 268–272. [CrossRef]
8. Devos, D.; Lebouvier, T.; Lardeux, B.; Biraud, M.; Rouaud, T.; Pouclet, H.; Coron, E.; Bruley des Varannes, S.; Naveilhan, P.; Nguyen, J.M.; et al. Colonic inflammation in Parkinson's disease. *Neurobiol. Dis.* **2013**, *50*, 42–48. [CrossRef]
9. Fasano, A.; Visanji, N.P.; Liu, L.W.; Lang, A.E.; Pfeiffer, R.F. Gastrointestinal dysfunction in Parkinson's disease. *Lancet Neurol.* **2015**, *14*, 625–639. [CrossRef]
10. Shannon, K.M.; Keshavarzian, A.; Dodiya, H.B.; Jakate, S.; Kordower, J.H. Is alpha-synuclein in the colon a biomarker for premotor Parkinson's disease? Evidence from 3 cases. *Mov. Disord.* **2012**, *27*, 716–719. [CrossRef]
11. Pouclet, H.; Lebouvier, T.; Coron, E.; des Varannes, S.B.; Rouaud, T.; Roy, M.; Neunlist, M.; Derkinderen, P. A comparison between rectal and colonic biopsies to detect Lewy pathology in Parkinson's disease. *Neurobiol. Dis.* **2012**, *45*, 305–309. [CrossRef]
12. Braak, H.; Del Tredici, K.; Rub, U.; de Vos, R.A.; Jansen Steur, E.N.; Braak, E. Staging of brain pathology related to sporadic Parkinson's disease. *Neurobiol. Aging* **2003**, *24*, 197–211. [CrossRef] [PubMed]
13. Visanji, N.P.; Marras, C.; Hazrati, L.N.; Liu, L.W.; Lang, A.E. Alimentary, my dear Watson? The challenges of enteric alpha-synuclein as a Parkinson's disease biomarker. *Mov. Disord.* **2014**, *29*, 444–450. [CrossRef] [PubMed]
14. Beach, T.G.; Corbille, A.G.; Letournel, F.; Kordower, J.H.; Kremer, T.; Munoz, D.G.; Intorcia, A.; Hentz, J.; Adler, C.H.; Sue, L.I.; et al. Multicenter Assessment of Immunohistochemical Methods for Pathological Alpha-Synuclein in Sigmoid Colon of Autopsied Parkinson's Disease and Control Subjects. *J. Parkinson's Dis.* **2016**, *6*, 761–770. [CrossRef] [PubMed]
15. Gold, A.; Turkalp, Z.T.; Munoz, D.G. Enteric alpha-synuclein expression is increased in Parkinson's disease but not Alzheimer's disease. *Mov. Disord.* **2013**, *28*, 237–240. [CrossRef] [PubMed]
16. Yu, S.; Zuo, X.; Li, Y.; Zhang, C.; Zhou, M.; Zhang, Y.A.; Ueda, K.; Chan, P. Inhibition of tyrosine hydroxylase expression in alpha-synuclein-transfected dopaminergic neuronal cells. *Neurosci. Lett.* **2004**, *367*, 34–39. [CrossRef]
17. Beekes, M. The Neural Gut-Brain Axis of Pathological Protein Aggregation in Parkinson's Disease and Its Counterpart in Peroral Prion Infections. *Viruses* **2021**, *13*, 1394. [CrossRef]
18. Cockfield, J.A.; Schafer, Z.T. SGK1-regulated metabolism: Key for the survival of cells detached from the extracellular matrix. *Mol. Cell Oncol.* **2021**, *8*, 1976583. [CrossRef]
19. Mercado, E.; Paniagua, N.; Sanchez-Robles, E.M.; Giron, R.; Alvarez de la Rosa, D.; Giraldez, T.; Goicoechea, C. SGK1.1 isoform is involved in nociceptive modulation, offering a protective effect against noxious cold stimulus in a sexually dimorphic manner. *Pharmacol. Biochem. Behav.* **2022**, *212*, 173302. [CrossRef]
20. Park, J.C.; Jeon, Y.J.; Jang, Y.S.; Cho, J.; Choi, D.H.; Han, J.S. SGK1 knockdown in the medial prefrontal cortex reduces resistance to stress-induced memory impairment. *Eur. Neuropsychopharmacol.* **2021**, *45*, 29–34. [CrossRef]
21. Lang, F.; Strutz-Seebohm, N.; Seebohm, G.; Lang, U.E. Significance of SGK1 in the regulation of neuronal function. *J. Physiol.* **2010**, *588*, 3349–3354. [CrossRef] [PubMed]
22. Schoenebeck, B.; Bader, V.; Zhu, X.R.; Schmitz, B.; Lubbert, H.; Stichel, C.C. Sgk1, a cell survival response in neurodegenerative diseases. *Mol. Cell Neurosci.* **2005**, *30*, 249–264. [CrossRef] [PubMed]
23. Hollister, R.D.; Page, K.J.; Hyman, B.T. Distribution of the messenger RNA for the extracellularly regulated kinases 1, 2 and 3 in rat brain: Effects of excitotoxic hippocampal lesions. *Neuroscience* **1997**, *79*, 1111–1119. [CrossRef] [PubMed]
24. Lang, F.; Cohen, P. Regulation and physiological roles of serum- and glucocorticoid-induced protein kinase isoforms. *Sci. STKE* **2001**, *2001*, re17. [CrossRef]
25. Rangone, H.; Poizat, G.; Troncoso, J.; Ross, C.A.; MacDonald, M.E.; Saudou, F.; Humbert, S. The serum- and glucocorticoid-induced kinase SGK inhibits mutant huntingtin-induced toxicity by phosphorylating serine 421 of huntingtin. *Eur. J. Neurosci.* **2004**, *19*, 273–279. [CrossRef]
26. Radi, E.; Formichi, P.; Battisti, C.; Federico, A. Apoptosis and oxidative stress in neurodegenerative diseases. *J. Alzheimer's Dis.* **2014**, *42* (Suppl. S3), S125–S152. [CrossRef]
27. Heo, E.J.; Lee, Y.; Hyung Seo, M.; Yeo, S. Association between SGK1 and alpha-synuclein in skeletal muscle in an MPTP-induced Parkinson's disease model. *Neurosci. Lett.* **2023**, *814*, 137464. [CrossRef]
28. Yeo, S.; Sung, B.; Hong, Y.M.; van den Noort, M.; Bosch, P.; Lee, S.H.; Song, J.; Park, S.K.; Lim, S. Decreased expression of serum- and glucocorticoid-inducible kinase 1 (SGK1) promotes alpha-synuclein increase related with down-regulation of dopaminergic cell in the Substantia Nigra of chronic MPTP-induced Parkinsonism mice and in SH-SY5Y cells. *Gene* **2018**, *661*, 189–195. [CrossRef]

29. Alvarez de la Rosa, D.; Gimenez, I.; Forbush, B.; Canessa, C.M. SGK1 activates Na+-K+-ATPase in amphibian renal epithelial cells. *Am. J. Physiol. Cell Physiol.* **2006**, *290*, C492–C498. [CrossRef]
30. Henke, G.; Setiawan, I.; Bohmer, C.; Lang, F. Activation of Na+/K+-ATPase by the serum and glucocorticoid-dependent kinase isoforms. *Kidney Blood Press. Res.* **2002**, *25*, 370–374. [CrossRef]
31. Koumangoye, R.; Omer, S.; Kabeer, M.H.; Delpire, E. Novel Human NKCC1 Mutations Cause Defects in Goblet Cell Mucus Secretion and Chronic Inflammation. *Cell Mol. Gastroenterol. Hepatol.* **2020**, *9*, 239–255. [CrossRef] [PubMed]
32. Zecevic, M.; Heitzmann, D.; Camargo, S.M.; Verrey, F. SGK1 increases Na,K-ATP cell-surface expression and function in Xenopus laevis oocytes. *Pflug. Arch.* **2004**, *448*, 29–35. [CrossRef]
33. Markossian, S.; Kreydiyyeh, S.I. TNF-alpha down-regulates the Na+-K+ ATPase and the Na+-K+-2Cl-cotransporter in the rat colon via PGE2. *Cytokine* **2005**, *30*, 319–327. [CrossRef] [PubMed]
34. Allgayer, H.; Brown, L.; Kruis, W.; Erdmann, E.; Paumgartner, G. Inhibition of human colonic (Na+ + K+)-ATPase by arachidonic and linoleic acid. *Naunyn Schmiedeberg's Arch. Pharmacol.* **1986**, *332*, 398–402. [CrossRef] [PubMed]
35. Wen, X.; Wan, Q. Research advances on the role of Na(+)-K(+)-ATPase regulation in pulmonary edema clearance of acute respiratory distress syndrome. *Zhonghua Wei Zhong Bing Ji Jiu Yi Xue* **2021**, *33*, 1011–1016. [PubMed]
36. Geering, K.; Beggah, A.; Good, P.; Girardet, S.; Roy, S.; Schaer, D.; Jaunin, P. Oligomerization and maturation of Na,K-ATPase: Functional interaction of the cytoplasmic NH2 terminus of the beta subunit with the alpha subunit. *J. Cell Biol.* **1996**, *133*, 1193–1204. [CrossRef] [PubMed]
37. Geering, K. The functional role of beta subunits in oligomeric P-type ATPases. *J. Bioenerg. Biomembr.* **2001**, *33*, 425–438. [CrossRef]
38. Ashmore, L.J.; Hrizo, S.L.; Paul, S.M.; Van Voorhies, W.A.; Beitel, G.J.; Palladino, M.J. Novel mutations affecting the Na, K ATPase alpha model complex neurological diseases and implicate the sodium pump in increased longevity. *Hum. Genet.* **2009**, *126*, 431–447. [CrossRef]
39. Holm, T.H.; Lykke-Hartmann, K. Insights into the Pathology of the alpha3 Na(+)/K(+)-ATPase Ion Pump in Neurological Disorders; Lessons from Animal Models. *Front. Physiol.* **2016**, *7*, 209. [CrossRef]
40. Lang, F.; Stournaras, C.; Alesutan, I. Regulation of transport across cell membranes by the serum- and glucocorticoid-inducible kinase SGK1. *Mol. Membr. Biol.* **2014**, *31*, 29–36. [CrossRef]
41. Setiawan, I.; Henke, G.; Feng, Y.; Bohmer, C.; Vasilets, L.A.; Schwarz, W.; Lang, F. Stimulation of Xenopus oocyte Na(+),K(+)ATPase by the serum and glucocorticoid-dependent kinase sgk1. *Pflug. Arch.* **2002**, *444*, 426–431. [CrossRef]
42. Kwon, O.C.; Song, J.J.; Yang, Y.; Kim, S.H.; Kim, J.Y.; Seok, M.J.; Hwang, I.; Yu, J.W.; Karmacharya, J.; Maeng, H.J.; et al. SGK1 inhibition in glia ameliorates pathologies and symptoms in Parkinson disease animal models. *EMBO Mol. Med.* **2021**, *13*, e13076. [CrossRef] [PubMed]
43. Lang, F.; Shumilina, E. Regulation of ion channels by the serum- and glucocorticoid-inducible kinase SGK1. *FASEB J.* **2013**, *27*, 3–12. [CrossRef] [PubMed]
44. Iqbal, S.; Howard, S.; LoGrasso, P.V. Serum- and Glucocorticoid-Inducible Kinase 1 Confers Protection in Cell-Based and in In Vivo Neurotoxin Models via the c-Jun N-Terminal Kinase Signaling Pathway. *Mol. Cell Biol.* **2015**, *35*, 1992–2006. [CrossRef] [PubMed]
45. Stichel, C.C.; Schoenebeck, B.; Foguet, M.; Siebertz, B.; Bader, V.; Zhu, X.R.; Lubbert, H. sgk1, a member of an RNA cluster associated with cell death in a model of Parkinson's disease. *Eur. J. Neurosci.* **2005**, *21*, 301–316. [CrossRef]
46. Chen, X.; Tagliaferro, P.; Kareva, T.; Yarygina, O.; Kholodilov, N.; Burke, R.E. Neurotrophic effects of serum- and glucocorticoid-inducible kinase on adult murine mesencephalic dopamine neurons. *J. Neurosci.* **2012**, *32*, 11299–11308. [CrossRef]

International Journal of
Molecular Sciences

MDPI

Article

YKL-40 and the Cellular Metabolic Profile in Parkinson's Disease

Maria Gevezova [1,2,†], Maria Kazakova [1,2,*,†], Anastasia Trenova [3,4] and Victoria Sarafian [1,2]

1 Department of Medical Biology, Medical University, 4000 Plovdiv, Bulgaria; mariya.gevezova@mu-plovdiv.bg (M.G.); victoria.sarafian@mu-plovdiv.bg (V.S.)
2 Research Institute at MU-Plovdiv, 4000 Plovdiv, Bulgaria
3 Department of Neurology, Medical University, 4000 Plovdiv, Bulgaria; anastasiya.trenova@mu-plovdiv.bg
4 University Hospital "Kaspela", 4000 Plovdiv, Bulgaria
* Correspondence: mariya.kazakova@mu-plovdiv.bg; Tel.: +897-571-982
† These authors contributed equally to this work.

Abstract: Parkinson's disease (PD) is the second most common neurodegenerative disease worldwide. A growing body of evidence suggests that mitochondrial dysfunction and inflammation play a crucial role as a pathogenetic mechanism in PD. The glycoprotein YKL-40 (CHI3L1) is a potential biomarker involved in inflammation and tumor processes. The aim of the present study was to investigate the metabolic profile of PBMCs from PD patients and to search for a possible relationship between cellular bioenergetics and YKL-40. The study included 18 naïve PD patients and an age-matched control group (HC, n = 7). Patients were diagnosed according to the MDS-PD, the UPDRS, and the Hoen–Yahr scales. Mitochondrial activity was measured by a metabolic analyzer on isolated PBMCs from PD patients. Gene (qPCR) and protein (ELISA) expression levels of YKL40 were investigated. New data are reported revealing changes in the mitochondrial activity and YKL-40 levels in PD patients. Bioenergetic parameters showed increased respiratory reserve capacity in PD compared to HC. The protein levels of YKL-40 were threefold higher in PD. We found a correlation between the YKL-40 protein levels and basal respiration and between YKL-40 and ATP production. These observations suggest an interplay between YKL-40 and mitochondrial function in PD. We assume that the YKL-40 gene and protein levels in combination with changes in mitochondrial function might serve as an additional tool to monitor the clinical course of PD.

Keywords: Parkinson's disease; YKL-40; mitochondrial dysfunction

1. Introduction

Parkinson's disease (PD) is defined as a complex neurodegenerative disorder characterized by dopaminergic neuronal cell death in the substantia nigra [1]. It is the second most common neuropathological condition after Alzheimer's disease, associated with significant disability and poor quality of life [2]. Many factors such as aging, smoking, and alcohol consumption are thought to be related to the initiation and progression of PD [3,4].

It has been suggested that the effect of the aging process on the pathogenesis of PD is partially mediated by mitochondrial dysfunction [5]. With aging, mitochondria become sensitive to environmental factors, thus increasing the possibility of mitochondrial dysfunction. A variety of mtDNA mutations lead to a reduction in ATP production, augment free radicals and oxidative stress, and cause mitochondrial dysfunction [6]. As mitochondrial respiration is linked to ATP production, neurons and differentiated neuronal cells are dependent on oxidative phosphorylation and display a considerable mitochondrial respiratory capacity to meet the requirements of increased functional demands and stress [7]. Strong evidence for altered mitochondrial metabolism in human neurodegenerative diseases, especially in PD patients, is still missing.

The implementation of reliable secreted biomarkers associated with motor or cognitive PD subtypes could allow accurate prediction of clinical outcomes. Recently, the research on novel biomarkers in PD is expanding, and data about prognostic cerebrospinal fluid (CSF) and blood biomarkers have been reported. They could help clinicians reliably predict patients' survival and determine the best choice of therapy. Furthermore, the potential advantage of biomarkers as drug targets in clinical trials has also been discussed.

Chitinase-3-like protein 1 (CHI3L1/YKL-40) or human cartilage glycoprotein-39 is a 40-kDa glycoprotein, a member of the mammalian chitinase-like protein family. It is considered as a marker for macrophage and microglial differentiation and activation [8]. It is an extracellular matrix glycoprotein belonging to glycoside hydrolase family 18 [9]. Because of the amino acid substitutions in the active site of chitinases, YKL-40 has no catalytic activity against chitinase substrates [10]. The protein is normally expressed by multiple cell types such as macrophages, chondrocytes, and vascular smooth muscle cells. However, increased YKL-40 levels have been detected in several solid tumors and chronic inflammatory conditions [9,11,12].

Elevated CSF YKL-40 levels were reported in different infectious and noninfectious diseases of the CNS [13]. Previous studies found that the YKL-40 protein may have a potential role as a promising biomarker reflecting the severity of inflammation in PD [14]. There are no data available on YKL-40 levels and mitochondrial metabolism in PD patients.

Due to the complex nature of this disease, biomarkers with diagnostic and prognostic value are needed to reliably identify the early stages of PD and to stratify patients into clinical subtypes, to allow individualization and the more effective application of therapy. In this context, the present study aimed to search for a possible relationship between cellular bioenergetics and YKL-40 gene and protein levels.

2. Results

2.1. Patients

Eighteen, newly diagnosed according to MDS-PD criteria, treatment naïve patients (nine women, nine men), aged between 42 and 79 years (mean age 65.5 $\pm$ 11.87) were included in this study. The duration between the starting point of the symptoms, reported by the patient, and the time point of diagnosis varied between 3 months and 15 months (mean 9 $\pm$ 6.43 months). The UPDRS score was between 20 and 48 (mean 33.39 $\pm$ 8.261). According to the Hoen–Yahr scale, the degree of disability in the majority of the patients (15/18, 83.33%) was 1.5 (unilateral signs and involvement of the axial muscles), and 3 out of 18 (16.67%) had 1.0 (unilateral Parkinsonian signs only). The results from both scales indicated that all patients were in the early stage of the disease.

2.2. Mitochondrial Activity

After measuring the mitochondrial function in PD patients and healthy controls (HC), we obtained data on five metabolic parameters that best described different factors influencing the ability of PBMCs to meet cell energy needs.

Our results showed an increased spare respiratory capacity in PD patients compared to HC. This difference was statistically significant (p = 0.038) and depicted the characteristic respiration pattern of PD PBMC (Figure 1). The studied indicator in patients averaged 243% $\pm$ 1.15, while for HC, it was 168% $\pm$ 0.4. However, the rest of the bioenergetic data showed a tendency for decreased values in PD. The difference in ATP-coupled respiration (p = 0.194), coupling efficiency (p = 0.161), maximal respiration (p = 0.680), and proton leak (p = 0.445), compared to HC, was not statistically significant. In addition, both PD PBMCs and HC PBMCs had a similar mean basal respiration without any statistical significance (p = 0.794), indicating the homogenous onset and the same conditions in studying the target groups. The results are shown in Figure 1.

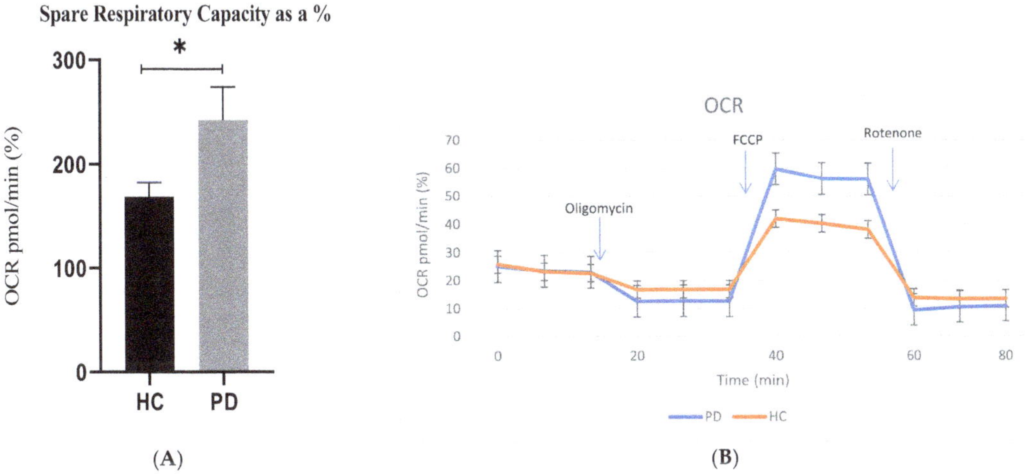

Figure 1. PBMC bioenergetic data in real time (Seahorse Xp analyzer). (**A**) The graph shows the differences between Parkinson's disease (PD) and healthy controls (HC). Values are expressed as means ± S.E.M; * = $p < 0.05$. (**B**) Profiles of Mito Stress Test data for OCR with arrows indicating injections into the media of specific inhibitors: oligomycin, carbonyl cyanite-4 (trifluoromethoxy) phenylhydrazone (FCCP), and Rotenone.

2.3. Gene and Protein Expression of YKL-40

The expression levels of YKL-40 showed only a trend for decreased mRNA levels in patients compared to the controls without any statistical significance ($p = 0.651$) in contrast to the protein levels ($p = 0.0001$) (Figure 2). The mean YKL-40 mRNA values in PD were 1.125 ± 0.6, while in HC they were 1.317 ± 0.4. Interestingly, YKL-40 plasma levels in patients with PD were three times higher than those of the HC. The reported mean values for protein levels in PD were 164.70 ng/mL ± 79.7, compared to the control group—46.06 ng/mL ± 19.1.

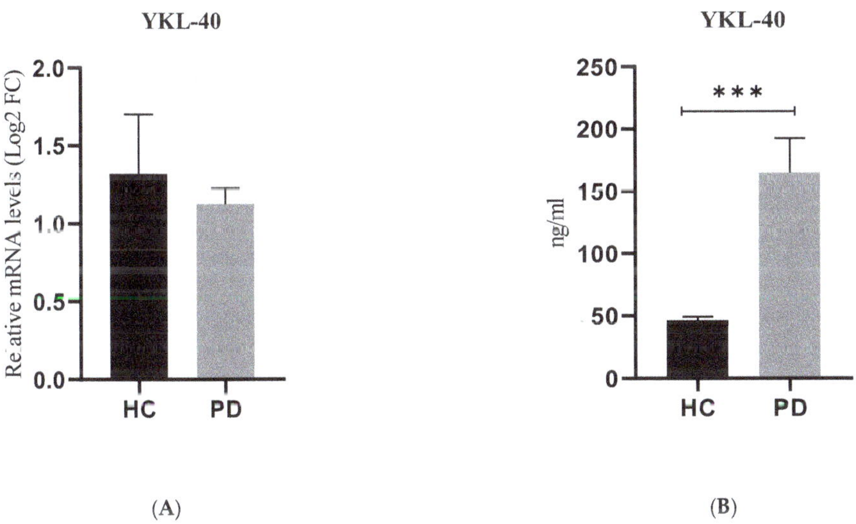

Figure 2. Gene (**A**) and protein (**B**) expression of YKL-40 in Parkinson's disease (PD) and healthy controls (HC). Values are expressed as means ± S.E.M; *** $p < 0.001$.

2.4. Correlation Analysis

A possible relationship between YKL-40 protein levels and bioenergetic parameters was investigated. The glycoprotein was shown to be directly related to ATP production (R = 0.9, $p = 0.012$) and to basal respiration (R = 0.92, $p = 0.008$).

2.5. Receiver Operating Characteristic (ROC) Analysis

To evaluate the predictive value of the two statistically significant indicators—the spare respiratory capacity and the YKL-40 protein levels, ROC analysis was performed (Figure 3). Our results revealed Sensitivity%—75.08 and Specificity%—58.33 (Area—0.77, STDEV—0.10, $p = 0.03$) for the reserve respiratory capacity and Sensitivity%—80.0 and Specificity%—94.12 (Area-0.92, STDEV-0.04, $p < 0.0001$) for the plasma levels of YKL-40. The data show that YKL-40 plasma levels appear to be an excellent marker for the discrimination of PD from the HC. The spare respiratory capacity also differentiates the studied groups but is less important than the plasma levels of the glycoprotein.

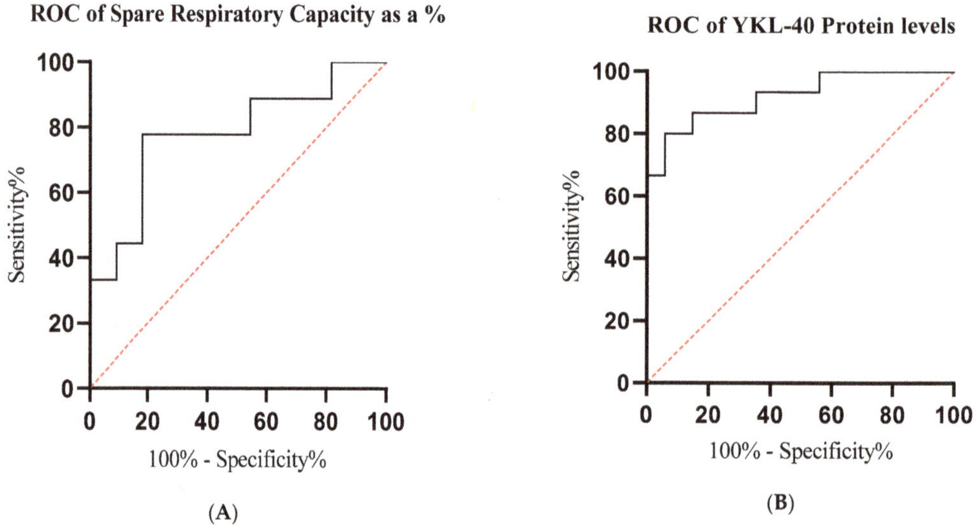

Figure 3. ROC analysis of Parkinson's disease patients (n = 18). (**A**) Spare respiratory capacity as %; (**B**) protein levels of YKL-40.

3. Discussion

In recent years, many reports have assumed mitochondrial dysfunction as a key factor in dopaminergic neuronal sensitivity in PD in both idiopathic and toxin-induced Parkinsonism [15–20]. This is explained by the fact that dopaminergic neurons have a high energy demand, and organelle dysfunction in patients promotes the loss of dopaminergic neurons in the substantia nigra pars compacta (SNpc). Gonzalez-Rodriguez et al. reported that the impairment of mitochondrial function simulates the pathogenesis of PD in animal models [15]. This evidence suggests that mitochondrial dysfunction occurs before the onset of PD. Therefore, one of the key challenges is the discovery of biomarkers that identify the early stages of the disease. The deeper understanding of the clinical–pathological mechanisms underlying the disease is fundamental for designing new strategies for therapeutic intervention in PD.

In the present study, we report high concentrations of plasma YKL-40 related to bioenergetic parameters. Data from previous investigations show a characteristic PD bioenergetic profile [21,22] with an increased reserve respiratory capacity compared to HC. The results obtained are consistent with other studies reporting changes in mitochondrial activity in PD [21]. The difference between the examined groups is statistically significant and shows that the reserve capacity is sensitive to oxidative stress and is therefore a suitable marker for the detection of physiological abnormalities in neurodegenerative diseases [23] and in particular to dopaminergic disorders such as schizophrenia, restless legs syndrome (RLS), and PD [16].

Schirinzi et al. (2022) also reported increased reserve capacity and maximal respiration in PD that correlated with the severity of motor impairment and with CSF Aβ42 levels [21]. The team hypothesized that mitochondrial respiration capacity may increase with PD progression. It is interpreted as a compensatory bioenergetic activation as a result of abnormal mitochondrial function [21]. However, we detected that the other bioenergetic parameters, which were reduced in the patient group (ATP-coupled respiration, coupling efficiency, maximal respiration and proton leak), were not statistically significant. This suggests that mitochondrial function is suppressed, as observed by Walter J et al. as well [23]. The possible explanation is that the observed abnormalities in the investigated parameters are intrinsic to PD mitochondria or that their activity is related to other pathogenic processes leading to PD. It is known that mitochondrial dysfunction can provoke deposition of α-synuclein and formation of Lewy bodies, as well as a neuronal loss [24]. Altered astrocytic function also leads to α-synuclein accumulation, neuroinflammation, impaired mitochondrial metabolism, and oxidative stress [25]. Metabolic reprogramming results in less efficient production of ATP and ROS generation [26]. Neurons and differentiated neuronal cells are largely dependent upon oxidative phosphorylation and usually exhibit a substantial mitochondrial spare respiratory capacity, to rapidly meet the requirements of increased functional demands and stress [7]. Furthermore, respiratory capacity features the grade of toxicity, ischemia, and oxidative stress [7,27]. Schapira and Tolosa concluded that mitochondrial dysfunction in the brain of PD patients may exist before the point of diagnosis during the prodromal period [28]. As we expected, mitochondrial function was altered in our PD patients' group. These findings are in accordance with previous studies describing fluctuation in the OCR parameters in different neurological pathologies. It was found that the spare respiratory capacity in fibroblasts correlated with early neuropsychiatric symptoms in Alzheimer's disease [29]. This finding together with our data suggest that moderate changes in the activity of the electron transport chain in mitochondria can significantly affect the disease phenotype. Reduced mitochondrial ATP production was also found to correlate with the downregulation of the PD-related genes in human neurons [30]. Another factor involved in PD development, α-synuclein, was proven to modulate the activity of ATP synthase through interaction with its α-subunit [31].

Our data add new evidence to the existing literature and show the association of mitochondria with inflammation. Previous studies on skin fibroblasts and prodromal and early Parkinson's blood cells have reported the presence of compromised mitochondrial structure and function, oxidative stress, and increased levels of reactive oxygen species (ROS) coupled with decreased levels of the antioxidant superoxide dismutase [32].

In addition to mitochondrial changes, we report a correlation of bioenergetic indices and YKL-40 protein levels. This may be explained by the close relationship between inflammation and metabolism [33]. During the immune response, immune cells switch from a metabolic quiescence to an active phase. This transition is associated with metabolic reprograming from a catabolic to anabolic state and a concerted balance between ATP generation and its utilization [34–36]. We assume that YKL-40 might reflect different sides of the response to brain injury, such as neuroinflammation and brain damage. Many studies have reported that the levels of YKL-40 were associated with the number of cells involved in neurodegeneration and glial activation [37]. It was found that the concentration of YKL-40 was lower in patients with PD compared to those with atypical Parkinson's syndrome but still higher than the levels in the control group. No correlations with disease stages or severity were observed in this study [38].

Interestingly, the data regarding YKL-40 in PD remain controversial. Our results showed significantly increased plasma YKl-40 levels compared to HC without any important change at the mRNA level. We could propose that YKL-40 expression might be regulated posttranscriptionally. In previous studies, a possible regulatory axis lncRNAs/miR-30e/YKL-40 was proven in systemic sclerosis [39]. It is quite possible that this glycoprotein is posttranscriptionally regulated in this pathology too.

High plasma levels of YKL-40 were detected in neurodegenerative dementias. Significantly elevated plasma YKL-40 in Creutzfeldt–Jakob disease (CJD) with a moderate potential to discriminate CJD cases from controls was determined. Additionally, YKL-40 concentrations appear significantly higher at late disease stages [40]. Furthermore, higher YKL-40 concentrations were related to the deterioration of cognitive abilities [37]. It was found that the high expression levels of YKL-40 were associated with an inflammatory brain profile in Alzheimer's disease [41]. Another research group determined that YKL-40 had the highest sensitivity and specificity in differentiating between healthy subjects and patients with mild dementia. It was concluded that only YKL-40 proved to be indicative for dementia progression [42]. On the contrary, other authors revealed lower levels of YKL-40 in patients with PD than in healthy controls or in other neurodegenerative conditions [43].

Along with the original results presented herein, however, there are some limitations in this study that could be addressed in future research. First, the number of patients and healthy individuals is insufficient; therefore, the data need further confirmation on a larger patient cohort. Second, a followup of the patients subtypes and the assessment of therapy is required. Although we used innovative and advanced technology to determine the mitochondrial status of living cells in real time, it is still an expensive approach, and it could be a limitation in routine clinical practice.

4. Materials and Methods

4.1. Patients and Controls

After signing a written informed consent in accordance with the instructions of the Ethics Committee at the Medical University of Plovdiv (Protocol № 6/2021) and following the principles of the Declaration of Helsinki, patients and healthy controls were recruited. It is a cross-sectional study that included 18 naïve PD patients and a control group (n = 7) of age-matched healthy individuals from a Bulgarian cohort. Venous blood was collected in an EDTA-Vacutainer monovette. The diagnosis was verified by a team of certified professionals from the Department of Neurology according to the internationally accepted criteria MDS-PD (Movement Disorder Society—Parkinson's Disease, 2015) (Postuma et al.; 2015). The severity of the symptoms was assessed by the UPDRS (Unified Parkinson's Disease Rating Scale) and the Hoen–Yahr scale. The full revised version of UPDRS contains 50 questions, divided into four parts: nonmotor aspects of experiences of daily living, motor experiences of daily living, motor examination, and motor complications. Each question is anchored with five responses. The total UPDRS score ranges between 0 (normal neurological examination) and 199 (the highest severity of the symptoms). The Hoen–Yahr scale is a short five-point scale that emphasizes the lateralization of symptoms and the patients' need for assistance in performing activities of daily living. The cases were recruited from the Department of Neurology, University Hospital "Kaspela", Medical University of Plovdiv, and amongst the patients referred to the Expert committee for diagnosis and treatment of Parkinson's disease at the University Hospital "Kaspela".

The control group included volunteers without a history or clinical data of neurological diseases. According to the exclusion criteria, subjects with acute/chronic infectious/inflammatory and autoimmune diseases were not involved in the study. Demographic and medical history information was also collected.

4.2. Isolation of Peripheral Blood Mononuclear Cells (PBMCs)

The peripheral blood was processed immediately, and PBMCs were isolated using Pancoll (Pan Biotech Cat # P04-60500) according to the protocol established by the manufacturer. Cells were cultured overnight in RPMI-1640 medium (Pan Biotech Cat # P04-22100) with 10% FBS, 1% penicillin/streptomycin in a cell culture incubator at 37 °C, 5% CO_2, and high humidity. The number of cells and their viability were determined using an automatic counter "LUNA" (Logos Biosystems, Anyang, South Korea). Cells were diluted with RPMI-1640 to a final concentration of 2×10^5 cells per well and then plated in Seahorse microplates for analysis. Separated plasma was frozen and stored at −80 °C for subsequent analysis.

4.3. Metabolic Analysis in Real Time

Mitochondrial activity was measured using a Seahorse XFp analyzer (Agilent, Santa Clara, CA, USA). The advantage of our study is that the cells are processed immediately after their isolation, which limits the negative effect of freezing. PBMCs were cultured on poly-D-lysine coated wells at a density of 2×10^5 cells/well. To verify that the cells were evenly distributed in a monolayer, the wells were visualized using an inverted microscope (Nikon eclipse TS 100, NIKON, Amstelveen, The Netherlands).

4.4. Mito Stress Test

A mitochondrial stress test was performed according to recommendations of Agilent (Agilent, Santa Clara, CA, USA) and following the protocol established by the manufacturer. Before analysis, the growth medium was replaced with Seahorse XF medium (Agilent Seahorse, pH = 7.4), and PBMCs were incubated at 37 °C without CO_2. Each sample was tested in triplicate, and the results were averaged. The Seahorse analyzer measures the oxygen consumption rate (OCR) as an indicator of mitochondrial respiration. The first measurement determines the baseline OCR followed by OCR after injection of the following inhibitors: oligomycin (1.5 μM), carbonyl cyanide 4-(trifluoromethoxy) phenylhydrazone (1 μM) (FCCP), and rotenone (0.5 μM), according to the manufacturer's instructions.

The use of the indicated inhibitors allows measurement of ATP-coupled respiration (oligomycin inhibits ATP synthase), maximal respiration (FCCP, reveals maximal ETC capacity), reserve respiratory capacity (difference between maximal respiration and basal respiration), and non-mitochondrial respiration (rotenone inhibits complex I).

4.5. YKL-40 Gene Expression by qPCR

Total RNA from white blood cells was extracted using Trizol (Invitrogen, No. 15596026) according to the manufacturer's instructions. The concentration of RNA was measured on a NanoDrop Nucleic Acid Quantification (Thermo Fisher Scientific, Waltham, MA, USA). cDNA was transcribed from 2 μg RNA using Genaxon GreenMasterMix (2×) (Genaxxon bioscienceGmbH, Ulm, Germany, Lot. No. M3023.0500) according to the kit's instructions. The primer combinations used for the longest RNA transcripts of CHI3L1 (YKL-40) were synthesized by Integrated DNA Technologies, (Leuven, Belgium). The sequences were as follows: Fw 5′-CTGCTCCAGTGCTGCTCT-3′, Rev 5′-TACAGAGGAAGCGGTCAAGG-3′. RNA normalization was performed using several internal controls: GAPDH (Fw 5′-AGG TCCACCACTGACACGTTG-3′, Rw5′-AGCTGAACGGGATGCTCACT-3′), ACTINβ (Fw 5′-AGTGTGACG TGGACATCCGGA-3′, Rev 5′-GCC AGGGCAGTGATCTCCTCCT-3′), and hUBC (Fw 5′TCCTCAGGCAGAG-GTT GATCTT-3′, Rev 5′-GCACCAAGTGCAGAGG TGGACTCTT-3′) (Integrated DNA Technologies, Leuven, Belgium). The qPCR reactions were run on Rotor-Gene Q 600 (Qiagen, Hilden, Germany), and quantitative evaluations were performed by the $2^{-\Delta\Delta Ct}$ method. Each sample was analyzed in duplicate.

4.6. Detection of YKL-40 in Plasma

Plasma was isolated from peripheral blood after centrifugation; then, the supernatant was frozen at −80 °C. It was later used to measure the YKL-40 levels with an ELISA kit, specific for the protein encoded by the longer isoform (383 amino acids) (MicroVue YKL-40, Quidel, 9975 Summers Ridge Road, San Diego, CA 92121, USA, Lot. No. 088337), following the manufacturer's instructions. The detection limit was 10 μg/L. The intra-assay coefficient of variation (CV) was 5%, and the inter-assay CV was <6%. YKL-40 is stable for at least 15 years in plasma samples stored at −80 °C [44]. The assay employs the sandwich-based ELISA method with optical density measured at 405 nm.

4.7. Statistical Analysis

Preprocessing of OCR values was performed using Wave Software2.6.3. (Seahorse Bioscience, Agilent, (available at https://www.agilent.com/en/products/cell-analysis/software-download-for-wave-desktop, accessed on 4 October 2023). Statistical analyses of data were performed with Graphpad Prism. Differences between normally distributed variables were evaluated for significance using the Welch's *t*-test for independent samples or the paired *t*-test for dependent samples and the Wilcoxon–Mann–Whitney test. The significance threshold was set at p-value < 0.05.

5. Conclusions

In conclusion, our study reveals new aspects of the relationship between the glycoprotein YKL-40 and mitochondrial function in PD. The bioenergetic and inflammatory changes affecting immune cells can illustrate the development of neuroinflammation and neurodegeneration. We consider that YKL-40 protein levels in combination with the dynamics in mitochondrial metabolism might serve as an additional tool to evaluate inflammatory activity and the clinical course of PD.

Author Contributions: Conceptualization, V.S. and M.K.; methodology, M.G.; software, M.G.; formal analysis, M.G.; clinical investigation, A.T.; investigation, M.K. and M.G.; writing—original draft preparation, M.G. and M.K.; writing—review and editing, V.S.; supervision, V.S. and A.T. All authors have read and agreed to the published version of the manuscript.

Funding: This research was financed by European Union-NextGenerationEU, through the National Recovery and Resilience Plan of the Republic of Bulgaria, project № BG-RRP-2.004-0007-C01.

Institutional Review Board Statement: The study was conducted in accordance with the Declaration of Helsinki and approved by the of the Ethics Committee at the Medical University of Plovdiv (Protocol № 6/2021).

Informed Consent Statement: Written informed consent was obtained from all subjects involved in the study.

Data Availability Statement: Data are contained within the article.

Acknowledgments: The authors acknowledge the work of Danail Minchev and Valentin Dichev in sample collection and processing.

Conflicts of Interest: The authors declare no conflict of interest.

References

1. Kalia, V.; Lang, E. Parkinson's disease. *Lancet* **2015**, *386*, 896–912. [CrossRef] [PubMed]
2. Váradi, C. Clinical features of Parkinson's disease: The evolution of critical symptoms. *Biology.* **2020**, *9*, 103. [CrossRef] [PubMed]
3. Pringsheim, T.; Jette, N.; Frolkis, A.; Steeves, T. The prevalence of Parkinson's disease: A systematic review and metaanalysis. *Mov. Disord.* **2014**, *29*, 1583–1590. [CrossRef]
4. Collier, J.; Kanaan, M.; Kordower, H. Ageing as a primary risk factor for Parkinson's disease: Evidence from studies of non-human primates. *Nat. Rev. Neurosci.* **2011**, *12*, 359–366. [CrossRef] [PubMed]
5. Park, S.; Davis, L.; Sue, M. Mitochondrial dysfunction in Parkinson's disease: New mechanistic insights and therapeutic perspectives. *Curr. Neurol. Neurosci.* **2018**, *18*, 18–21. [CrossRef]
6. Rango, M.; Bresolin, N. Brain mitochondria, aging, and Parkinson's disease. *Genes* **2018**, *9*, 250. [CrossRef]
7. Yadava, N.; Nicholls, G. Spare respiratory capacity rather than oxidative stress regulates glutamate excitotoxicity after partial respiratory inhibition of mitochondrial complex I with rotenone. *J. Neurosci.* **2007**, *27*, 7310–7317. [CrossRef]
8. Renkema, G. Chitotriosidase, a chitinase, and the 39-kDa human cartilage glycoprotein, a chitin-binding lectin, are homologues of family 18 glycosyl hydrolases secreted by human macrophages. *Eur. J. Biochem.* **1998**, *251*, 504–509. [CrossRef]
9. Jensen, B.V.; Johansen, J.S.; Price, P.A. High levels of serum HER-2/neu and YKL-40 independently reflect aggressiveness of metastatic breast cancer. *Clin. Cancer Res.* **2003**, *9*, 4423–4434.
10. Hakala, B.E.; White, C.; Recklies, A.D. Human cartilage gp-39, a major secretory product of articular chondrocytes and synovial cells, is a mammalian member of the chitinase protein family. *J. Biol. Chem.* **1993**, *268*, 25803–25810. [CrossRef]

11. Brasso, K. Prognostic value of PINP, bone alkaline phosphatase, CTX-I, and YKL-40 in patients with metastatic prostate carcinoma. *Prostate* **2006**, *66*, 503–513. [CrossRef] [PubMed]
12. Shao, R. YKL-40 acts as an angiogenic factor to promote tumor angiogenesis. *Front. Physiol.* **2013**, *4*, 122. [CrossRef] [PubMed]
13. Prakash, M.; Bodas, M.; Prakash, D.; Nawani, N.; Khetmalas, M.; Mandal, A. Diverse pathological implications of YKL-40: Answers may lie in 'outside-in' signaling. *Cell Signal.* **2013**, *25*, 1567–1573. [CrossRef]
14. Riabov, V.; Gudima, A.; Wang, N.; Mickley, A.; Orekhov, A.; Kzhyshkowska, J. Role of tumor associated macrophages in tumor angiogenesis and lymphangiogenesis. *Front. Physiol.* **2014**, *5*, 75. [CrossRef]
15. González-Rodríguez, P.; Zampese, E.; Stout, K.; Guzman, N.; Ilijic, E.; Yang, B.; Tkatch, T.; Stavarache, A.; Wokosin, L.; Gao, L.; et al. Disruption of mitochondrial complex I induces progressive parkinsonism. *Nature* **2021**, *599*, 650–656. [CrossRef] [PubMed]
16. Smith, A.; Depp, C.; Ryan, J.; Johnston, I.; Alegre-Abarrategui, J.; Evetts, S.; Rolinski, M.; Baig, F.; Ruffmann, C.; Simon, K.; et al. Mitochondrial dysfunction and increased glycolysis in prodromal and early Parkinson's blood cells. *Mov. Disord.* **2018**, *33*, 1580–1590. [CrossRef] [PubMed]
17. Gao, Y.; Yang, T.; Gu, Y.; Sun, X. Mitochondrial dysfunction in Parkinson's disease: From mechanistic insights to therapy. *Front. Aging Neurosci.* **2022**, *14*, 885500. [CrossRef] [PubMed]
18. Malpartida, A.; Williamson, M.; Narendra, D.; Wade-Martins, R.; Ryan, B. Mitochondrial dysfunction and mitophagy in Parkinson's disease: From mechanism to therapy. *Trends Biochem. Sci.* **2021**, *46*, 329–343. [CrossRef]
19. Betarbet, R.; Sherer, T.; MacKenzie, G.; Garcia-Osuna, M.; Panov, A.; Greenamyre, J. Chronic systemic pesticide exposure reproduces features of Parkinson's disease. *Nat. Neurosci.* **2000**, *3*, 1301–1306. [CrossRef]
20. Tanner, C.; Kamel, F.; Ross, G. Rotenone, paraquat, and Parkinson's disease. *Environ. Health Perspect.* **2011**, *119*, 866–872. [CrossRef] [PubMed]
21. Schirinzi, T.; Salvatori, I.; Zenuni, H.; Grillo, P.; Valle, C.; Martella, G.; Mercuri, N.; Ferri, A. Pattern of Mitochondrial Respiration in Peripheral Blood Cells of Patients with Parkinson's Disease. *Int. J. Mol. Sci.* **2022**, *23*, 10863. [CrossRef] [PubMed]
22. Minchev, D.; Kazakova, M.; Sarafian, V. Neuroinflammation and Autophagy in Parkinson's Disease—Novel Perspectives. *Int. J. Mol. Sci.* **2022**, *23*, 14997. [CrossRef] [PubMed]
23. Walter, J.; Bolognin, S.; Antony, P.; Nickels, S.; Poovathingal, S.; Salamanca, L.; Magni, S.; Perfeito, R.; Hoel, F.; Qing, X.; et al. Neural stem cells of Parkinson's disease patients exhibit aberrant mitochondrial morphology and functionality. *Stem Cell Rep.* **2019**, *12*, 878–889. [CrossRef] [PubMed]
24. Valdinocci, D.; Simões, R.; Kovarova, J.; Cunha-Oliveira, T.; Neuzil, J.; Pountney, D. Intracellular and intercellular mitochondrial dynamics in Parkinson's disease. *Front. Neurosci.* **2019**, *13*, 930. [CrossRef] [PubMed]
25. Lauro, C.; Limatola, C. Metabolic reprograming of microglia in the regulation of the innate inflammatory response. *Front. Immunol.* **2020**, *11*, 493. [CrossRef] [PubMed]
26. Booth, H.D.; Hirst, W.D.; Wade-Martins, R. The Role of Astrocyte Dysfunction in Parkinson's Disease Pathogenesis. *Trends Neurosci.* **2017**, *40*, 358–370. [CrossRef]
27. Flynn, M.; Choi, S.; Day, N.; Gerencser, A.; Hubbard, A.; Melov, S. Impaired spare respiratory capacity in cortical synaptosomes from Sod2 null mice. *Free. Radic. Biol. Med.* **2011**, *50*, 866–873. [CrossRef]
28. Schapira, H.; Tolosa, E. Molecular and clinical prodrome of Parkinson disease: Implications for treatment. *Nat. Rev. Neurol.* **2010**, *6*, 309–317. [CrossRef]
29. Bell, S.; De Marco, M.; Barnes, K.; Shaw, P.; Ferraiuolo, L.; Blackburn, D.; Mortiboys, H.; Venneri, A. Deficits in mitochondrial spare respiratory capacity contribute to the neuropsychological changes of Alzheimer's disease. *J. Pers. Med.* **2020**, *10*, 32. [CrossRef]
30. Gandhi, P.; Chen, S.; Wilson-Delfosse, A. Leucine-rich repeat kinase 2 (LRRK2): A key player in the pathogenesis of Parkinson's disease. *J. Neurosci. Res.* **2009**, *87*, 1283–1295. [CrossRef]
31. Ludtmann, M.; Angelova, P.; Ninkina, N.; Gandhi, S.; Buchman, V.; Abramov, A. Monomeric alpha-synuclein exerts a physiological role on brain ATP synthase. *J. Neurosci.* **2016**, *36*, 10510–10521. [CrossRef] [PubMed]
32. Teves, Y.; Bhargava, V. Parkinson's disease skin fibroblasts display signature alterations in growth, redox homeostasis, mitochondrial function, and autophagy. *Front. Neurosci.* **2018**, *11*, 737. [CrossRef] [PubMed]
33. Angajala, A.; Lim, S. Diverse roles of mitochondria in immune responses: Novel insights into immuno-metabolism. *Front. Immunol.* **2018**, *9*, 1605. [CrossRef] [PubMed]
34. Pearce, L.; Pearce, J. Metabolic pathways in immune cell activation and quiescence. *Immunity* **2013**, *38*, 633–643. [CrossRef]
35. Carr, L.; Kelman, A.; Wu, G.; Gopaul, R.; Senkevitch, E.; Aghvanyan, A. Glutamine uptake and metabolism are coordinately regulated by ERK/MAPK during T lymphocyte activation. *J. Immunol.* **2010**, *185*, 1037–1044. [CrossRef] [PubMed]
36. Le, A.; Lane, N.; Hamaker, M.; Bose, S.; Gouw, A.; Barbi, J. Glucose-independent glutamine metabolism via TCA cycling for proliferation and survival in B cells. *Cell Metab.* **2012**, *15*, 110–121. [CrossRef]
37. Hall, S.; Janelidze, S.; Surova, Y.; Hansson, O. Cerebrospinal fluid concentrations of inflammatory markers in Parkinson's disease and atypical parkinsonian disorders. *Sci. Rep.* **2018**, *8*, 13276. [CrossRef]
38. Magdalinou, N.; Paterson, W.; Schott, J.; Fox, N.; Mummery, C.; Blennow, K.; Bhatia, K.; Morris, H.; Giunti, P.; Warner, T. A panel of nine cerebrospinal fluid biomarkers may identify patients with atypical parkinsonian syndromes. *J. Neurol. Neurosurg. Psychiatry* **2015**, *86*, 1240–1247. [CrossRef]
39. Dichev, V.; Mehterov, N.; Kazakova, M.; Karalilova, R.; Batalov, A.; Sarafian, V. The lncRNAs/miR-30e/CHI3L1 axis is dysregulated in systemic sclerosis. *Biomedicines* **2022**, *10*, 496. [CrossRef]

40. Villar-Piqué, A.; Schmitz, M.; Hermann, P.; Goebel, S.; Bunck, T.; Varges, D.; Ferrer, I.; Riggert, J.; Llorens, F.; Zerr, I. Plasma YKL-40 in the spectrum of neurodegenerative dementia. *J. Neuroinflamm.* **2019**, *16*, 1–5. [CrossRef]
41. Sanfilippo, C.; Castrogiovanni, P.; Imbesi, R.; Musumeci, G.; Vecchio, M.; Volti, G.L.; Tibullo, D.; Broggi, G.; Caltabiano, R.; Ulivieri, M.; et al. Sex-dependent neuro-deconvolution analysis of Alzheimer's disease brain transcriptomes according to CHI3L1 expression levels. *J. Neuroimmunol.* **2022**, *373*, 577977. [CrossRef] [PubMed]
42. Wilczyńska, K.; Maciejczyk, M.; Zalewska, A.; Waszkiewicz, N. Serum amyloid biomarkers, tau protein and YKL-40 utility in detection, differential diagnosing, and monitoring of dementia. *Front. Psychiatry* **2021**, *12*, 725511. [CrossRef] [PubMed]
43. Olsson, B.; Hertze, J.; Lautner, R.; Zetterberg, H.; Nägga, K.; Höglund, K.; Basun, H.; Annas, P.; Lannfelt, L.; Andreasen, N.; et al. Microglial markers are elevated in the prodromal phase of Alzheimer's disease and vascular dementia. *J. Alzheimer's Dis.* **2013**, *33*, 45–53. [CrossRef] [PubMed]
44. Bojesen, E.; Johansen, S.; Nordestgaard, G. Plasma YKL-40 levels in healthy subjects from the general population. *Clin. Chim. Acta* **2011**, *412*, 709–712. [CrossRef] [PubMed]

International Journal of
Molecular Sciences

MDPI

Article

Alpha-Synuclein-Specific Regulatory T Cells Ameliorate Parkinson's Disease Progression in Mice

Seon-Young Park [1], HyeJin Yang [1], Soyoung Kim [1], Juwon Yang [2], Hyemin Go [2] and Hyunsu Bae [1,2,*]

[1] Department of Science in Korean Medicine, College of Korean Medicine, Graduate School, Kyung Hee University, 26 Kyungheedae-ro, Dongdaemun-gu, Seoul 02447, Republic of Korea; psys12@naver.com (S.-Y.P.); emilly86@naver.com (H.Y.); samanda0@nate.com (S.K.)

[2] Department of Korean Medicine, College of Korean Medicine, Graduate School, Kyung Hee University, 26 Kyungheedae-ro, Dongdaemun-gu, Seoul 02447, Republic of Korea; jwyang@khu.ac.kr (J.Y.); comrp1521@hanmail.net (H.G.)

* Correspondence: hbae@khu.ac.kr; Tel.: +82-2-961-9316; Fax: +82-2-962-9316

Abstract: Parkinson's disease (PD) is a long-term neurodegenerative disease characterized by dopaminergic neuronal loss and the aggregation of alpha-synuclein (α-syn) in the brain. Cell therapy using regulatory T cells (Tregs) has therapeutic potential on PD progression in a mouse model; however, several challenges were associated with its applications. Here, we propose a strategy for α-syn specific Treg expansion (α-syn Treg). We presented α-syn to T cells via dendritic cells. This method increased the mobility of Tregs towards the site of abundant α-syn in vitro ($p < 0.01$; α-syn Tregs versus polyclonal Tregs (poly Tregs)) and in vivo. Consequently, α-syn Tregs showed noteworthy neuroprotective effects against motor function deficits ($p < 0.05$, $p < 0.01$; α-syn Tregs versus poly Tregs), dopaminergic neuronal loss ($p < 0.001$; α-syn Tregs versus poly Tregs), and α-syn accumulation ($p < 0.05$; α-syn Tregs versus poly Tregs) in MPTP-induced PD mice. Furthermore, the adoptive transfer of α-syn Tregs exerted immunosuppressive effects on activated microglia, especially pro-inflammatory microglia, in PD mice. Our findings suggest that α-syn presentation may provide a significant improvement in neuroprotective activities of Tregs and suggest the effective clinical application of Treg therapy in PD.

Keywords: Parkinson's disease; regulatory T cell; alpha-synuclein

1. Introduction

Parkinson's disease (PD) is a progressive neurodegenerative disorder that leads to motor deficits, including tremors, rigidity, and postural instability [1]. The current therapy for Parkinson's is aimed only at relieving the disease symptoms. None of the prescribed drugs, such as dopamine agonists and levodopa, have been proven disease-modifying [2]. Additionally, there are many side effects, such as drug-related motor fluctuations, on–off effects, and dyskinesia [3].

A characteristic feature of PD is the presence of soluble or insoluble aggregates of alpha-synuclein (α-syn) in the brain. Though α-syn is a synaptic protein present in the brain, it accompanies neurotoxicity and ultimately leads to neuronal dysfunction [4]. Aggregated α-syn is phagocytosed by microglia and induces microglial activation with ROS production. Activated microglia are found in specific brain regions in patients with PD and are considered influential in the degeneration of dopaminergic neurons by amplifying neuronal inflammation [5,6]. The 1-methyl-4-phenyl-1,2,3,6-tetrahydropyridine (MPTP) model is the most widely used experimental model for studying PD in mice. MPTP causes α-syn accumulation in the substantia nigra (SN) and PD-like motor impairment [7,8]. In the MPTP model, as in patients with PD, microglial activation is induced before the dopaminergic neuronal loss and is implicated in PD pathology. Therefore, activated microglia are a promising therapeutic target for PD [9].

CD4$^+$CD25$^+$ regulatory T cells (Tregs) play a crucial role in controlling immune balance in the central nervous system (CNS). However, their suppressive activity is dysregulated during inflammation, aggravating the damage caused by autoreactive T effector cells [10,11]. In line with this, Tregs have been considered an attractive therapeutic modality that induces neuroprotective activity [12]. Adoptive transfer of Tregs has been attempted as a potential treatment for neurodegenerative diseases such as Alzheimer's disease (AD), amyotrophic lateral sclerosis (ALS), and even PD [13–15]. However, these studies have some limitations. A major problem with polyclonal Treg (poly Treg) therapy is the suppression of off-target immune responses, which may increase susceptibility to opportunistic infections or inhibit antitumor activity. As a solution to this problem, the development of antigen-specific Tregs has been attempted. It has been reported that antigen-specific Tregs are more effective and safer than polyclonal Tregs in a mouse model of non-obese diabetes and graft-versus-host disease (GVHD) [16–19]. Therefore, α-syn, a PD-specific antigen, could increase the effectiveness of Treg therapy targeting PD.

The aim of the current study is to establish the effects of α-syn presentation on the ex vivo expansion of Tregs (α-syn Tregs). Thus, we adoptively transferred expanded Tregs into an MPTP-induced mouse model of PD and assessed the disease pathology. The advantage of α-syn presentation showed the enhancement of the neuroprotective effects of Tregs on motor function, neuronal loss, and pro-inflammatory microglial activation. In conclusion, our findings provide insight into the clinical application of Treg therapy which has great potential for PD.

2. Results

2.1. α-syn Tregs Maintain the Suppressive Phenotype during Ex Vivo Expansion

To prepare antigen-specific suppressive Tregs, CD4+ T cells were co-cultured with poly- or α-syn-presenting DCs for four days. After 4 days, poly- or α-syn-specific CD4$^+$CD25$^+$ Tregs were isolated and expanded for 14 days (Figure 1a). The expression of CD62L in Tregs was analyzed using flow cytometry (Figure 1b). At the start (day 0), there were no differences between poly and α-syn Tregs. However, on day 14, a higher expression of CD62L was observed in α-syn Tregs (Figure 1c). After expansion, mRNAs were extracted from Tregs, and the expression of Treg markers, such as FoxP3, GATA3, IL-10, and Amphiregulin (Areg), was analyzed (Figure 1d). The expression of these markers was increased in α-syn Tregs. These results suggest that α-syn Tregs maintain the immunosuppressive phenotype during ex vivo expansion.

2.2. Antigen Presentation Increases the Mobility of Tregs toward Disease Site

To determine whether antigen presentation affected TCR Vβ usage, the TCR repertoire was assessed using the Mouse Vβ TCR Screening Panel (Figure 2a). The usage of some Vβ, particularly Vβ 8.1/2, 8.3, and 13, increased slightly after antigen presentation.

It is presumed that antigen-specific Tregs are more efficient than poly Tregs because they migrate towards the specific area of the brain where the antigen exists [20]. Therefore, we performed an in vitro migration assay to assess the mobility of Tregs towards α-syn (Figure 2b). After 4 h of seeding Tregs, approximately 50% of α-syn Tregs moved to the bottom where α-syn existed, whereas less than 10% of poly Tregs moved. To determine whether these in vitro results were reproduced in vivo, we produced poly and α-syn Tregs expressing Thy1.1 from Thy1.1$^+$ mice (Figure 2c). These Tregs were adoptively transferred to Thy1.2$^+$ MPTP-intoxicated mice. Their distributions were detected with Thy1.1 expression in CD4$^+$7AAD$^-$ cells. The results showed that α-syn Tregs accounted for a higher proportion in the brain, while there were no differences between the groups in other tissues such as lung, spleen, kidney, and lymph node. These results support that α-syn presentation improves the mobility of Tregs to α-syn-rich environments in vitro as well as in vivo.

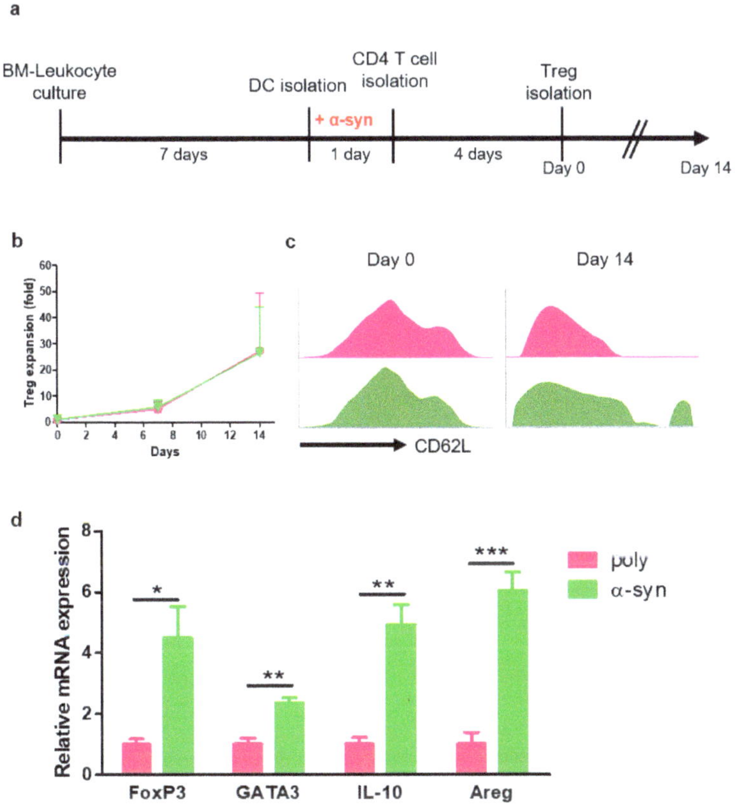

Figure 1. α-syn Tregs have a more suppressive phenotype. (**a**) Schematic diagram of Treg preparation and expansion. (**b**,**c**) On days 0, 7, and 14 after Treg isolation, the phenotypes of Tregs were analyzed using flow cytometry. (**d**) The relative mRNA expressions of FoxP3, GATA3, IL-10, and Areg in expanded Tregs were analyzed. Data are presented as mean ± SEM. Statistical analyses were conducted with two-tailed t-test; * $p < 0.05$, ** $p < 0.01$, and *** $p < 0.001$. $n = 3$–4.

2.3. α-syn Tregs Improve Motor Function in MPTP-Induced Parkinson's Disease Mice

To investigate the effect of Tregs on MPTP-induced neurodegeneration, 5×10^5 Tregs were adoptively transferred into MPTP-treated mice (12 mg/kg) (Figure 3a). A pole test was carried out to examine PD-related motor deficits on day 6, and mice were sacrificed on day 7. MPTP treatment causes deficits in motor function resembling those observed in human PD [21]. As expected, MPTP treatment significantly extended the time taken to turn downward compared to the wild type (WT) (Figure 3b). This time was significantly shortened in α-syn Tregs but not in poly Tregs. In particular, α-syn Tregs significantly improved motor function compared with poly Tregs in both indices. These results suggest that α-syn Tregs significantly improve motor function.

2.4. α-syn Tregs Reduce Parkinson's Disease Pathology in MPTP Induced Parkinson's Disease Mice

PD is characterized by the loss of dopaminergic neurons in the SN [22]. To confirm the loss of dopaminergic neurons, TH was assessed in the SN (Figure 4a). MPTP treatment significantly reduced the number of TH-positive dopaminergic neurons compared with the control. α-syn Tregs significantly inhibited the loss of dopaminergic neurons compared with both MPTP and poly Tregs.

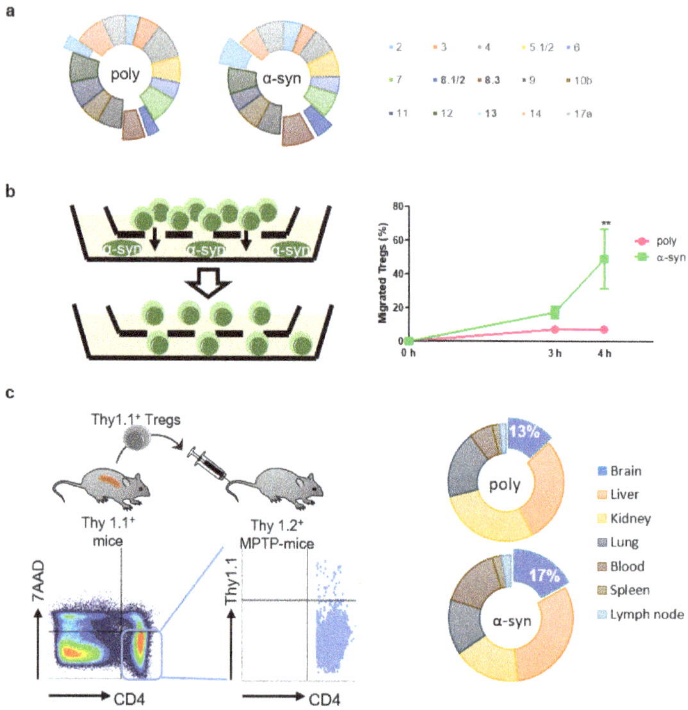

Figure 2. α-syn Tregs move more to site with α-syn. (**a**) TCR repertoire was assessed using Mouse Vb TCR Screening Panel. (**b**) Tregs and α-syn were seeded in the upper and bottom chambers of transwell, respectively. Migration of Tregs was assessed after 3 and 4 h. (**c**) Thy1.1$^+$ Tregs were isolated from splenocytes of Thy1.1$^+$ mice. Thy1.1$^+$ Tregs were adoptively transferred to MPTP-intoxicated mice and detected in the lymph node, spleen, blood, lung, kidney, liver, and brain after 7 days. Data are presented as mean ± SEM. Statistical analyses were conducted with two-way ANOVA; ** $p < 0.01$. n = 3–4.

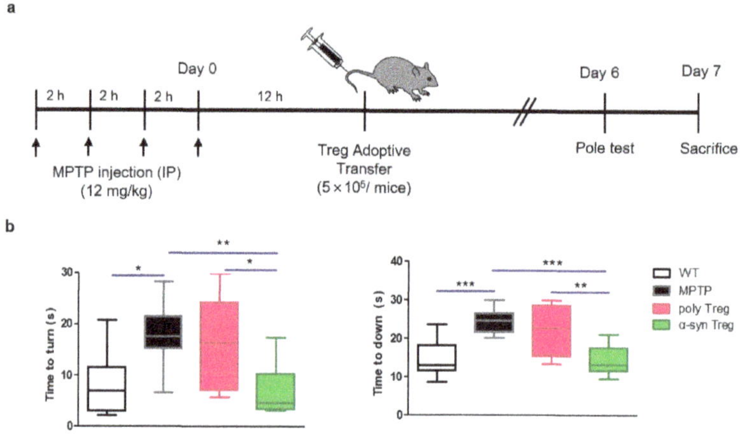

Figure 3. α-syn Tregs moderate motor dysfunction of MPTP-induced mice model of PD. (**a**) Schematic diagram of Treg transfer in MPTP-induced PD mice. (**b**) Time to turn downward and descend the pole was measured by pole test on day 6. Data are presented as mean ± SEM. Statistical analyses were conducted with one-way ANOVA; * $p < 0.05$, ** $p < 0.01$, and *** $p < 0.001$. n = 8–10.

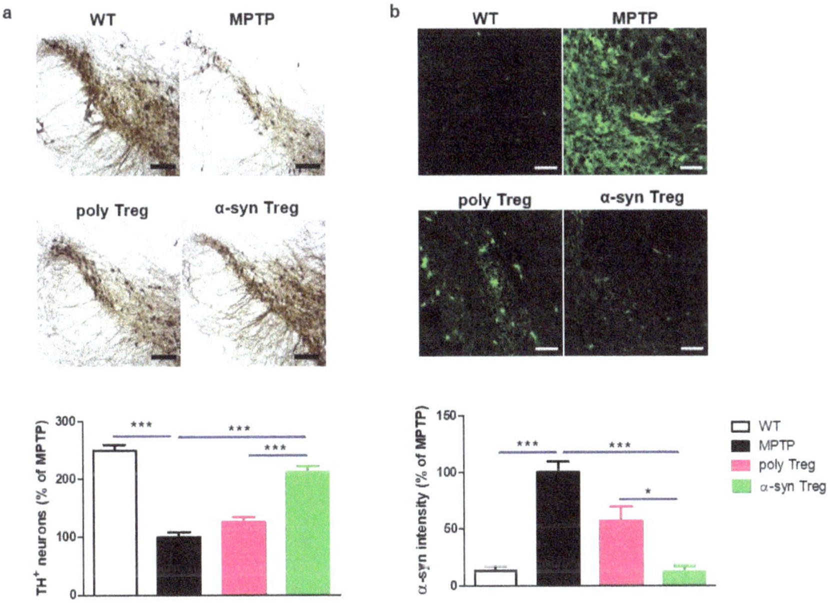

Figure 4. Immunohistochemical staining of TH and α-syn in the SN of MPTP-induced mice model of PD. (**a**) Immunohistochemistry was performed for expression of TH, a dopaminergic neuron marker, in SN of MPTP-induced mice model of PD. (**b**) To measure α-syn accumulation, brain sections were stained with α-syn. The intensity was calculated and normalized with MPTP. Data are presented as mean ± SEM. Statistical analyses were conducted with one-way ANOVA; * $p < 0.05$, and *** $p < 0.001$. $n = 4$–6. Scale bars: 50 μm.

To investigate the effects of Tregs on the upregulation of α-syn in the MPTP-induced mouse model of PD, SNs were stained with α-syn (Figure 4b). The α-syn intensity in the SN was increased in MPTP-intoxicated mice; α-syn Tregs decreased the accumulation of α-syn in the SN compared to MPTP and poly Tregs. These results suggest that antigen presentation improves the effects of Tregs on PD pathology.

2.5. α-syn Tregs Modulate Microglial Polarization in MPTP-Induced Parkinson's Disease Mice

Activated microglia secrete M1-associated pro-inflammatory cytokines that aggravate neurodegeneration in the PD brain, while M2 microglia participate in tissue repair by producing anti-inflammatory cytokines [23]. To assess the activation of pro-inflammatory M1 microglia, SNs were immunostained with Iba1 as a marker of activated microglia and NOS2 as a marker of pro-inflammatory microglia (Figure 5A). The increased intensities of NOS2 and Iba1 in MPTP-treated mice were significantly decreased only by the adoptive transfer of α-syn Tregs. To further confirm the effects of Tregs on microglial polarization, the mRNA expression of M1 and M2 microglial markers was evaluated with quantitative real-time PCR (Figure 5B,C). The relative expression of NOS2, an M1 phenotypic marker, was significantly reduced only by the α-syn Treg transfer. Interestingly, the expression levels of the pro-inflammatory cytokines TNF-α, IL-1β, and IL-6 were significantly decreased with any Tregs. However, the adoptive transfer of Tregs has no significant effect on the relative mRNA levels of Arg1 and IL-4, an M2 phenotypic marker. Additionally, the mRNA levels of FoxP3 and IL-10 were significantly increased in the α-syn Treg-transferred mice brains (Figure 5D). Collectively, these data show that α-syn Tregs exert a better therapeutic effect than polyclonal Tregs.

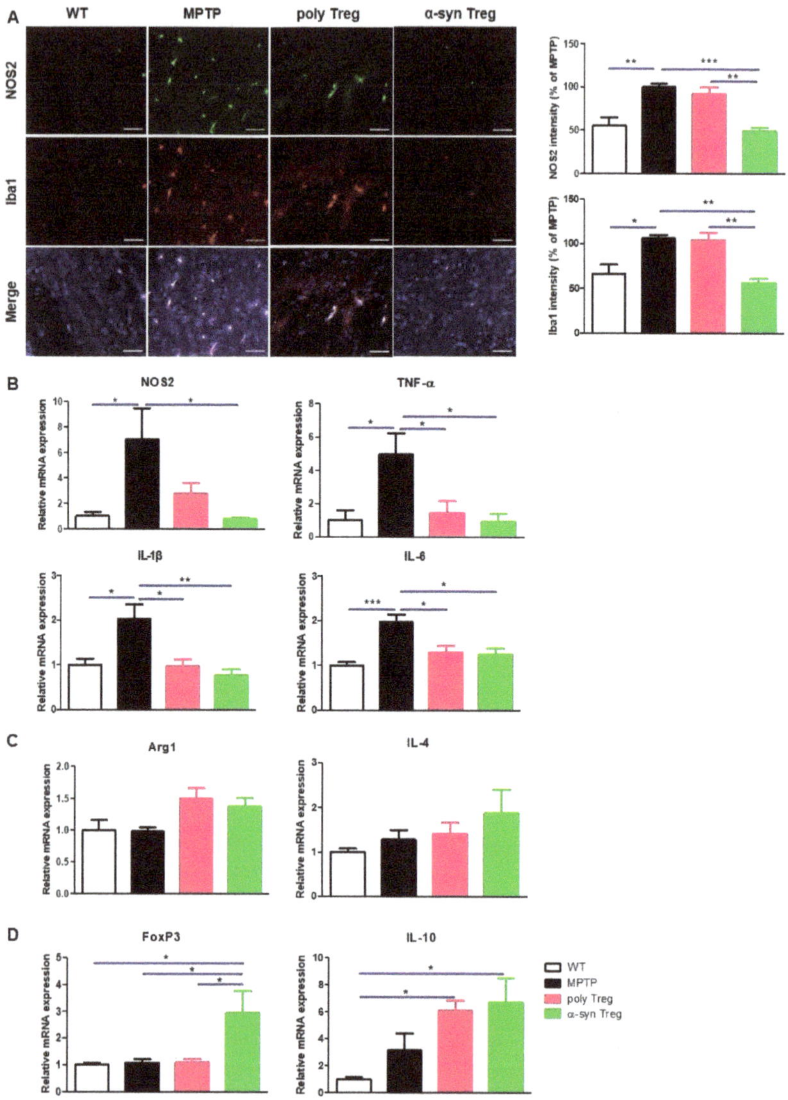

Figure 5. α-syn Tregs modulate microglial polarization. (**A**) The expression of NOS2 and Iba1 were detected in the SN. (**B–D**) The relative mRNA expressions of pro-inflammatory markers (NOS2, TNF-α, IL-1β, and IL-6), anti-inflammatory markers (Arg1, IL-4), and Treg markers (IL-10 and FoxP3) were analyzed in the brain. Data are presented as mean ± SEM. Statistical analyses were conducted with one-way ANOVA; * $p < 0.05$, ** $p < 0.01$, and *** $p < 0.001$. n = 4–6. Scale bars: 50 μm.

3. Discussion

Our present study proposes α-syn presentation as a method for enhancing the therapeutic effects of Tregs in PD. To prepare Tregs, $CD4^+CD25^+$ Tregs were isolated and expanded ex vivo after α-syn presentation via DCs. The α-syn Tregs showed the following characteristics. First, the expression of suppressive markers was steadily maintained. Second, these Tregs moved more frequently towards α-syn both in vitro and in vivo. Finally, Tregs showed better neuroprotective effects on PD pathology with improved motor deficits, dopaminergic neuronal loss, and α-syn accumulation.

Tregs tightly regulate immune balance in the CNS as immune suppressors. In neurodegenerative diseases, Treg dysfunction leads to neuroinflammation. Many studies have focused on the importance of Tregs in the pathology of neurodegenerative diseases [10,24]. Tregs are significantly reduced or show poor suppressive activity in mild AD, ALS, and multiple sclerosis (MS) [25–27]. In line with this, Treg therapy has been attempted, and its neuroprotective effects have been demonstrated in neurodegenerative diseases, including AD, ALS, and PD [13–15]. In patients and mouse models with PD, $CD4^+$ and $CD8^+$ T cells were demonstrated to infiltrate the brain [28,29]. Other studies have reported that $CD3^+$ T cells were accumulated in the brains of AD and MS patients [30,31]. These studies suggest the possibility of infiltration of adoptively transferred cells into the brain with neurodegenerative diseases and its potential as a therapeutic strategy for these diseases targeting adaptive immunity.

For efficient Treg therapy, it is important to diminish the number of Tregs to be administrated. Previously, Ashley et al. demonstrated that adoptively transfer of higher than 3.5×10^6 Tregs was necessary to elicit sufficient neuroprotective effects in a behavior test and TH immunoreactivity MPTP-mice model [13]. Similarly, in AD and ALS models, the neuroprotective effects of Tregs have been shown by transferring 1×10^6 or more Tregs which were cultured with stimulation for 4 days [14,15]. To obtain such an amount of Tregs in humans, ex vivo expansion is essential for patients and many protocols for the ex vivo expansion of Tregs have been developed [18,32,33]. In this study, we adoptively transferred only 5×10^5 ex vivo expanded Tregs to MPTP-intoxicated mice and demonstrated that they could show sufficient neuroprotective effects with α-syn presentation. However, since the Tregs should be isolated from peripheral blood mononuclear cells (PBMC) in humans for clinical Treg therapy, there is a need for adequate modifications with antigen presentation and expansion.

For the application of Treg therapy to actual patients, it is also desirable to increase their suppressive potential. Several studies suggest the importance of CD62L expression in Tregs for clinical manipulation in GVHD [34,35]. Highly expressed CD62L is rapidly lost after antigen experience [36]. However, we maintain the expression of CD62L using bee venom phospholipase A2 (bvPLA2), a Treg inducer maintaining CD62L expression during α-syn presentation [37]. Finally, we also examined the expression of Treg-specific genes. Areg is a critical factor for Treg suppressive function in vivo and in vitro [38]. Tregs produce Areg and IL-10 for tissue repair in damaged tissues, regulated by GATA3 [39]. These results demonstrate that our method of preparation of α-syn Treg enhances the efficiency of Treg therapy. A previous study has reported that in PD, peripheral DCs could migrate across the blood–brain barrier and uptake α-syn aggregates in the brain and then proceed to the cervical lymph nodes in which they enhance the MHC-II expression and α-syn presentation to activate antigen-specific T cells [40,41]. Ng et al. found that treatment of α-syn with monocyte-derived DCs stimulated robust inflammatory α-syn specific CD4+ T cell responses. They also found that both endo lysosomal and autophagic pathways are associated with α-syn, and treatment of DCs with α-syn appeared to downregulate the expression of Rab and autophagosomal proteins, suggesting the possibility that α-syn regulated its interaction with antigen processing components which limited its degradation and allowed the activation of T cells [40]. Alam et al. confirmed that the expression of co-stimulatory (CD80, CD83, and CD86) and MHC (HLA-ABC) molecules upregulated in DCs after treating α-syn to DCs. It has also been confirmed that α-syn can stimulate phenotypic and functional maturation of both DCs, endowing them with increased antigen-presenting capacity [42]. Like these studies, additional studies about antigen presentation of DCs need to be added in future experiments, although we presented α-syn to T cells via DCs and found that antigen presentation increases Tregs migration toward the brain region where the antigen exists (Figure 2).

Microglia, the primary immune cells in the CNS, play crucial roles in orchestrating brain inflammation [43]. Microglia may perform pathogenic or neuroprotective functions, depending on their phenotype. M1 microglia express NOS2 as a phenotypic marker and secrete pro-inflammatory cytokines such as IL-1β, IL-6, and TNF-α. M2 microglia, which express Arg1, are associated with tissue repair by producing IL-4 and IL-10 [23]. Activated microglia are involved in the pathogenesis of certain neurological diseases such as PD [44]. In the brains of patients with PD, an increased level of activated microglia is representative, and it causes neurodegeneration. In animal models, activation of microglia and secretion of M1-associated pro-inflammatory cytokines are induced by MPTP [23]. Additionally, α-syn acts as a stimulus that activates M1 microglia in the PD brain [45]. Therefore, targeting the balance of impaired pro- and anti-inflammatory microglia can be a strategy for disease modification [46]. Tregs ameliorate neuroinflammatory injury by regulating microglial polarization. In addition, loss of Tregs induced pro-inflammatory microglial activation in ICH and stroke models [47,48]. Indeed, studies attempting Treg therapy in neurodegenerative diseases have commonly observed the inhibition of microglial inflammation after Treg transfer [13–15]. Our findings are consistent with those of previous studies showing that Tregs inhibit neuroinflammation by modulating microglial polarization. Since Th1 and Th17 cells contribute to neuroinflammation by accompanying proinflammatory cytokines to induce neurotoxicity by microglia in PD [49], we additionally measured the expression level of IFN-g and IL-17 with qPCR. However, there were no significant effects of Tregs.

The current data showed that α-syn presentation improved the neuroprotective effects of Tregs in a PD mouse model by ameliorating neurodegeneration, α-syn accumulation, and microglial inflammation. Our data suggest that α-syn presentation may provide a significant improvement in the neuroprotective activities of Tregs, indicating a potential therapeutic value. Ultimately, our strategy may provide a promising avenue for immunotherapy with clinical relevance and cost-effectiveness for treating PD. However, there are certain limitations. It remains unclear how α-syn Tregs improved motor function, and unlike α-syn Tregs, why poly Tregs failed to do so. It also remains unclear how distinct α-syn specific Tregs may reduce inflammation and neurodegeneration. In this regard, further studies are required for a better understanding of the mechanisms of Treg therapy. Resolving this issue will enable the development of a more efficient protocol for Treg expansion.

The implication that α-syn presentation improved the suppressive potential and mobility of Tregs towards the disease site in a mouse model of PD is clear. These improvements increase the therapeutic effects of Tregs on the amelioration of PD progression by modulating microglial polarization (Figure 6).

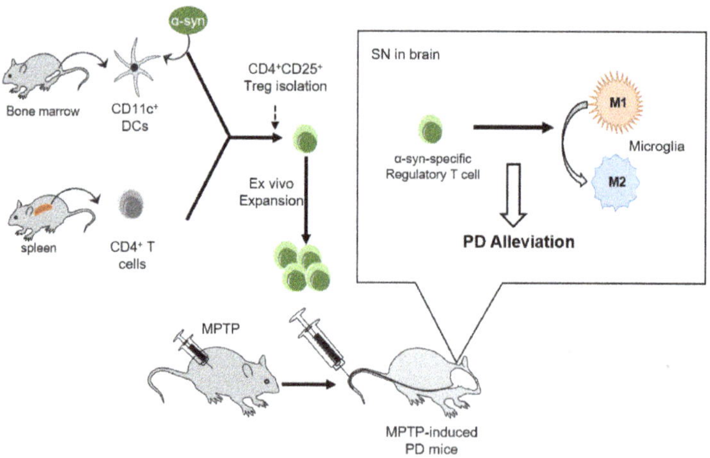

Figure 6. Adoptive transfer of α-syn-specific Tregs alleviated PD in MPTP-intoxicated mice. $CD11C^+$ DCs and $CD4^+$ T cells were isolated from the bone marrow and spleen, respectively. $CD4^+CD25^+$ Tregs were isolated and expanded ex vivo after α-syn presentation. Adoptive transfer of α-syn specific Tregs inhibited pro-inflammatory M1 microglial activation and alleviated PD pathologies such as motor dysfunction and dopaminergic neuronal loss in MPTP-induced PD mice.

4. Materials and Methods

4.1. Instrumentation

The instruments used in the experiment are listed as follows (Table 1).

Table 1. Instruments used in the study.

Instrumentation	Manufacturing Company/City, Country
LUNA-II automated cell counter	Logos Biosystems/Anyang, Republic of Korea
FACSLyric™ flow cytometer	BD Biosciences/San Jose, CA, USA
Cryomicrotome	Thermo Fisher Scientific/Waltham, MA, USA
Bright-field microscope	Nikon/Tokyo, Japan
LSM 800 confocal laser-scanning microscope	Carl Zeiss/Oberkochen, Germany
T100™ thermal cycler	BIORAD/Hercules, CA, USA
CFX connect Real-time PCR system	BIORAD/Hercules, CA, USA

4.2. Reagents

Alpha-synuclein (α-syn) was purchased from Prospec (cat #PRO-393). Granulocyte-macrophage colony-stimulating factor was purchased from R&D systems (Minneapolis, MN, USA) (cat #415-ML-010-CF). CD11c (cat #130-052-001) and $CD4^+CD25^+$ Regulatory T Cell Isolation Kit (cat #130-091-041) were purchased from Miltenyi Biotec Inc. (Auburn, CA, USA). 1-Methyl-4-phenyl-1,2,3,6-tetrahydropyridine (MPTP)-hydrochloride (HCl) was obtained from Sigma-Aldrich (St. Louis, MO, USA) (cat #M0896). PE-Cy7-CD4 (cat #25-0041-82), APC-CD62L (cat #17-0621-82), PE-Cy5-CD4 (cat #15-0041-82), and 7AAD (cat #00-6993-50) were obtained from Invitrogen (Waltham, MA, USA); the APC-Cy7-CD25 (cat #557658), mouse Vβ TCR Screening Panel (cat #557004), PE-cy7-CD4 (cat #552775), and BV421-Thy1.1 (cat #740044) were from BD Bioscience. To detect dopaminergic neurons, primary antibodies directed against tyrosine hydroxylase (TH, cat #P40101-150) and an avidin-biotin complex kit (Vectastain ABC kit, cat #PK-6102) were purchased from Pel-Freez Clinical System (Rogers, AR, USA) and Vector Laboratories (Newark, CA, USA), respectively. For immunofluorescence, the primary antibodies against ionized calcium-binding adaptor molecule 1 (Iba1 cat #ab178846) were from Abcam (Cambridge, UK); the primary antibodies against nitric oxide synthase 2 (NOS2 cat #sc-7271) and α-syn were from Santa Cruz Biotechnology (Dallas, TX, USA). Alexa 488- or 594-conjugated IgG secondary antibodies were from Invitrogen.

4.3. Animals

Male C57BL/6 mice and Thy1.1 (B6.PL-Thy1<a>/CyJ) mice were obtained from Jackson Laboratory (Bar Harbor, ME, USA). The mice were maintained on a 12 h light/dark cycle and under temperature-controlled conditions, with food and water ad libitum. All experiments were performed in accordance with approved animal protocols and guidelines established by Kyung Hee University (KHUASP(SE)-22-148 and 20-240).

4.4. Cell Preparation

To present the Parkinson's disease-specific antigen to Tregs via dendritic cells (DCs), α-syn (Prospec, East Brunswick, NJ, USA) was aggregated with incubation in a shaker at 37 °C for 7 days. To obtain bone marrow (BM)-DCs, BM-leukocytes were isolated from the femurs and tibiae of mice and resuspended in a medium containing 20 ng/mL granulocyte-macrophage colony-stimulating factor (R&D Systems, Minneapolis, MN, USA) [50]. After 7 days, $CD11c^+$ DCs were isolated using CD11c MicroBeads (Miltenyi Biotec Inc., Auburn, CA, USA) and seeded at a density of 2×10^5/mL in 96-well U-bottom plates. For antigen presentation, the samples were treated with (for α-syn Tregs) or without (for poly Treg) 2 μg/mL aggregated α-syn for 24 h. CD4 T cells were isolated from splenocytes using CD4 (L3T4) microbeads (Miltenyi Biotec Inc.). The cells were added to the DC at a ratio of 10:1 (CD4 T: DC) with 0.4 μg/mL bvPLA2 (Sigma-Aldrich, St. Louis, MO, USA). Four days after CD4 T cells-DC co-culture, $CD4^+CD25^+$ T cells (Tregs) were isolated using magnetic-activated cell sorting, according to the manufacturer's protocol ($CD4^+CD25^+$ Regulatory T Cell Isolation Kit; Miltenyi Biotec Inc.). The $CD4^+CD25^+$ T cells were stimulated using the Treg Expansion Kit (Miltenyi Biotec Inc.) for 14 days.

4.5. *Treg Migration Assay*

Treg migration was assayed using 24-well Transwell chambers with 5 μm pores. A total of 10^5 cells in 100 μL of media (RPMI 1640 with 0.5% BSA) were placed in the upper chambers. Aggregated α-syn, diluted in 600 μL media, was placed in the lower wells, and the chambers were incubated at 37 °C. After 3 and 4 h, the migrated cells present in the bottom wells were counted using a LUNA-II automated cell counter (Logos Biosystems, Anyang, Republic of Korea).

4.6. *Animal Experiment*

For MPTP (1-methyl-4-phenyl-1,2,3,6-tetrahydropyridine) intoxication, seven-week-old male mice received four i.p. injections of MPTP-HCl (12 mg/kg free base in saline; Sigma-Aldrich) at 2 h intervals as previously described [13]. A wild-type mouse, which did not receive the injections of MPTP-HCl, was prepared for the control group. Twelve hours after the last MPTP injection, the MPTP-intoxicated mice were randomly divided and received adoptive transfers of 5×10^5 Tregs (N = 8–10, total N = 37). The checklist for the guidelines for animal experiments is uploaded as a Supplementary File.

4.7. *Trafficking*

For trafficking of adoptively transferred Tregs, poly and α-syn Tregs were prepared from Thy 1.1-mice as described in the cell preparation section. MPTP-treated mice (Thy 1.2^+) were divided randomly into four groups and received an adoptive transfer of 1×10^6 Thy 1.1^+ Tregs. After 7 days, the mice were euthanized, and the inguinal lymph nodes, spleen, blood, lung, kidney, liver, and brain were harvested. T cells were enriched with 30–70% Percoll (Cytiva, Marlborough, MA, USA) density gradient centrifugation and debris removal solution (Miltenyi) from the lung, kidney, liver, and brain.

4.8. *Flow Cytometry*

For flow cytometry, cells were washed with BD FACS Stain buffer (BD Bioscience, San Jose, CA, USA) and stained with fluorescently labeled antibodies for 30 min at 4 °C in the dark. The following antibodies were used: PE-Cy7-CD4 (Invitrogen, Waltham, MA, USA), APC-CD62L (Invitrogen), and APC-Cy7-CD25 (BD Pharmingen, San Diego, CA, USA) for the Treg phenotype, PE-Cy5-CD4 (Invitrogen) and mouse Vβ TCR Screening Panel (BD Pharmingen) for the T cell receptor (TCR) repertoire, and PE-cy7-CD4, BV421-Thy1.1 (BD OptiBuild, San Jose, CA, USA), and 7AAD (Invitrogen) for Treg trafficking. All data were acquired using FACSLyric™ (BD Biosciences) and analyzed using the FACSuite software v1.2 (BD Biosciences).

4.9. *Pole Test*

Six days after the adoptive transfer of Tregs, a pole test was performed to determine forelimb and hindlimb motor coordination and balance. Briefly, the mice were placed on top of a gauze-banded wooden pole (50 cm in length and 0.8 cm in diameter) facing upward. The animals were allowed to climb down to the base of the pole. The time taken to turn completely downward and the total time taken for the mouse to reach the floor (time to down) were recorded. The maximum cut-off time to stop was 30 s.

4.10. *Tissue Processing and Immunohistochemistry*

Mice were anesthetized with isoflurane (Forane solution; ChoongWae Pharma, Seoul, Republic of Korea) and transcardially perfused with phosphate-buffered saline (PBS). Brains were dissected and divided in half. The brain halves were postfixed in 4% paraformaldehyde for 18 h at 4 °C, transferred to 30% sucrose solution, and subsequently frozen. Tissues were serially cut on a cryostat into 30-μm-thick coronal sections using a cryomicrotome (HM525 NX; Thermo Fisher Sientific, Inc., Waltham, MA, USA).

To detect dopaminergic neurons, primary antibodies were directed against tyrosine hydroxylase (TH; 1:2000, Pel-Freez Clinical System, Rogers, AR, USA). The sections were washed with PBS, incubated with the appropriate biotinylated secondary antibody, and processed using an avidin-biotin complex kit (Vectastain ABC kit; Vector Laboratories, Newark, CA, USA) for 1 h at 25 °C. The reaction product was visualized with 0.05% diaminobenzidine-HCl and 0.003% hydrogen peroxide in 0.1 M phosphate buffer. The labeled tissue sections were subsequently mounted and analyzed under a bright-field microscope (Nikon, Tokyo, Japan). An unbiased stereological estimation of the total number of TH-positive dopaminergic neurons in the SN was performed using the optical fractionator method on an Olympus computer-assisted stereological toolbox system version 2.1.4. (Olympus, Tokyo, Japan) as previously described [51]. The sections used for counting covered the entire SN, from the rostral tip of the pars compacta to the caudal end of the pars reticulata.

For immunofluorescence, brain sections were incubated with 50% formic acid for 10 min at 25 °C and heated with 10 mM sodium citrate buffer (pH 6.0) for epitope retrieval. After washing with cold PBS, nonspecific binding was reduced by blocking the sections with 5% BSA in 0.2% Triton X-100 in TBS for 30 min at 25 °C. The sections were incubated with antibodies against ionized calcium-binding adaptor molecule 1 (Iba1; 1:1000, Abcam, Cambridge, UK), nitric oxide synthase 2 (NOS2; 1:500, Santa Cruz Biotechnology, Dallas, TX, USA), and α-syn (1:100, Santa Cruz Biotechnology) O/N at 4 °C. Brain sections were washed with TBSTr, incubated for 2 h at 25 °C with Alexa 488- or 594-conjugated IgG secondary antibodies, and then counterstained with DAPI. The tissues were examined using an LSM 800 confocal laser-scanning microscope (Carl Zeiss, Oberkochen, Germany). The staining intensity was quantified by measuring the integral density of the region of interest from monochromatic images using the ImageJ software v1.52a. The percentage of staining intensity was calculated relative to the MPTP group and multiplied by 100.

4.11. RNA Extraction and Quantitative Real-Time PCR (qRT-PCR)

Total RNA was isolated from Tregs or unfixed brain halves using the easy-BLUE RNA extraction kit (iNtRON Biotechnology, Seongnam, Republic of Korea), and cDNA was synthesized using Cyclescript reverse transcriptase (Bioneer, Daejeon, Republic of Korea). Samples were prepared for real-time PCR using the SensiFAST SYBR no-Rox kit (Bioline, Randolph, MA, USA). The cycling conditions were: 1 cycle at 95 °C for 30 s, 49 cycles at 95 °C for 10 s, 55 °C for 30 s, followed by a melting curve at 9 5 °C for 10 s, 50 °C for 5 s, and then a gradual increase until 95 °C was reached. The base sequences of the primers were as follows: β-actin, forward: 5′-GTG CTA TGT TGC TCT AGA CTT CG-3′ and reverse: 5′-ATG CCA CAG GAT TCC ATA CC-3′; FoxP3: forward: 5′ CTG CTC CTC CTA TTC CCG TAA C-3′ and reverse: 5′-AGC TAG AGG CTT TGC CTT CG-3′; GATA3: forward: 5′-GAA GGC ATC CAG ACC CGA AAC 3′ and reverse: 5′-ACC CAT GGC GGT GAC CAT GC-3′; IL-10: forward: 5′-CAG CCG GGA AGA CAA TAA CTG-3′ and reverse: 5′-CCG CAG CTC TAG GAG CAT GT-3′; Areg: forward: 5′-ACT GTG CAT GCC ATT GCC TA-3′ and reverse: 5′-ACT GGG CAT CTG GAA CCA TC-3′; NOS2: forward: 5′-AGG ACA TCC TGC GGC AGC-3′ and reverse: 5′-GCT TTA ACC CCT CCT GTA-3′; TNF-α: forward: 5′-GGC AGG TTC TGT CCC TTT CAC-3′ and reverse: 5′-TTC TGT GCT CAT GGT GTC TTT TCT-3′; IL-1β: forward: 5′-AAG CCT CCT CCT CTC CCA CC 3′ and reverse: 5′-TGA GGC CCA AGG CCA CAG G-3′; IL-6: forward: 5′-TTC CAT CCA GTT GCC TTC TTG-3′ and reverse: 5′-GGG AGT GGT ATC CTC TGT GAA GTC-3′; Arg1: forward: 5′-CTC CAA GCC AAA GTC CTT AGA G-3′ and reverse: 5′-AGG AGC TGT CAT TAG GGA CAT C-3′; IL-4: forward: 5′-ATC CTG CTC TTC TTT CTC GAA TGT-3′ and reverse: 5′-GCC GAT GAT CTC TCT CAA GTG ATT-3′.

4.12. Statistical Analysis

All data were analyzed using GraphPad Prism 5.01 (GraphPad Software Inc., San Diego, CA, USA). The data are presented as the mean and standard error of the mean (SEM), where indicated. The statistical significance of each variable was evaluated with a two-tailed *t*-test for single comparison and one-way analysis of variance (ANOVA), followed by Tukey's multiple comparison test for multiple comparisons. Multiple comparisons within groups were analyzed using a two-way ANOVA, followed by Bonferroni post-hoc tests. All experiments were performed in a blind manner and were repeated independently under identical conditions. Statistical significance was set at $p < 0.05$.

Supplementary Materials: The following supporting information can be downloaded at: https://www.mdpi.com/article/10.3390/ijms242015237/s1.

Author Contributions: Conceptualization, H.B.; methodology, S.-Y.P.; formal analysis, S.K., J.Y. and H.G.; writing—original draft preparation, S.-Y.P.; writing—review and editing, S.-Y.P., H.Y. and H.B.; supervision, H.B.; funding acquisition, H.B. All authors have read and agreed to the published version of the manuscript.

Funding: This work was supported by the National Research Foundation of Korea (NRF) grant, funded by the Korean government (MSIT) (RS-2023-00208347).

Institutional Review Board Statement: The animal study protocol was approved by the Institutional Animal Care and Use Committee of Kyung Hee University (KHUASP(SE)-22-148 and 20-240).

Informed Consent Statement: Not applicable.

Data Availability Statement: All data generated or analyzed during this study are included in this article.

Conflicts of Interest: The authors declare no conflict of interest.

References

1. Jankovic, J. Parkinson's disease: Clinical features and diagnosis. *J. Neurol. Neurosurg. Psychiatry* **2008**, *79*, 368–376. [CrossRef]
2. De Virgilio, A.; Greco, A.; Fabbrini, G.; Inghilleri, M.; Rizzo, M.I.; Gallo, A.; Conte, M.; Rosato, C.; Ciniglio Appiani, M.; de Vincentiis, M. Parkinson's disease: Autoimmunity and neuroinflammation. *Autoimmun. Rev.* **2016**, *15*, 1005–1011. [CrossRef]
3. Bezard, E.; Brotchie, J.M.; Gross, C.E. Pathophysiology of levodopa-induced dyskinesia: Potential for new therapies. *Nat. Rev. Neurosci.* **2001**, *2*, 577–588. [CrossRef]
4. Jiang, T.; Sun, Q.; Chen, S. Oxidative stress: A major pathogenesis and potential therapeutic target of antioxidative agents in Parkinson's disease and Alzheimer's disease. *Prog. Neurobiol.* **2016**, *147*, 1–19. [CrossRef]
5. Su, R.; Zhou, T. Alpha-Synuclein Induced Immune Cells Activation and Associated Therapy in Parkinson's Disease. *Front. Aging Neurosci.* **2021**, *13*, 769506. [CrossRef]
6. Glass, C.K.; Saijo, K.; Winner, B.; Marchetto, M.C.; Gage, F.H. Mechanisms underlying inflammation in neurodegeneration. *Cell* **2010**, *140*, 918–934. [CrossRef]
7. D'Amico, R.; Gugliandolo, E.; Siracusa, R.; Cordaro, M.; Genovese, T.; Peritore, A.F.; Crupi, R.; Interdonato, L.; Di Paola, D.; Cuzzocrea, S.; et al. Toxic Exposure to Endocrine Disruptors Worsens Parkinson's Disease Progression through NRF2/HO-1 Alteration. *Biomedicines* **2022**, *10*, 1073. [CrossRef]
8. Cordaro, M.; Siracusa, R.; Crupi, R.; Impellizzeri, D.; Peritore, A.F.; D'Amico, R.; Gugliandolo, E.; Di Paola, R.; Cuzzocrea, S. 2-Pentadecyl-2-Oxazoline Reduces Neuroinflammatory Environment in the MPTP Model of Parkinson Disease. *Mol. Neurobiol.* **2018**, *55*, 9251–9266. [CrossRef]
9. Czlonkowska, A.; Kohutnicka, M.; Kurkowska-Jastrzebska, I.; Czlonkowski, A. Microglial reaction in MPTP (1-methyl-4-phenyl-1,2,3,6-tetrahydropyridine) induced Parkinson's disease mice model. *Neurodegeneration* **1996**, *5*, 137–143. [CrossRef]
10. He, F.; Balling, R. The role of regulatory T cells in neurodegenerative diseases. *Wiley Interdiscip. Rev. Syst. Biol. Med.* **2013**, *5*, 153–180. [CrossRef]
11. Bailey-Bucktrout, S.L.; Bluestone, J.A. Regulatory T cells: Stability revisited. *Trends Immunol.* **2011**, *32*, 301–306. [CrossRef]
12. Gendelman, H.E.; Appel, S.H. Neuroprotective activities of regulatory T cells. *Trends Mol. Med.* **2011**, *17*, 687–688. [CrossRef]
13. Reynolds, A.D.; Banerjee, R.; Liu, J.; Gendelman, H.E.; Mosley, R.L. Neuroprotective activities of $CD4^+CD25^+$ regulatory T cells in an animal model of Parkinson's disease. *J. Leukoc. Biol.* **2007**, *82*, 1083–1094. [CrossRef]
14. Baek, H.; Ye, M.; Kang, G.H.; Lee, C.; Lee, G.; Choi, D.B.; Jung, J.; Kim, H.; Lee, S.; Kim, J.S.; et al. Neuroprotective effects of $CD4^+CD25^+Foxp3^+$ regulatory T cells in a 3xTg-AD Alzheimer's disease model. *Oncotarget* **2016**, *7*, 69347–69357. [CrossRef]

15. Beers, D.R.; Henkel, J.S.; Zhao, W.; Wang, J.; Huang, A.; Wen, S.; Liao, B.; Appel, S.H. Endogenous regulatory T lymphocytes ameliorate amyotrophic lateral sclerosis in mice and correlate with disease progression in patients with amyotrophic lateral sclerosis. *Brain* **2011**, *134*, 1293–1314. [CrossRef]
16. Bluestone, J.A. Regulatory T-cell therapy: Is it ready for the clinic? *Nat. Rev. Immunol.* **2005**, *5*, 343–349. [CrossRef]
17. Rana, J.; Biswas, M. Regulatory T cell therapy: Current and future design perspectives. *Cell. Immunol.* **2020**, *356*, 104193. [CrossRef]
18. Tang, Q.; Henriksen, K.J.; Bi, M.; Finger, E.B.; Szot, G.; Ye, J.; Masteller, E.L.; McDevitt, H.; Bonyhadi, M.; Bluestone, J.A. In vitro-expanded antigen-specific regulatory T cells suppress autoimmune diabetes. *J. Exp. Med.* **2004**, *199*, 1455–1465. [CrossRef]
19. Trenado, A.; Sudres, M.; Tang, Q.; Maury, S.; Charlotte, F.; Gregoire, S.; Bonyhadi, M.; Klatzmann, D.; Salomon, B.L.; Cohen, J.L. Ex vivo-expanded $CD4^{+}CD25^{+}$ immunoregulatory T cells prevent graft-versus-host-disease by inhibiting activation/differentiation of pathogenic T cells. *J. Immunol.* **2006**, *176*, 1266–1273. [CrossRef]
20. Selck, C.; Dominguez-Villar, M. Antigen-Specific Regulatory T Cell Therapy in Autoimmune Diseases and Transplantation. *Front. Immunol.* **2021**, *12*, 661875. [CrossRef]
21. Sedelis, M.; Schwarting, R.K.; Huston, J.P. Behavioral phenotyping of the MPTP mouse model of Parkinson's disease. *Behav. Brain Res.* **2001**, *125*, 109–125. [CrossRef]
22. Bertram, L.; Tanzi, R.E. The genetic epidemiology of neurodegenerative disease. *J. Clin. Investig.* **2005**, *115*, 1449–1457. [CrossRef]
23. Subramaniam, S.R.; Federoff, H.J. Targeting Microglial Activation States as a Therapeutic Avenue in Parkinson's Disease. *Front. Aging Neurosci.* **2017**, *9*, 176. [CrossRef]
24. Kipnis, J.; Avidan, H.; Caspi, R.R.; Schwartz, M. Dual effect of $CD4^{+}CD25^{+}$ regulatory T cells in neurodegeneration: A dialogue with microglia. *Proc. Natl. Acad. Sci. USA* **2004**, *101* (Suppl. 2), 14663–14669. [CrossRef]
25. Larbi, A.; Pawelec, G.; Witkowski, J.M.; Schipper, H.M.; Derhovanessian, E.; Goldeck, D.; Fulop, T. Dramatic shifts in circulating CD4 but not CD8 T cell subsets in mild Alzheimer's disease. *J. Alzheimer's Dis.* **2009**, *17*, 91–103. [CrossRef]
26. Rentzos, M.; Evangelopoulos, E.; Sereti, E.; Zouvelou, V.; Marmara, S.; Alexakis, T.; Evdokimidis, I. Alterations of T cell subsets in ALS: A systemic immune activation? *Acta Neurol. Scand.* **2012**, *125*, 260–264. [CrossRef]
27. Bhela, S.; Kempsell, C.; Manohar, M.; Dominguez-Villar, M.; Griffin, R.; Bhatt, P.; Kivisakk-Webb, P.; Fuhlbrigge, R.; Kupper, T.; Weiner, H.; et al. Nonapoptotic and extracellular activity of granzyme B mediates resistance to regulatory T cell (Treg) suppression by HLA-DR-CD25hiCD127lo Tregs in multiple sclerosis and in response to IL-6. *J. Immunol.* **2015**, *194*, 2180–2189. [CrossRef]
28. Brochard, V.; Combadiere, B.; Prigent, A.; Laouar, Y.; Perrin, A.; Beray-Berthat, V.; Bonduelle, O.; Alvarez-Fischer, D.; Callebert, J.; Launay, J.M.; et al. Infiltration of $CD4^{+}$ lymphocytes into the brain contributes to neurodegeneration in a mouse model of Parkinson disease. *J. Clin. Investig.* **2009**, *119*, 182–192. [CrossRef]
29. Williams, G.P.; Schonhoff, A.M.; Jurkuvenaite, A.; Gallups, N.J.; Standaert, D.G.; Harms, A.S. CD4 T cells mediate brain inflammation and neurodegeneration in a mouse model of Parkinson's disease. *Brain* **2021**, *144*, 2047–2059. [CrossRef]
30. Togo, T.; Akiyama, H.; Iseki, E.; Kondo, H.; Ikeda, K.; Kato, M.; Oda, T.; Tsuchiya, K.; Kosaka, K. Occurrence of T cells in the brain of Alzheimer's disease and other neurological diseases. *J. Neuroimmunol.* **2002**, *124*, 83–92. [CrossRef]
31. Babbe, H.; Roers, A.; Waisman, A.; Lassmann, H.; Goebels, N.; Hohlfeld, R.; Friese, M.; Schroder, R.; Deckert, M.; Schmidt, S.; et al. Clonal expansions of $CD8^{+}$ T cells dominate the T cell infiltrate in active multiple sclerosis lesions as shown by micromanipulation and single cell polymerase chain reaction. *J. Exp. Med.* **2000**, *192*, 393–404. [CrossRef]
32. Godfrey, W.R.; Ge, Y.G.; Spoden, D.J.; Levine, B.L.; June, C.H.; Blazar, B.R.; Porter, S.B. In vitro-expanded human $CD4^{+}CD25^{+}$ T-regulatory cells can markedly inhibit allogeneic dendritic cell-stimulated MLR cultures. *Blood* **2004**, *104*, 453–461. [CrossRef]
33. Hoffmann, P.; Eder, R.; Kunz-Schughart, L.A.; Andreesen, R.; Edinger, M. Large-scale in vitro expansion of polyclonal human $CD4^{+}CD25^{high}$ regulatory T cells. *Blood* **2004**, *104*, 895–903. [CrossRef]
34. Fu, S.; Yopp, A.C.; Mao, X.; Chen, D.; Zhang, N.; Chen, D.; Mao, M.; Ding, Y.; Bromberg, J.S. $CD4^{+}$ $CD25^{+}$ $CD62^{+}$ T-regulatory cell subset has optimal suppressive and proliferative potential. *Am. J. Transplant.* **2004**, *4*, 65–78. [CrossRef]
35. Ermann, J.; Hoffmann, P.; Edinger, M.; Dutt, S.; Blankenberg, F.G.; Higgins, J.P.; Negrin, R.S.; Fathman, C.G.; Strober, S. Only the $CD62L^{+}$ subpopulation of $CD4^{+}CD25^{+}$ regulatory T cells protects from lethal acute GVHD. *Blood* **2005**, *105*, 2220–2226. [CrossRef]
36. Anderson, B.E.; McNiff, J.; Yan, J.; Doyle, H.; Mamula, M.; Shlomchik, M.J.; Shlomchik, W.D. Memory $CD4^{+}$ T cells do not induce graft-versus-host disease. *J. Clin. Investig.* **2003**, *112*, 101–108. [CrossRef]
37. Baek, H.; Park, S.Y.; Ku, S.J.; Ryu, K.; Kim, Y.; Bae, H.; Lee, Y.S. Bee Venom Phospholipase A2 Induces Regulatory T Cell Populations by Suppressing Apoptotic Signaling Pathway. *Toxins* **2020**, *12*, 198. [CrossRef]
38. Zaiss, D.M.; van Loosdregt, J.; Gorlani, A.; Bekker, C.P.; Grone, A.; Sibilia, M.; van Bergen en Henegouwen, P.M.; Roovers, R.C.; Coffer, P.J.; Sijts, A.J. Amphiregulin enhances regulatory T cell-suppressive function via the epidermal growth factor receptor. *Immunity* **2013**, *38*, 275–284. [CrossRef]
39. Delacher, M.; Imbusch, C.D.; Weichenhan, D.; Breiling, A.; Hotz-Wagenblatt, A.; Trager, U.; Hofer, A.C.; Kagebein, D.; Wang, Q.; Frauhammer, F.; et al. Genome-wide DNA-methylation landscape defines specialization of regulatory T cells in tissues. *Nat. Immunol.* **2017**, *18*, 1160–1172. [CrossRef]
40. Ng, L.; Wang, X.; Yang, C.; Su, C.; Li, M.; Cheung, A.K.L. Corrigendum: Celastrol Downmodulates Alpha-Synuclein-Specific T Cell Responses by Mediating Antigen Trafficking in Dendritic Cells. *Front. Immunol.* **2022**, *13*, 907993. [CrossRef]
41. Garretti, F.; Monahan, C.; Sette, A.; Agalliu, D.; Sulzer, D. T cells, alpha-synuclein and Parkinson disease. *Handb. Clin. Neurol.* **2022**, *184*, 439–455. [CrossRef]

42. Alam, M.M.; Yang, D.; Li, X.Q.; Liu, J.; Back, T.C.; Trivett, A.; Karim, B.; Barbut, D.; Zasloff, M.; Oppenheim, J.J. Alpha synuclein, the culprit in Parkinson disease, is required for normal immune function. *Cell Rep.* **2022**, *38*, 110090. [CrossRef]
43. Hansen, D.V.; Hanson, J.E.; Sheng, M. Microglia in Alzheimer's disease. *J. Cell Biol.* **2018**, *217*, 459–472. [CrossRef]
44. McGeer, P.L.; Itagaki, S.; Boyes, B.E.; McGeer, E.G. Reactive microglia are positive for HLA-DR in the substantia nigra of Parkinson's and Alzheimer's disease brains. *Neurology* **1988**, *38*, 1285–1291. [CrossRef]
45. Benner, E.J.; Banerjee, R.; Reynolds, A.D.; Sherman, S.; Pisarev, V.M.; Tsiperson, V.; Nemachek, C.; Ciborowski, P.; Przedborski, S.; Mosley, R.L.; et al. Nitrated alpha-synuclein immunity accelerates degeneration of nigral dopaminergic neurons. *PLoS ONE* **2008**, *3*, e1376. [CrossRef]
46. Pisanu, A.; Lecca, D.; Mulas, G.; Wardas, J.; Simbula, G.; Spiga, S.; Carta, A.R. Dynamic changes in pro- and anti-inflammatory cytokines in microglia after PPAR-gamma agonist neuroprotective treatment in the MPTPp mouse model of progressive Parkinson's disease. *Neurobiol. Dis.* **2014**, *71*, 280–291. [CrossRef]
47. Zhou, K.; Zhong, Q.; Wang, Y.C.; Xiong, X.Y.; Meng, Z.Y.; Zhao, T.; Zhu, W.Y.; Liao, M.F.; Wu, L.R.; Yang, Y.R.; et al. Regulatory T cells ameliorate intracerebral hemorrhage-induced inflammatory injury by modulating microglia/macrophage polarization through the IL-10/GSK3beta/PTEN axis. *J. Cereb. Blood Flow Metab.* **2017**, *37*, 967–979. [CrossRef]
48. Liesz, A.; Suri-Payer, E.; Veltkamp, C.; Doerr, H.; Sommer, C.; Rivest, S.; Giese, T.; Veltkamp, R. Regulatory T cells are key cerebroprotective immunomodulators in acute experimental stroke. *Nat. Med.* **2009**, *15*, 192–199. [CrossRef]
49. Appel, S.H.; Beers, D.R.; Henkel, J.S. T cell-microglial dialogue in Parkinson's disease and amyotrophic lateral sclerosis: Are we listening? *Trends Immunol.* **2010**, *31*, 7–17. [CrossRef]
50. Lutz, M.B.; Kukutsch, N.; Ogilvie, A.L.; Rossner, S.; Koch, F.; Romani, N.; Schuler, G. An advanced culture method for generating large quantities of highly pure dendritic cells from mouse bone marrow. *J. Immunol. Methods* **1999**, *223*, 77–92. [CrossRef]
51. Chung, E.S.; Kim, H.; Lee, G.; Park, S.; Kim, H.; Bae, H. Neuro-protective effects of bee venom by suppression of neuroinflammatory responses in a mouse model of Parkinson's disease: Role of regulatory T cells. *Brain Behav. Immun.* **2012**, *26*, 1322–1330. [CrossRef]

International Journal of
Molecular Sciences

MDPI

Review

Biological Function Analysis of MicroRNAs and Proteins in the Cerebrospinal Fluid of Patients with Parkinson's Disease

Ji Su Hwang [1,2,†], Seok Gi Kim [1,2,†], Nimisha Pradeep George [1,2], Minjun Kwon [1,2], Yong Eun Jang [1,2], Sang Seop Lee [3] and Gwang Lee [1,2,*]

1 Department of Molecular Science and Technology, Ajou University, Suwon 16499, Republic of Korea; js3004@ajou.ac.kr (J.S.H.); rlatjrrl9977@ajou.ac.kr (S.G.K.); nimishapgeorge@ajou.ac.kr (N.P.G.); kmj936@ajou.ac.kr (M.K.); jye120@ajou.ac.kr (Y.E.J.)
2 Department of Physiology, Ajou University School of Medicine, Suwon 16499, Republic of Korea
3 Department of Pharmacology, Inje University College of Medicine, Busan 47392, Republic of Korea; leess@inje.ac.kr
* Correspondence: glee@ajou.ac.kr; Tel.: +82-31-219-4554
† These authors contributed equally to this work.

Abstract: Parkinson's disease (PD) is a progressive neurodegenerative disorder characterized by alpha-synuclein aggregation into Lewy bodies in the neurons. Cerebrospinal fluid (CSF) is considered the most suited source for investigating PD pathogenesis and identifying biomarkers. While microRNA (miRNA) profiling can aid in the investigation of post-transcriptional regulation in neurodegenerative diseases, information on miRNAs in the CSF of patients with PD remains limited. This review combines miRNA analysis with proteomic profiling to explore the collective impact of CSF miRNAs on the neurodegenerative mechanisms in PD. We constructed separate networks for altered miRNAs and proteomes using a bioinformatics method. Altered miRNAs were poorly linked to biological functions owing to limited information; however, changes in protein expression were strongly associated with biological functions. Subsequently, the networks were integrated for further analysis. In silico prediction from the integrated network revealed relationships between miRNAs and proteins, highlighting increased reactive oxygen species generation, neuronal loss, and neurodegeneration and suppressed ATP synthesis, mitochondrial function, and neurotransmitter release in PD. The approach suggests the potential of miRNAs as biomarkers for critical mechanisms underlying PD. The combined strategy could enhance our understanding of the complex biochemical networks of miRNAs in PD and support the development of diagnostic and therapeutic strategies for precision medicine.

Keywords: cerebrospinal fluid; integrated omics; microRNA; Parkinson's disease; proteomes

1. Introduction

Parkinson's disease (PD) is a chronic and progressive neurodegenerative disorder ranking second after Alzheimer's disease (AD) [1]. PD is marked by the formation of Lewy bodies (cytoplasmic inclusions containing alpha-synuclein (α-syn)) along with the degeneration of dopaminergic (DA) neurons in the substantia nigra (SN) [2–4]. Post-mortem studies of PD brains have shown a reduction in DA neuronal viability in the SN by approximately $\geq$30%, leading to motor impairments, such as tremors, rigidity, and bradykinesia, as well as non-motor features, including reduced olfactory ability, cognitive impairment, and psychiatric symptoms [5]. However, such hallmark symptoms of PD often overlap with those of other neurodegenerative diseases [6], posing a challenge for its early diagnosis. Moreover, current diagnostic practices primarily rely on clinical assessments [7], underscoring the necessity for refined diagnostic criteria.

Given the dopamine deficiency following DA neuronal death, medications such as levodopa, dopamine agonists, and monoamine oxidase B inhibitors have been developed to supplement dopamine action and have shown beneficial effects in the early stages of the disease [8–10]. However, unwanted side effects, such as hallucinations and impulse control disorders, remain a concern for patients [11,12]. Additionally, multiple system atrophy (MSA) shares several common impairments with PD but differs from it in terms of disease progression and response to dopaminergic treatment, hindering the fast and accurate diagnosis of PD [13,14]. Therefore, elucidation of the physiological mechanisms and development of critical biomarkers for discriminating PD from other neurodegenerative disorders are urgently required.

Numerous studies on the pathology of PD have identified several contributing factors, including genetic mutations, protein misfolding, oxidative stress, and mitochondrial dysfunction [15–18]. However, no definitive biomarker has been identified yet. To this end, a substantial number of studies have focused on changes in biomolecules in the brain and/or body fluids, including saliva, urine, blood, and cerebrospinal fluid (CSF) [19–22].

Among the fluids, CSF is increasingly regarded as a valuable source for research on PD owing to its proximity to the central nervous system (CNS) [23]. Unlike peripheral fluids, CSF directly interfaces with the extracellular space of the brain, allowing for an unrestricted two-way exchange of molecules between these regions [7]. In contrast, brain-derived proteins are typically not detected in blood-derived samples [24]. Approximately 20% of the proteins in CSF are derived specifically from brain cells, while the remainder originate from peripheral blood filtration [25]. Despite the relatively low proportion of brain-derived components, CSF is preferred over other sources to accurately mirror pathophysiological conditions [7]. In addition, CSF from PD patients was found to include toxic factors for dopaminergic neurons [26], highlighting its usefulness for PD research. Consequently, alterations in biomolecules, including transcripts, microRNAs (miRNAs), proteins, and metabolites, in the CSF are actively being evaluated using both experimental and bioinformatics methods [27–30].

Among the biomolecules of interest, miRNAs have emerged as important regulators of gene expression and translation [31]. These small non-coding RNAs, approximately 22 nucleotides in length [32], were first identified in *Caenorhabditis elegans* and regulate gene expression by interacting with complementary sequences on target mRNA through antisense RNA–RNA interactions [33]. miRNAs are implicated in various biological processes from the development, differentiation, proliferation, cell death, and immune systems [34]. In the context of neurodegenerative diseases, dysregulated miRNAs influence key pathways, such as neuroinflammation [35–37], oxidative stress [38–40], α-syn aggregation [41–43], and mitochondrial dysfunction [44–46]. These disruptions are particularly critical in PD, highlighting their significant role in disease onset and progression [47–51]. Consequently, significant efforts have been made to analyze various miRNAs in the CSF of patients with PD to identify potential miRNA biomarkers associated with these pathways [52–57]. However, the physiological roles of altered CSF miRNAs in patients with PD remain poorly understood, indicating the need for further research to elucidate their functions and diagnostic potential.

In this review, we aimed to employ an integrative strategy that combines miRNA and proteomic analyses to overcome the limitations of miRNAs and their unclear biological roles. Datasets from both miRNAomic and proteomic studies were collected from the published literature and analyzed to predict the biological functions related to PD using the bioinformatics tool Ingenuity Pathway Analysis (IPA, http://www.ingenuity.com, accessed on 28 October 2024) [58]. This integrative approach enabled the simultaneous examination of miRNAs and proteins, offering a more detailed understanding of the molecular mechanisms and biological functions underlying PD. Moreover, this analysis could uncover potential therapeutic targets by identifying miRNAs that modulate neuroprotective or neurodegenerative processes, thereby providing valuable insights into biomarker discovery and therapeutic development for PD. Herein, we divide this review into four sections as follows: (1) the advantages of CSF in PD diagnosis, (2) CSF miRNA in PD,

(3) CSF protein in PD, and (4) the integration of miRNA and protein of the CSF of patients with PD.

2. Advantages of CSF in PD Diagnosis

Several studies have aimed to identify feasible biomarkers for the early diagnosis of PD by analyzing biomolecules, such as transcripts, miRNAs, proteins, and metabolites, in the blood of patients [20,59–64] and animal models [29,65–67]. Although blood-derived signatures are relatively easy to obtain, and associations be-tween molecular changes in the blood and PD continue to emerge [68], these studies are generally deemed less effective than CSF-derived studies in capturing the intricate mechanisms underlying PD [69].

CSF is primarily secreted by the choroid plexus [70,71]; its close contact with the extracellular space of the brain provides it distinct advantages for reflecting the state of the CNS [72]. Critical proteins involved in the pathology of PD and related disorders, such as tau, neuron-specific enolase, deglycase-1, chitinase-3-like protein 1, and α-syn, have been detected in CSF, underscoring the necessity of CSF analysis [44,73]. Notably, as aging progresses, the turnover and exchange of CSF components decreases, leading to the accumulation of proteins and other molecules, which may serve as potential biomarkers for PD [70,74]. Leveraging CSF analysis to elucidate PD mechanisms, discover biomarkers, and aid in the early diagnosis of PD may significantly enhance our understanding of its pathophysiology and lead to the development of more precise diagnostic tools, ultimately improving the outcomes of patients with PD.

3. CSF miRNA in PD

Emerging evidence suggests that dysregulated miRNAs, such as miR-7 and miR-153, are closely related to PD pathogenesis [75,76], and their presence in various body fluids underscores the potential of miRNAs as biomarkers of PD [77]. Dos Santos et al. identified three promising CSF miRNAs, namely miR-10b-5p, miR-22-3p, and miR-151a-3p, which are involved in PD pathogenesis, using small RNA sequencing and biomarker panel identification via machine learning techniques [78]. These findings suggest the diagnostic applicability of CSF miRNAs with their notable stability, quantifiability, and cost-effectiveness in the detection of neurodegenerative diseases, such as PD [79,80], MSA [56,81], amyotrophic lateral sclerosis (ALS) [82], and AD [83,84].

Building on this, miRNAs associated with PD pathology show limited overlap and different expression patterns in various biofluids [49,85]. Specifically, the miRNAs let-7g-3p, miR-10a-5p, miR-409-3p, miR-324-3p, and miR-205-5p were found to be upregulated in the CSF but not in other samples, such as whole blood, plasma, serum, and peripheral blood mononuclear cells [49,85]. Moreover, miR-7-5p was found to be downregulated in CSF but upregulated in serum [49]. This feature underscores the importance of analyzing miRNAs specifically in the CSF, which is structurally adjacent to the brain and more directly representative of the brain's environment, to identify disease-specific biomarkers rather than relying on blood or peripheral fluids.

Subsequently, we compiled a list of miRNAs from CSF that were significantly altered and validated using various analytical methods to investigate PD pathologies based on the published literature (Table 1). A total of 77 miRNAs were collected; 54 were upregulated and 23 were downregulated in the CSF of patients with PD. These changes in miRNAs were analyzed, and the miRNAomic network was constructed using IPA to explore their associations with biological functions and diseases related to PD.

Table 1. List of miRNAs that are altered in the cerebrospinal fluid of patients with Parkinson's disease.

miRNA Symbol	miRBase ID	Expression Pattern	Analysis Method [a]	Reference
miR-106b-5p	MIMAT0000680	Increased	RT-qPCR	[54]
miR-218-5p	MIMAT0000275	Increased	RT-qPCR	[54]
miR-7-5p	MIMAT0000252	Increased	RT-qPCR	[54]
miR-331-5p	MIMAT0004700	Increased	RT-qPCR, TaqMan low-density array	[54,86,87]
miR-34c-3p	MIMAT0004677	Increased	RT-qPCR	[54]
miR-30b-5p	MIMAT0000420	Increased	RT-qPCR	[54]
miR-30c-5p	MIMAT0000244	Increased	RT-qPCR	[54]
miR-19a-3p	MIMAT0000073	Increased	NGS	[55]
let-7g-3p	MIMAT0004584	Increased	NGS, TaqMan low-density array	[55,86]
mir-30b	MI0000441	Increased	RT-qPCR, TaqMan miRNA array	[86,87]
mir-16-2	MI0000115	Increased	RT-qPCR, TaqMan miRNA array	[86,87]
miR-205-5p	MIMAT0000266	Increased	RT-qPCR	[56]
miR-144-5p	MIMAT0004600	Increased	High-throughput deep sequencing, RT-qPCR	[88]
miR-200a-3p	MIMAT0000682	Increased	High-throughput deep sequencing, RT-qPCR	[87,88]
miR-542-3p	MIMAT0003389	Increased	High-throughput deep sequencing, RT-qPCR	[88]
miR-126-5p	MIMAT0000444	Increased	NGS	[89]
miR-126-3p	MIMAT0000445	Increased	NGS	[89]
miR-138-5p	MIMAT0000430	Increased	NGS	[89]
miR-9-5p	MIMAT0000441	Increased	NGS	[89]
miR-219a-2-3p	MIMAT0004675	Increased	NGS	[89]
miR-181a-5p	MIMAT0000256	Increased	NGS	[89]
miR-181b-5p	MIMAT0000257	Increased	NGS	[89]
miR-451a	MIMAT0001631	Increased	NGS	[52,89]
miR-486-5p	MIMAT0002177	Increased	NGS	[52,89]
miR-98-5p	MIMAT0000096	Increased	NGS	[89]
miR-144-3p	MIMAT0000436	Increased	NGS	[52,89]
miR-769-5p	MIMAT0003886	Increased	NGS	[89]
miR-211-5p	MIMAT0000268	Increased	NGS	[89]
miR-129-5p	MIMAT0000242	Increased	NGS	[89]
miR-16-5p	MIMAT0000069	Increased	NGS	[89]
miR-151a-5p	MIMAT0004697	Increased	RT-qPCR	[87]
miR-24	MI0000080, MI0000081	Increased	RT-qPCR	[87]
mir-214	MI0000290	Increased	RT-qPCR	[87]
let-7b	MI0000063	Increased	RT-qPCR	[87]
let-7f-1-3p	MIMAT0004486	Increased	RT-qPCR	[87]
miR-16	MI0000070	Increased	RT-qPCR	[87]
miR-200a-5p	MIMAT0001620	Increased	RT-qPCR	[87]
miR-26a-5p	MIMAT0000082	Increased	RT-qPCR	[87]
miR-29b-1	MI0000105	Increased	RT-qPCR	[87]
miR-122-5p	MIMAT0000421	Increased	NGS	[52]
miR-423-5p	MIMAT0004748	Increased	NGS	[52]
miR-151a-3p	MIMAT0000757	Increased	NGS	[52]
miR-320a	MI0000542	Increased	NGS	[52]
miR-320b	MIMAT0005792	Increased	NGS	[52]
miR-574-5p	MIMAT0004795	Increased	NGS	[52]
miR-206	MIMAT0000462	Increased	NGS	[52]
miR-1298-5p	MIMAT0005800	Increased	NGS	[52]
miR-1246	MIMAT0005898	Increased	NGS	[52]
miR-1307-3p	MIMAT0005951	Increased	NGS	[52]
miR-128-3p	MIMAT0000424	Increased	NGS	[52]
let-7a-5p	MIMAT0000062	Increased	NGS	[52]
let-7d-3p	MIMAT0004484	Increased	NGS	[52]
miR-4508	MIMAT0019045	Increased	NGS	[52]
miR-155-5p	MIMAT0000646	Increased	NGS	[52]
miR-99a-5p	MIMAT0000097	Decreased	RT-qPCR, NGS	[54,89]
miR-99b-5p	MIMAT0000689	Decreased	NGS	[89]
miR-100-5p	MIMAT0000098	Decreased	RT-qPCR	[54]
miR-145-5p	MIMAT0000437	Decreased	RT-qPCR	[54]
miR-92a-3p	MIMAT0000092	Decreased	RT-qPCR	[54]
miR-106a-5p	MIMAT0000103	Decreased	RT-qPCR	[54]
miR-128	MI0000727	Decreased	NGS	[55]
miR-431-3p	MIMAT0004757	Decreased	NGS	[55]
miR-212-3p	MIMAT0000269	Decreased	NGS	[55]
miR-1224-5p	MIMAT0005458	Decreased	NGS	[55]
miR-4448	MIMAT0018967	Decreased	NGS	[55]
mir-1	MI0000651	Decreased	TaqMan miRNA array, RT-qPCR	[86,87]
mir-126	MI0000471	Decreased	TaqMan miRNA array	[86]
mir-28	MI0000086	Decreased	TaqMan miRNA array	[86]
mir-301a	MI0000745	Decreased	TaqMan miRNA array	[86]
mir-29c	MI0000735	Decreased	TaqMan miRNA array	[86]
miR-24-3p	MIMAT0000080	Decreased	RT-qPCR	[56]
miR-626	MIMAT0003295	Decreased	RT-qPCR	[53]
miR-501-3p	MIMAT0004774	Decreased	NGS	[89]
miR-186-5p	MIMAT0000456	Decreased	NGS	[89]
miR-331-3p	MIMAT0000760	Decreased	RT-qPCR	[87]
miR-485-3p	MIMAT0002176	Decreased	RT-qPCR	[87]
miR-125b	MI0000446, MI0000470	Decreased	RT-qPCR	[90]

[a] Abbreviations: RT-qPCR, reverse transcription-quantitative polymerase chain reaction; NGS, next-generation sequencing.

Of these 77 miRNAs, 16 miRNAs were identified as contributors to seven different biological functions, including reactive oxygen species (ROS) production, ATP synthesis, mitochondrial function, the release of neurotransmitters, the loss of neurons, neurodegeneration, and PD (Figure 1A). Since the symbols for each miRNA that IPA software automatically recognized are different from Table 1, the details of 16 miRNAs in the miRNAomic network are provided in Supplementary Table S1. Then, based on the observation of 12 upregulated miRNAs and 4 downregulated miRNAs, the activation or inhibition of biological functions was predicted (Figure 1B). Notably, "synthesis of ATP" and "function of mitochondria" were predicted to be inhibited owing to the upregulation of miR-7a-5p, miR-16-5p, and mir-15, which directly inhibit these functions. Additionally, "neurodegeneration" was predicted to be activated by the upregulation of miR-17-5p. These findings are supported by previous studies. Specifically, miR-7a-5p, which is associated with ATP synthesis, has been reported to suppress cell proliferation upon upregulation, whereas its downregulation reduced apoptosis in non-small-cell lung cancer [91]. Similarly, miR-16-5p was shown to enhance mitochondrial function when its expression level was reduced in bladder cancer [92]. Furthermore, overexpression of mir-15 decreased ATP levels in rat ventricular myocytes [93]. Finally, miR-17-5p overexpression impaired TGF-beta signaling, leading to neurodegeneration in SH-SY5Y cells [94]. Taken together, these altered levels in the CSF are indicative of biological processes underlying PD. However, unfortunately, "generation of reactive oxygen species," "loss of neurons," and "Parkinson's disease" were not predicted, and "release of neurotransmitter" was not linked to any miRNA. Although a thorough analysis of CSF miRNAs in PD pathology is still insufficient owing to their limited information, altered levels of CSF miRNAs revealed a strong association with PD, underscoring their potential as diagnostic biomarkers and therapeutic targets.

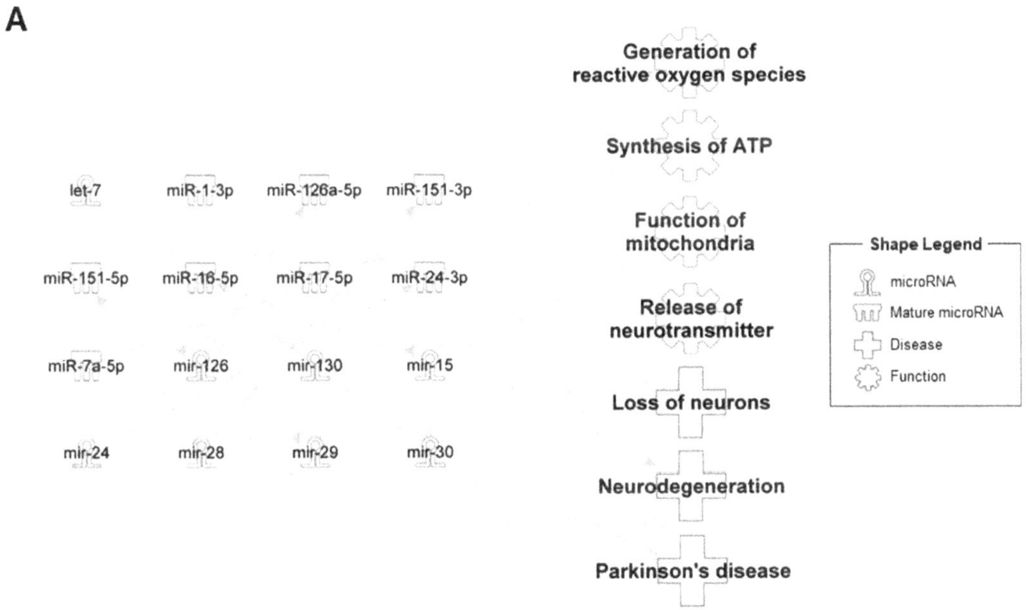

Figure 1. *Cont.*

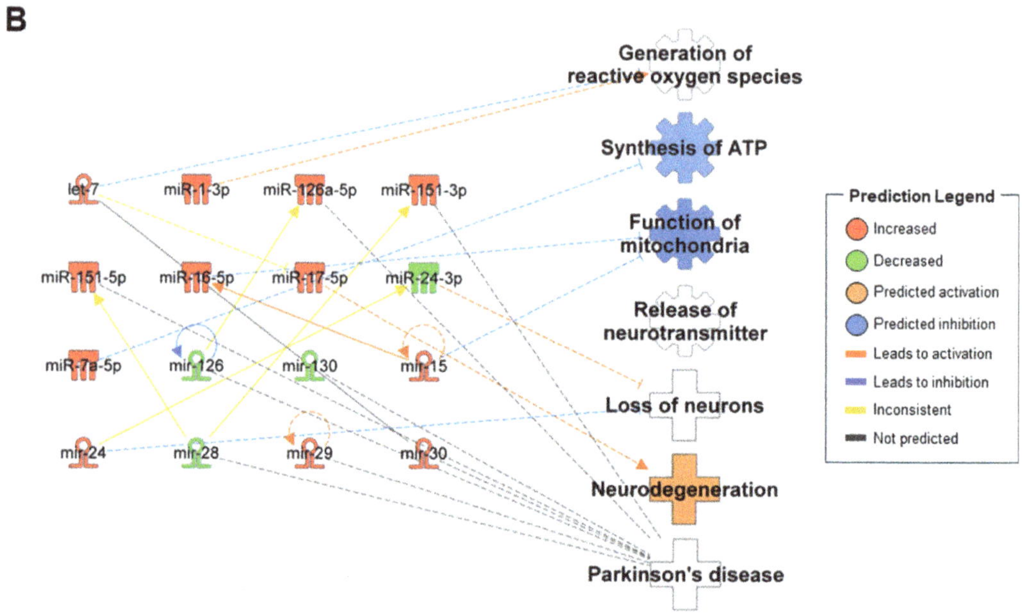

Figure 1. Biological function- and disease-related miRNAomic network from ingenuity pathway analysis. (**A**) Network of miRNAs (associated with pathological functions in Parkinson's disease) from the cerebrospinal fluid (CSF) of patients. (**B**) In silico prediction of the miRNAomic network based on alterations of the miRNAs from CSF. The color intensity reflects the confidence of the prediction.

4. CSF Protein in PD

Proteomics research enables both quantitative and qualitative analyses of neurodegenerative diseases, particularly focusing on proteins expressed in the human brain in cases with AD, PD, frontotemporal dementia, and ALS [95–97]. This approach is particularly advantageous for identifying pathological alterations in proteins. Indeed, α-syn, a key protein in PD pathology, has been extensively studied using proteomics techniques, such as mass spectrometry, gel electrophoresis, and chromatography, revealing its crucial role in PD [98]. Advanced mass spectrometry technology has greatly improved sensitivity for detecting the molecules while reducing the sample size needed for high-throughput analysis, thereby facilitating not only the analysis of simple proteins but also the identification of post-translational modifications [99] that play a crucial role in regulating protein function in neuroscience [100]. Hondius et al. utilized laser microdissection combined with liquid chromatography–tandem mass spectrometry to analyze postmortem brain tissue [101], shedding light on the understanding of protein pathways involved in PD mechanisms.

To address the shortcomings in miRNAomic analysis, we obtained proteomic profiles of the CSF of patients with PD and identified 92 altered proteins (Table 2). This set, which includes 16 more proteins than our previous study of 76 biologically significant proteins [102], enabled a more comprehensive mechanistic analysis and enhanced the investigation of biological functions related to PD [103–110]. Of these, 43 were upregulated and 49 were downregulated.

Table 2. List of proteins altered in the cerebrospinal fluid of patients with Parkinson's disease.

Protein Symbol [a]	Uniprot ID	Expression Pattern	Analysis Method [b]	Reference
APOE	P02649	Increased	LC-MS/MS	[111]
ENPP2	Q13822	Increased	LC-MS/MS	[111]
CNDP1	Q96KN2	Increased	LC-MS/MS	[111]
LDHB	P07195	Increased	LC-MS/MS	[111]
PEDF	P36955	Increased	LC-MS/MS	[111]
ALBU	P02768	Increased	LC-MS/MS	[111]
APLD1	Q96LR9	Increased	LC-MS/MS	[89]
DNS2A	O00115	Increased	LC-MS/MS	[89]
KV401	P06312	Increased	LC-MS/MS	[89]
IPSP	P05154	Increased	LC-MS/MS	[89]
APOA4	P06727	Increased	LC-MS/MS	[89]
KAIN	P29622	Increased	LC-MS/MS	[89]
HV102	P23083	Increased	LC-MS/MS	[89]
IGHG4	P01861	Increased	LC-MS/MS	[89]
LV147	P01700	Increased	LC-MS/MS	[89]
KLKB1	P03952	Increased	LC-MS/MS	[89]
TM198	Q66K66	Increased	LC-MS/MS	[89]
CBPB2	Q96IY4	Increased	LC-MS/MS	[89]
RET4	P02753	Increased	LC-MS/MS	[89]
LSAMP	Q13449	Increased	MRM-LC-MS/MS	[112]
APOH	P02749	Increased	MRM-LC-MS/MS	[112]
C1QC	P02747	Increased	LC-MS/MS	[106]
SIAL	P21815	Increased	Protein microarray	[113]
CCL2	P13500	Increased	Protein microarray, Meta-analysis, Bead-based cytokine array	[110,113,114]
SV2A	Q7L0J3	Increased	Protein microarray	[113]
CCL14	Q16627	Increased	Protein microarray, LC-MS/MS	[115]
SCUBE1	Q8IWY4	Increased	Protein microarray, LC-MS/MS	[115]
OMD	Q99983	Increased	Protein microarray, LC-MS/MS	[115,116]
CLN5	O75503	Increased	Protein microarray, LC-MS/MS	[115]
MA1C1	Q9NR34	Increased	Protein microarray, LC-MS/MS	[115]
CD44	P16070	Increased	LC-MS/MS	[116]
IL6	P05231	Increased	Meta-analysis	[114]
TNFA	P01375	Increased	Meta-analysis	[114]
IL1B	P01584	Increased	Meta-analysis	[114]
CRP	P02741	Increased	Meta-analysis	[114]
CCL28	Q9NRJ3	Increased	Systematic review	[114]
IL4	P05112	Increased	Meta-analysis (ELISA)	[114]
TGFB1	P01137	Increased	Meta-analysis (ELISA)	[114]
A2GL	P02750	Increased	LC-IMS-MS, LC-MRM	[103]
POTEE	Q6S8J3	Increased	LC-IMS-MS	[103]
FGFR2	P21802	Increased	LC-IMS-MS	[103]
ANT3	P01008	Increased	LC-MS/MS	[106]
UCHL1	P09936	Increased	Human Neurology 4-Plex A Advantage Kit	[107]
SCG2	P13521	Decreased	LC-MS/MS	[89]
CLUS	P10909	Decreased	LC-MS/MS	[111]
CO3	P01024	Decreased	LC-MS/MS	[111]
CO4A	P0C0L4	Decreased	LC-MS/MS	[111]
CO4B	P0C0L5	Decreased	LC-MS/MS	[111]
DCD	P81605	Decreased	LC-MS/MS	[111]
HPT	P00738	Decreased	LC-MS/MS	[111]
TTHY	P02766	Decreased	LC-MS/MS	[111]
GOGA3	Q08378	Decreased	LC-MS/MS	[117]
APOB	P04114	Decreased	LC-MS/MS	[117]
VGF	O15240	Decreased	LC-MS/MS, HPLC-MS/MS	[89,106]
CADM2	Q8N3J6	Decreased	LC-MS/MS	[89]
PTPR2	Q92932	Decreased	LC-MS/MS, HPLC-MS/MS	[89,106]
NEC1	P29120	Decreased	LC-MS/MS	[89]
CMGA	P10645	Decreased	LC-MS/MS	[89]
VTM2A	Q8TAG5	Decreased	LC-MS/MS, HPLC-MS/MS	[89,106]
SLIK1	Q96PX8	Decreased	LC-MS/MS	[89]
NPY	P01303	Decreased	LC-MS/MS	[89]
7B2	P05408	Decreased	LC-MS/MS	[89]
APOC2	P02655	Decreased	LC-MS/MS	[89]
AP2B1	P63010	Decreased	LC-PRM-MS	[118]
CATF	Q9UBX1	Decreased	LC-PRM-MS	[118]
SAP3	P17900	Decreased	LC-PRM-MS	[118]

Table 2. *Cont.*

Protein Symbol [a]	Uniprot ID	Expression Pattern	Analysis Method [b]	Reference
CGRE1	Q99674	Decreased	HPLC-MS/MS	[106]
TGON2	O43493	Decreased	HPLC-MS/MS	[106]
PARK7	Q99497	Decreased	Meta-analysis	[119]
CH3L1	P36222	Decreased	Meta-analysis	[114,119]
IL16	Q14005	Decreased	Systematic review	[114]
IL17	Q16552	Decreased	Systematic review	[114]
CCL8	P80075	Decreased	Systematic review	[114]
CCL23	P55773	Decreased	Systematic review	[114]
GROA	P09341	Decreased	Systematic review	[114]
NGF	P01138	Decreased	Systematic review	[114]
FGF19	O95750	Decreased	Systematic review	[114]
SCF	P21583	Decreased	Systematic review	[114]
CSF1	P09603	Decreased	Systematic review	[114]
PD1L1	Q9NZQ7	Decreased	Systematic review	[114]
VEGFA	P15692	Decreased	Systematic review	[114]
TGFA	P01135	Decreased	Meta-analysis (multiplex cytokine)	[114]
SAP	P07602	Decreased	LC-IMS-MS	[103]
NPTX2	P47972	Decreased	LC-MS/MS	[104]
NFL	P07196	Decreased	SIMOA	[105]
CADH2	P19022	Decreased	LC-MS/MS	[106]
EPHA4	P54764	Decreased	LC-MS/MS	[106]
QSOX1	O00391	Decreased	LC-MS/MS	[106]
MP3B2	A6NCE7	Decreased	ELISA	[108]
BECN1	Q14457	Decreased	ELISA	[108]
LAMP2	P13473	Decreased	ELISA	[108]
LAMP1	P11279	Decreased	Immunoblotting	[109]

[a] The protein symbol indicates the human protein designated by UniProtKB. [b] Abbreviations: LC-MS/MS, liquid chromatography–tandem mass spectrometry; MRM-LC-MS/MS, multiple reaction monitoring—liquid chromatography—tandem mass spectrometry; ELISA, enzyme-linked immunosorbent assay; LC-IMS-MS, liquid chromatography—ion mobility spectrometry—mass spectrometry; LC-MRM, liquid chromatography–multiple reaction monitoring; LC-PRM-MS, liquid chromatography–parallel reaction monitoring–mass spectrometry; SIMOA, single-molecule array.

Based on the identified profiles of proteins, a proteomic network was constructed with the same biological functions and diseases predicted in the miRNA network (Figure 2A). From this proteomic dataset, 40 proteins were identified as contributors in the network. Of these, 17 proteins showed increased levels, while 23 exhibited decreased levels. With these changes, "generation of reactive oxygen species," "loss of neurons," "neurodegeneration," and "Parkinson's disease" were predicted to be activated, while "synthesis of ATP," "function of mitochondria," and "release of neurotransmitter" were inhibited (Figure 2B). In particular, CO3, IL1B, PARK7, NGF, TGFA, TGFB1, and TNFA affected more than two functions, highlighting their significant roles in the pathogenesis and progression of PD. This proteomic network analysis indicated that the protein changes reflect PD conditions and that the interconnected proteins underlie the relationship between functions and diseases.

Notably, "release of neurotransmitter" was connected and predicted to be suppressed only in proteomics, not in miRNAome. This finding indicates that proteomic analysis of CSF can compensate for the limitations of miRNAome, as it captures direct evidence of altered protein expressions. Therefore, we incorporated two different omics together for synergetic effect to understand PD mechanisms and identify key molecules.

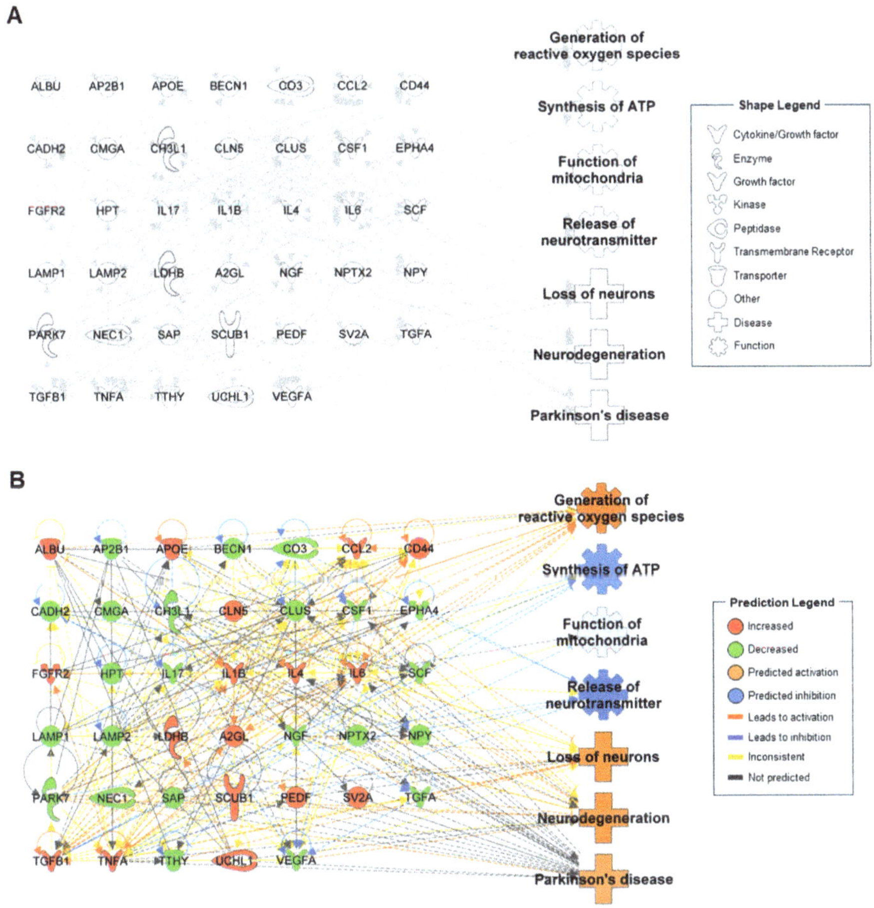

Figure 2. Biological function- and disease-related proteomic network from ingenuity pathway analysis. (**A**) Network of proteins associated with the pathological functions in Parkinson's disease in the cerebrospinal fluid (CSF) of patients. (**B**) In silico prediction of the proteomic network based on the alterations of proteins from CSF. The color intensity reflects the confidence of the prediction.

5. Integration of miRNA and Protein from the CSF of Patients with PD

The foundational concept of integrated omics was proposed by Dr. Leroy Hood [120,121], who proposed a systems biology approach that combines different types of omics data, such as genomics, epigenomics, transcriptomics, metabolomics, proteomics, and phosphoproteomics, to provide a comprehensive understanding of complex multifactorial biological systems [122]. The trans-omics concept of dynamic networks was proposed by Yugi et al. [123]. These approaches enable the identification of molecular signatures and potential therapeutic targets, facilitating precise diagnostics and advancing research across various fields of molecular biology, including neuroinflammation [124], nanotoxicity [125–128], mechanobiology [129], stem cell therapy [130–132], and neurodegenerative disease [133–135].

In this study, the miRNAome and proteome were integrated, and their relationships were considered using the latest IPA program, thereby identifying critical relationships between miRNAs and proteins (Figure 3A). Notably, the upregulation of miR-16-5p, which contributes to the inhibition of mitochondrial function, was associated with the regulation of AP2B1, CSF1, CLUS, SCF, LAMP2, IL6, and VEGFA, potentializing it as a biomarker with a significant contribution to PD pathology. Furthermore, upregulation of miR-7a-5p revealed an inhibitory effect on TGFA, which was found to activate "loss of neurons" and "neurodegeneration." Next, the trends in biological functions and diseases in the integrated network were similar to those observed in both the miRNAomic and proteomic networks (Figure 3B). However, inhibition of "synthesis of ATP" and "function of mitochondria" was predicted to be more severe than in miRNAomic and proteomic networks by connecting the additional relationships between miRNAs and proteins. Therefore, a more profound prediction in the integrated network could provide more accurate and comprehensive insights. Furthermore, this approach enabled the investigation of the contributions of molecular expression patterns to the pathogenesis of PD (Figure 3C,D). Adhering to specific criteria (categories associated with "neurological" and |activation z-score| > 1), 97 upregulated molecules were found to induce immune neurological disorders, including "autoimmune neurological disorder" and "experimental autoimmune encephalomyelitis" and activate the "damage of neurons" (Figure 3C). Moreover, 72 downregulated molecules were associated with the activation of 16 disease and biological functions, indicating neuronal degeneration, impaired movement, and non-motor symptoms (Figure 3D). Among them, the exact same functions, "neurodegeneration" and "loss of neurons," were predicted to be activated, highlighting the importance of CSF analysis in PD studies. Through this integrative approach, strong correlations between CSF components and biological functions were identified, underscoring the need for further research on the alterations in CSF to unravel PD-related mechanisms and disease progression.

In this review, we collected data on the changes in miRNAs and proteins in the CSF of patients with PD to analyze the molecular mechanisms associated with PD. Although CSF directly reflects the changes in the CNS, the discomfort and potential complications of the lumbar puncture process to obtain CSF samples limit its application in routine monitoring or large-scale diagnostic testing [136]. In contrast, peripheral blood sampling has the advantage of being relatively noninvasive and easily accessible. Despite obvious differences in the changes in miRNAs in the CSF and blood of patients with PD, certain miRNAs have been observed to exhibit identical patterns of increase or decrease in both the CSF and blood. For instance, Zhuang et al. reported that miR-125b levels are significantly higher in both CSF and plasma samples of patients with PD [90]. Tong et al. analyzed exosomes derived from the blood and CSF of patients with PD, highlighting five common miRNAs upregulated in both of the samples (miR-151a-5p, miR-24, mir-485-5p, mir-331-5p, and mir-214) [87]. These findings suggest that important miRNAs identified in both samples with the same pattern can be easily tested using blood-based assays, enabling early diagnosis and disease progression monitoring of PD. Therefore, comprehensive studies with miRNA levels in both CSF and blood-derived samples with integrative multi-omics analysis, such as transcriptomics, proteomics, phosphoproteomics, and metabolomics, with key miRNAs may elucidate the critical mechanisms of PD.

While this review highlights the potential of integrating multi-omics for PD analysis, it also has limitations. Data collection primarily relied on individual studies that were not corroborated by other research, emphasizing the need for follow-up studies. For instance, discrepancies in findings related to molecules such as miR-127-3p [55,89], miR-136-3p [86,87], and miR-433 [55,86] have hindered efforts to clarify mechanisms underlying PD. These inconsistencies suggest that more robust and reproducible studies are essential to resolve conflicting results and improve our understanding of molecular targets in PD.

Moreover, integrated omics approaches present unique challenges. Owing to the limited number of samples and the vast number of quantified molecules, they can often lead to false positives and negatives. Additionally, the analysis of time-series data and the distinction between true biological signals and noise in both targeted and untargeted approaches present significant hurdles. Therefore, effective data handling using multiple-omics data would be essential to ensure accurate reflection and reliability of biological phenotypes. Key steps include rigorous data filtering to remove noise and irrelevant features and thorough data cleaning to correct errors and manage missing values. This preprocessing ensures that downstream analyses, including statistical modeling and machine learning applications, yield meaningful biological insights and reliable predictions. For clustering multi-omics data, machine learning algorithms are useful in integrated omics research [124,133].

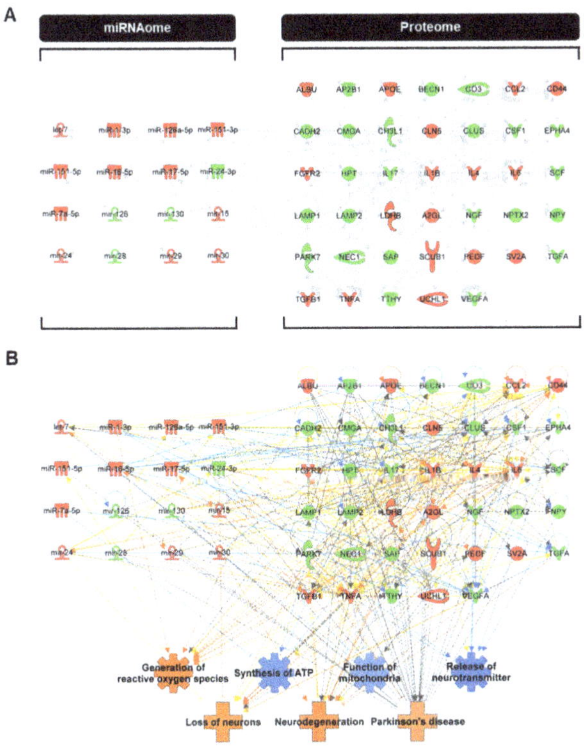

C

	Disease & Biological functions	Activation z-score
Upregulated molecules	Autoimmune neurological disorder	1.54
	Experimental autoimmune encephalomyelitis	1.53
	Cardiomyopathy	1.51
	Damage of neurons	1.48

D

	Disease & Biological functions	Activation z-score
Downregulated molecules	Degeneration of nervous system	1.75
	Gliosis	1.66
	Memory deficits	1.63
	Degeneration of neurons	1.55
	Neurodegeneration	1.50
	Loss of neurons in central nervous system	1.50
	Loss of neurons	1.43
	Movement Disorders	1.35
	Injury of nervous system	1.25
	Injury of central nervous system	1.24
	Brain damage	1.24
	Cell death of brain cells	1.15
	Ventricular dysfunction	1.11
	Progressive encephalopathy	1.07
	Progressive neurological disorder	1.07
	Neurological signs	1.03

Figure 3. Integrated omics network analysis and prediction from ingenuity pathway analysis. (**A**) Integrated network of the miRNAome and proteome. (**B**) In silico prediction of integrated network based on alterations of miRNAs and proteins. (**C**,**D**) Analyses of diseases and biological functions in upregulated molecules (**C**) and downregulated molecules (**D**) from the integrated omics dataset. A positive activation z-score indicates the activation of the function. The color intensity reflects the confidence of the prediction.

The application of artificial intelligence (AI) in integrated omics is a promising approach to address the challenges. Because integrated omics data are huge data, AI enhances their capabilities by processing vast amounts of complex biological time differences in omics data, non-obvious relationships between omics regimes, and high-dimensional omics data [137,138]. Additionally, AI can classify various omics data, uncover hidden patterns, perform feature selection, generate predictive models with high accuracy, accelerate biomarker discovery, and aid in clinical decision-making. In addition, the use of bioinformatics tools, such as IPA programs, can bring breakthroughs in this area. Identifying an apparent relationship between the different datasets within the biological context and independent statistical analysis of each omics regime is a challenge for researchers. In this context, we successfully integrated both datasets using the latest IPA program to construct a single and integrative network based on the pathological features of PD. By integrating omics data with AI approaches, researchers can derive deeper insights into the biological mechanisms in the CSF of patients with PD, develop more effective diagnostic tools, and investigate the pathophysiological mechanism of PD, thus advancing the field of neurodegenerative research.

6. Conclusions

This review highlights the importance of analyzing miRNAs and proteins in the CSF of patients with PD and proposes a powerful approach for understanding PD mechanisms by combining miRNAomic and proteomic data from the CSF using a computational method. Our functional network analyses of miRNAs and proteomes offer comprehensive insights into the post-transcriptional and translational regulations driving disease pathology including ROS production, ATP reduction, mitochondrial dysfunction, and inhibited release of neurotransmitters, resulting in the loss of neurons and neurodegeneration. This approach highlights key pathways involved in PD progression and identifies potential biomarkers and therapeutic targets in the CSF. As the field progresses, integrating multi-omics data and AI will provide a more in-depth understanding of PD and support the development of diagnostic and therapeutic strategies.

Supplementary Materials: The following supporting information can be downloaded at: https://www.mdpi.com/article/10.3390/ijms252413260/s1.

Author Contributions: Conceptualization and study design, G.L.; data collection and analysis, J.S.H., S.G.K., N.P.G., Y.E.J., M.K. and S.S.L.; manuscript writing, J.S.H., S.G.K., S.S.L. and G.L. All authors have read and agreed to the published version of the manuscript.

Funding: This research was funded by the National Research Foundation (NRF) Ministry of Science and ICT (MSIT) in Korea (grant numbers 2020M3E5D9080661, 2023R1A2C1004585, and RS-2024-00416536).

Conflicts of Interest: The authors declare no conflicts of interest.

References

1. Thomas, B.; Beal, M.F. Parkinson's disease. *Hum. Mol. Genet.* **2007**, *16*, R183–R194. [CrossRef] [PubMed]
2. Yi, S.; Wang, L. Pathogenesis of alpha-Synuclein in Parkinson's Disease: From a Neuron-Glia Crosstalk Perspective. *Int. J. Mol. Sci.* **2022**, *23*, 14753. [CrossRef] [PubMed]
3. Ryu, M.Y.; Kim, D.W. Localization of CKII beta subunits in Lewy bodies of Parkinson's disease. *J. Neurol. Sci.* **2008**, *266*, 9–12. [CrossRef] [PubMed]
4. Morris, H.R.; Spillantini, M.G. The pathogenesis of Parkinson's disease. *Lancet* **2024**, *403*, 293–304. [CrossRef] [PubMed]
5. Kouli, A.; Torsney, K.M.; Kuan, W.-L. *Parkinson's Disease: Etiology, Neuropathology, and Pathogenesis*; Exon Publications: Brisbane City, QLD, Australia, 2018; pp. 3–26.
6. Das, S.; Zhang, Z. Clinicopathological overlap of neurodegenerative diseases: A comprehensive review. *J. Clin. Neurosci.* **2020**, *78*, 30–33. [CrossRef]
7. Kwon, E.H.; Tennagels, S. Update on CSF biomarkers in Parkinson's disease. *Biomolecules* **2022**, *12*, 329. [CrossRef]
8. Zahoor, I.; Shafi, A. *Pharmacological Treatment of Parkinson's Disease*; Exon Publications: Brisbane City, QLD, Australia, 2018; pp. 129–144.
9. Olanow, C.W.; Stern, M.B.; Sethi, K. The scientific and clinical basis for the treatment of Parkinson disease (2009). *Neurology* **2009**, *72*, S1–S136. [CrossRef]
10. Caslake, R.; Macleod, A. Monoamine oxidase B inhibitors versus other dopaminergic agents in early Parkinson's disease. *Cochrane Database Syst. Rev.* **2009**, *4*, CD006661. [CrossRef]
11. Kataoka, H.; Sawa, N. Can dopamine agonists trigger tactile hallucinations in patients with Parkinson's disease? *J. Neurol. Sci.* **2014**, *347*, 361–363. [CrossRef] [PubMed]
12. Grall-Bronnec, M.; Victorri-Vigneau, C. Dopamine agonists and impulse control disorders: A complex association. *Drug Saf.* **2018**, *41*, 19–75. [CrossRef]
13. Lee, W.W.; Kim, H.J. Semiautomated Algorithm for the Diagnosis of Multiple System Atrophy With Predominant Parkinsonism. *J. Mov. Disord.* **2022**, *15*, 232–240. [CrossRef] [PubMed]
14. Hughes, A.J.; Colosimo, C. The dopaminergic response in multiple system atrophy. *J. Neurol. Neurosurg. Psychiatry* **1992**, *55*, 1009–1013. [CrossRef]
15. Nuytemans, K.; Theuns, J. Genetic etiology of Parkinson disease associated with mutations in the SNCA, PARK2, PINK1, PARK7, and LRRK2 genes: A mutation update. *Hum. Mutat.* **2010**, *31*, 763–780. [CrossRef]
16. Xiao, B.; Kuruvilla, J. Mitophagy and reactive oxygen species interplay in Parkinson's disease. *npj Park. Dis.* **2022**, *8*, 135. [CrossRef]
17. Guo, J.D.; Zhao, X. Damage to dopaminergic neurons by oxidative stress in Parkinson's disease. *Int. J. Mol. Med.* **2018**, *41*, 1817–1825. [CrossRef] [PubMed]
18. Tan, J.M.; Wong, E.S. Protein misfolding and aggregation in Parkinson's disease. *Antioxid. Redox Signal* **2009**, *11*, 2119–2134. [CrossRef] [PubMed]

19. Cocco, C.; Manai, A.L. Brain-biomarker changes in body fluids of patients with Parkinson's disease. *Int. J. Mol. Sci.* **2023**, *24*, 10932. [CrossRef]
20. Shamir, R.; Klein, C. Analysis of blood-based gene expression in idiopathic Parkinson disease. *Neurology* **2017**, *89*, 1676–1683. [CrossRef]
21. Seol, W.; Kim, H. Urinary biomarkers for neurodegenerative diseases. *Exp. Neurobiol.* **2020**, *29*, 325. [CrossRef]
22. Nijakowski, K.; Owecki, W. Salivary Biomarkers for Parkinson's Disease: A Systematic Review with Meta-Analysis. *Cells* **2024**, *13*, 340. [CrossRef]
23. Parnetti, L.; Gaetani, L. CSF and blood biomarkers for Parkinson's disease. *Lancet Neurol.* **2019**, *18*, 573–586. [CrossRef] [PubMed]
24. Constantinescu, R.; Mondello, S. Cerebrospinal fluid biomarker candidates for parkinsonian disorders. *Front. Neurol.* **2013**, *3*, 187. [CrossRef]
25. Reiber, H. Dynamics of brain-derived proteins in cerebrospinal fluid. *Clin. Chim. Acta* **2001**, *310*, 173–186. [CrossRef] [PubMed]
26. Le, W.-D.; Rowe, D.B. Effects of cerebrospinal fluid from patients with Parkinson disease on dopaminergic cells. *Arch. Neurol.* **1999**, *56*, 194–200. [CrossRef]
27. Hall, S.; Surova, Y. CSF biomarkers and clinical progression of Parkinson disease. *Neurology* **2015**, *84*, 57–63. [CrossRef]
28. Lerche, S.; Zimmermann, M. Inflammatory CSF profiles and longitudinal development of cognitive decline in sporadic and GBA-associated PD. *npj Park. Dis.* **2023**, *9*, 38. [CrossRef]
29. Weng, H.; Song, W. Proteomic profiling reveals the potential mechanisms and regulatory targets of sirtuin 4 in 1-methyl-4-phenyl-1, 2, 3, 6-tetrahydropyridine-induced Parkinson's mouse model. *Front. Neurosci.* **2023**, *16*, 1035444. [CrossRef] [PubMed]
30. Kwon, D.H.; Hwang, J.S. Cerebrospinal fluid metabolome in Parkinson's disease and multiple system atrophy. *Int. J. Mol. Sci.* **2022**, *23*, 1879. [CrossRef]
31. Gaudet, A.D.; Fonken, L.K. MicroRNAs: Roles in Regulating Neuroinflammation. *Neuroscientist* **2018**, *24*, 221–245. [CrossRef]
32. Ha, M.; Kim, V.N. Regulation of microRNA biogenesis. *Nat. Rev. Mol. Cell Biol.* **2014**, *15*, 509–524. [CrossRef]
33. Lee, R.C.; Feinbaum, R.L. The C. elegans heterochronic gene lin-4 encodes small RNAs with antisense complementarity to lin-14. *Cell* **1993**, *75*, 843–854. [CrossRef] [PubMed]
34. Tüfekci, K.U.; Meuwissen, R.L.J. The role of microRNAs in biological processes. *Methods Mol. Biol.* **2014**, *1107*, 15–31. [PubMed]
35. Shaheen, N.; Shaheen, A. MicroRNAs regulation in Parkinson's disease, and their potential role as diagnostic and therapeutic targets. *npj Park. Dis.* **2024**, *10*, 186. [CrossRef]
36. Wen, Q.; Wang, Y. MicroRNA-155-5p promotes neuroinflammation and central sensitization via inhibiting SIRT1 in a nitroglycerin-induced chronic migraine mouse model. *J. Neuroinflamm.* **2021**, *18*, 287. [CrossRef] [PubMed]
37. Zhao, W.; Spiers, J.G. microRNA-146a modulates behavioural activity, neuroinflammation, and oxidative stress in adult mice. *Mol. Cell Neurosci.* **2023**, *124*, 103820. [CrossRef]
38. Olufunmilayo, E.O.; Gerke-Duncan, M.B. Oxidative Stress and Antioxidants in Neurodegenerative Disorders. *Antioxidants* **2023**, *12*, 517. [CrossRef]
39. Banerjee, J.; Khanna, S. MicroRNA Regulation of Oxidative Stress. *Oxid. Med. Cell. Longev.* **2017**, *2017*, 2872156. [CrossRef]
40. Pan, K.; Chen, S. MicroRNA-23b attenuates tau pathology and inhibits oxidative stress by targeting GnT-III in Alzheimer's disease. *Neuropharmacology* **2021**, *196*, 108671. [CrossRef] [PubMed]
41. Li, R.; Huang, X. Novel C. elegans models of Lewy body disease reveal pathological protein interactions and widespread miRNA dysregulation. *Cell. Mol. Life Sci.* **2024**, *81*, 377. [CrossRef]
42. Liang, H.; Ding, B. MicroRNA-10a inhibits A30P alpha-synuclein aggregation and toxicity by targeting proapoptotic protein BCL2L11. *Int. J. Clin. Exp. Pathol.* **2018**, *11*, 624–633.
43. McMillan, K.J.; Murray, T.K. Loss of MicroRNA-7 Regulation Leads to alpha-Synuclein Accumulation and Dopaminergic Neuronal Loss In Vivo. *Mol. Ther.* **2017**, *25*, 2404–2414. [CrossRef] [PubMed]
44. Aaluri, G.R.; Choudhary, Y. Mitochondria-Associated MicroRNAs and Parkinson's Disease. *Neurosci. Insights* **2024**, *19*, 26331055241254846. [CrossRef] [PubMed]
45. Moradi Vastegani, S.; Nasrolahi, A. Mitochondrial Dysfunction and Parkinson's Disease: Pathogenesis and Therapeutic Strategies. *Neurochem. Res.* **2023**, *48*, 2285–2308. [CrossRef]
46. Catanesi, M.; d'Angelo, M. MicroRNAs Dysregulation and Mitochondrial Dysfunction in Neurodegenerative Diseases. *Int. J. Mol. Sci.* **2020**, *21*, 5986. [CrossRef]
47. Saadh, M.J.; Muhammad, F.A. MicroRNAs Modulating Neuroinflammation in Parkinson's disease. *Inflammation* **2024**, *47*, 2199–2219. [CrossRef]
48. Li, T.; Le, W. Biomarkers for Parkinson's Disease: How Good Are They? *Neurosci. Bull.* **2020**, *36*, 183–194. [CrossRef] [PubMed]
49. Guévremont, D.; Roy, J. MicroRNAs in Parkinson's disease: A systematic review and diagnostic accuracy meta-analysis. *Sci. Rep.* **2023**, *13*, 16272. [CrossRef]
50. Saghazadeh, A.; Rezaei, N. MicroRNA machinery in Parkinson's disease: A platform for neurodegenerative diseases. *Expert Rev. Neurother.* **2022**, *22*, 427–453. [CrossRef]
51. Angelopoulou, E.; Paudel, Y.N. miR-124 and Parkinson's disease: A biomarker with therapeutic potential. *Pharmacol. Res.* **2019**, *150*, 104515. [CrossRef]
52. Tan, X.; Hu, J. MicroRNA-409-3p Targeting at ATXN3 Reduces the Apoptosis of Dopamine Neurons Based on the Profile of miRNAs in the Cerebrospinal Fluid of Early Parkinson's Disease. *Front. Cell Dev. Biol.* **2021**, *9*, 755254. [CrossRef]

53. Qin, L.X.; Tan, J.Q. Preliminary study of hsa-miR-626 change in the cerebrospinal fluid of Parkinson's disease patients. *J. Clin. Neurosci.* **2019**, *70*, 198–201. [CrossRef] [PubMed]
54. Starhof, C.; Hejl, A.M. The biomarker potential of cell-free microRNA from cerebrospinal fluid in Parkinsonian Syndromes. *Mov. Disord.* **2019**, *34*, 246–254. [CrossRef]
55. Burgos, K.; Malenica, I. Profiles of extracellular miRNA in cerebrospinal fluid and serum from patients with Alzheimer's and Parkinson's diseases correlate with disease status and features of pathology. *PLoS ONE* **2014**, *9*, e94839. [CrossRef]
56. Marques, T.M.; Kuiperij, H.B. MicroRNAs in Cerebrospinal Fluid as Potential Biomarkers for Parkinson's Disease and Multiple System Atrophy. *Mol. Neurobiol.* **2017**, *54*, 7736–7745. [CrossRef] [PubMed]
57. Chandran, D.; Krishnan, S. Exosomal microRNAs in Parkinson's disease: Insights into biomarker potential and disease pathology. *Neurol. Sci.* **2024**, *45*, 3625–3639. [CrossRef]
58. Kramer, A.; Green, J. Causal analysis approaches in Ingenuity Pathway Analysis. *Bioinformatics* **2014**, *30*, 523–530. [CrossRef]
59. Zhang, W.T.; Wang, Y.J. Circulating microRNAs as potential biomarkers for the diagnosis of Parkinson's disease: A meta-analysis. *Neurologia* **2024**, *39*, 573–583. [CrossRef] [PubMed]
60. Yadav, S.K.; Jauhari, A. Transcriptomics and Proteomics Approach for the Identification of Altered Blood microRNAs and Plasma Proteins in Parkinson's Disease. *Cell. Mol. Neurobiol.* **2023**, *43*, 3527–3553. [CrossRef]
61. Pavelka, L.; Rauschenberger, A. Converging peripheral blood microRNA profiles in Parkinson's disease and progressive supranuclear palsy. *Brain Commun.* **2024**, *6*, fcae187. [CrossRef]
62. Augustine, J.; Jereesh, A.S. Blood-based gene-expression biomarkers identification for the non-invasive diagnosis of Parkinson's disease using two-layer hybrid feature selection. *Gene* **2022**, *823*, 146366. [CrossRef]
63. Posavi, M.; Diaz-Ortiz, M. Characterization of Parkinson's disease using blood-based biomarkers: A multicohort proteomic analysis. *PLoS Med.* **2019**, *16*, e1002931. [CrossRef] [PubMed]
64. D'Ascenzo, N.; Antonecchia, E. Metabolomics of blood reveals age-dependent pathways in Parkinson's Disease. *Cell Biosci.* **2022**, *12*, 102. [CrossRef] [PubMed]
65. Yadav, S.K.; Pandey, A. Identification of Altered Blood MicroRNAs and Plasma Proteins in a Rat Model of Parkinson's Disease. *Mol. Neurobiol.* **2022**, *59*, 1781–1798. [CrossRef] [PubMed]
66. Naidoo, D.; de Lencastre, A. Regulation of TIR-1/SARM-1 by miR-71 Protects Dopaminergic Neurons in a C. elegans Model of LRRK2-Induced Parkinson's Disease. *Int. J. Mol. Sci.* **2024**, *25*, 8795. [CrossRef]
67. Alieva, A.K.; Filatova, E.V. Transcriptome profile changes in mice with MPTP-induced early stages of Parkinson's disease. *Mol. Neurobiol.* **2017**, *54*, 6775–6784. [CrossRef]
68. Tönges, L.; Buhmann, C. Blood-based biomarker in Parkinson's disease: Potential for future applications in clinical research and practice. *J. Neural Transm.* **2022**, *129*, 1201–1217. [CrossRef]
69. Ma, Z.-L.; Wang, Z.-L. Biomarkers of Parkinson's Disease: From Basic Research to Clinical Practice. *Aging Dis.* **2024**, *15*, 1813–1830.
70. Sakka, L.; Coll, G. Anatomy and physiology of cerebrospinal fluid. *Eur. Ann. Otorhinolaryngol. Head Neck Dis.* **2011**, *128*, 309–316. [CrossRef]
71. Damkier, H.H.; Brown, P.D. Cerebrospinal fluid secretion by the choroid plexus. *Physiol. Rev.* **2013**, *93*, 1847–1892. [CrossRef]
72. Jiménez-Jiménez, F.J.; Alonso-Navarro, H. Cerebrospinal fluid biochemical studies in patients with Parkinson's disease: Toward a potential search for biomarkers for this disease. *Front. Cell. Neurosci.* **2014**, *8*, 369.
73. Faizan, M.; Sachan, N. Cerebrospinal fluid protein biomarkers in Parkinson's disease. *Clin. Chim. Acta* **2024**, *556*, 117848. [CrossRef] [PubMed]
74. Osher, D. The cerebrospinal fluid; anatomy and physiology. *Cinci J. Med.* **1949**, *30*, 379. [PubMed]
75. Je, G.; Kim, Y.-S. Mitochondrial ROS-mediated post-transcriptional regulation of α-synuclein through miR-7 and miR-153. *Neurosci. Lett.* **2017**, *661*, 132–136. [CrossRef]
76. Doxakis, E. Post-transcriptional regulation of α-synuclein expression by mir-7 and mir-153. *J. Biol. Chem.* **2010**, *285*, 12726–12734. [CrossRef] [PubMed]
77. Li, S.; Lei, Z. The role of microRNAs in neurodegenerative diseases: A review. *Cell Biol. Toxicol.* **2023**, *39*, 53–83. [CrossRef] [PubMed]
78. Dos Santos, M.C.T.; Barreto-Sanz, M.A. miRNA-based signatures in cerebrospinal fluid as potential diagnostic tools for early stage Parkinson's disease. *Oncotarget* **2018**, *9*, 17455. [CrossRef]
79. Wang, Y.; Yang, Z. Tiny But Mighty: Promising Roles of MicroRNAs in the Diagnosis and Treatment of Parkinson's Disease. *Neurosci. Bull.* **2017**, *33*, 543–551. [CrossRef]
80. Lu, X.; Cui, Z. MiRNAs participate in the diagnosis, pathogenesis and therapy of Parkinson's disease. *Histol. Histopathol.* **2018**, *33*, 447–453.
81. Kim, T.; Valera, E.; Desplats, P. Alterations in Striatal microRNA-mRNA Networks Contribute to Neuroinflammation in Multiple System Atrophy. *Mol. Neurobiol.* **2019**, *56*, 7003–7021. [CrossRef]
82. Casado Gama, H.; Amoros, M.A. Systematic review and meta-analysis of dysregulated microRNAs derived from liquid biopsies as biomarkers for amyotrophic lateral sclerosis. *Noncoding RNA Res.* **2024**, *9*, 523–535. [CrossRef]
83. Islam, M.A.; Sultana, O.F. MicroRNA-455-3P as a peripheral biomarker and therapeutic target for mild cognitive impairment and Alzheimer's disease. *Ageing Res. Rev.* **2024**, *100*, 102459. [CrossRef] [PubMed]

84. Alexandrov, P.N.; Dua, P. microRNA (miRNA) speciation in Alzheimer's disease (AD) cerebrospinal fluid (CSF) and extracellular fluid (ECF). *Int. J. Biochem. Mol. Biol.* **2012**, *3*, 365–373. [PubMed]
85. Roser, A.E.; Caldi Gomes, L. Circulating miRNAs as diagnostic biomarkers for Parkinson's disease. *Front. Neurosci.* **2018**, *12*, 625. [CrossRef] [PubMed]
86. Gui, Y.; Liu, H. Altered microRNA profiles in cerebrospinal fluid exosome in Parkinson disease and Alzheimer disease. *Oncotarget* **2015**, *6*, 37043–37053. [CrossRef] [PubMed]
87. Tong, G.; Zhang, P. Diagnostic test to Identify Parkinson's disease from the blood sera of Chinese population: A cross-sectional study. *Park. Dis.* **2022**, *2022*, 8683877. [CrossRef]
88. Mo, M.; Xiao, Y. MicroRNA expressing profiles in A53T mutant alpha-synuclein transgenic mice and Parkinsonian. *Oncotarget* **2017**, *8*, 15. [CrossRef]
89. Caldi Gomes, L.; Roser, A.E. MicroRNAs from extracellular vesicles as a signature for Parkinson's disease. *Clin. Transl. Med.* **2021**, *11*, e357. [CrossRef]
90. Zhuang, J.; Cai, P. Long noncoding RNA MALAT1 and its target microRNA-125b are potential biomarkers for Alzheimer's disease management via interactions with FOXQ1, PTGS2 and CDK5. *Am. J. Transl. Res.* **2020**, *12*, 5940.
91. Li, Q.; Wu, X. MicroRNA-7-5p induces cell growth inhibition, cell cycle arrest and apoptosis by targeting PAK2 in non-small cell lung cancer. *FEBS Open Bio* **2019**, *9*, 1983–1993. [CrossRef]
92. Li, H.-J.; Sun, X.-M. LncRNA UCA1 promotes mitochondrial function of bladder cancer via the MiR-195/ARL2 signaling pathway. *Cell Physiol. Biochem.* **2017**, *43*, 2548–2561. [CrossRef]
93. Nishi, H.; Ono, K. MicroRNA-15b Modulates Cellular ATP Levels and Degenerates Mitochondria via Arl2 in Neonatal Rat Cardiac Myocytes 2. *J. Biol. Chem.* **2010**, *285*, 1920–1930. [CrossRef] [PubMed]
94. Wang, H.; Liu, J. miR-106b aberrantly expressed in a double transgenic mouse model for Alzheimer's disease targets TGF-β type II receptor. *Brain Res.* **2010**, *1357*, 166–174. [CrossRef] [PubMed]
95. Li, K.W.; Ganz, A.B. Proteomics of neurodegenerative diseases: Analysis of human post-mortem brain. *J. Neurochem.* **2019**, *151*, 435–445. [CrossRef] [PubMed]
96. Demartini, D.R.; Schilling, L.P. Alzheimer's and Parkinson's diseases: An environmental proteomic point of view. *J. Proteom.* **2014**, *104*, 24–36. [CrossRef]
97. Vilkaite, G.; Vogel, J.; Mattsson-Carlgren, N. Integrating amyloid and tau imaging with proteomics and genomics in Alzheimer's disease. *Cell Rep. Med.* **2024**, *5*, 101735. [CrossRef]
98. Palanisamy, S.; Subramani, P. Advancement in Alpha-Synuclein Proteomics: Exploring Analytical Techniques and Their Clinical Implications in Parkinson's Disease. *Georgian Med. News* **2024**, *351*, 49–54.
99. Aebersold, R.; Mann, M. Mass-spectrometric exploration of proteome structure and function. *Nature* **2016**, *537*, 347–355. [CrossRef] [PubMed]
100. Hosp, F.; Mann, M. A Primer on Concepts and Applications of Proteomics in Neuroscience. *Neuron* **2017**, *96*, 558–571. [CrossRef]
101. Hondius, D.C.; Hoozemans, J.J.M. A Laser Microdissection-Liquid Chromatography-Tandem Mass Spectrometry Workflow for Post-mortem Analysis of Brain Tissue. *Methods Mol. Biol.* **2018**, *1723*, 371–383.
102. Kim, S.G.; Hwang, J.S. Integrative Metabolome and Proteome Analysis of Cerebrospinal Fluid in Parkinson's Disease. *Int. J. Mol. Sci.* **2024**, *25*, 11406. [CrossRef]
103. Zhu, S.; Bäckström, D. Alterations in self-aggregating neuropeptides in cerebrospinal fluid of patients with parkinsonian disorders. *J. Park. Dis.* **2022**, *12*, 1169–1189. [CrossRef]
104. Nilsson, J.; Constantinescu, J. Cerebrospinal fluid biomarkers of synaptic dysfunction are altered in Parkinson's disease and related disorders. *Mov. Disord.* **2023**, *38*, 267–277. [CrossRef]
105. Liu, Y.; Dou, K. Neurofilament light as a biomarker for motor decline in Parkinson's disease. *Front. Neurosci.* **2022**, *16*, 959261. [CrossRef]
106. Rotunno, M.S.; Lane, M. Cerebrospinal fluid proteomics implicates the granin family in Parkinson's disease. *Sci. Rep.* **2020**, *10*, 2479. [CrossRef]
107. Schulz, I.; Kruse, N. Systematic assessment of 10 biomarker candidates focusing on α-synuclein-related disorders. *Mov. Disord.* **2021**, *36*, 2874–2887. [CrossRef] [PubMed]
108. Youn, J.; Lee, S.-B. Cerebrospinal fluid levels of autophagy-related proteins represent potentially novel biomarkers of early-stage Parkinson's disease. *Sci. Rep.* **2018**, *8*, 16866. [CrossRef] [PubMed]
109. Boman, A.; Svensson, S. Distinct lysosomal network protein profiles in parkinsonian syndrome cerebrospinal fluid. *J. Park. Dis.* **2016**, *6*, 307–315. [CrossRef] [PubMed]
110. Schröder, J.B.; Pawlowski, M. Immune cell activation in the cerebrospinal fluid of patients with Parkinson's disease. *Front. Neurol.* **2018**, *9*, 1081. [CrossRef]
111. Guo, J.; Sun, Z. Proteomic analysis of the cerebrospinal fluid of Parkinson's disease patients. *Cell Res.* **2009**, *19*, 1401–1403. [CrossRef]
112. Heywood, W.E.; Galimberti, D. Identification of novel CSF biomarkers for neurodegeneration and their validation by a high-throughput multiplexed targeted proteomic assay. *Mol. Neurodegener.* **2015**, *10*, 64. [CrossRef]
113. Winchester, L.; Barber, I. Identification of a possible proteomic biomarker in Parkinson's disease: Discovery and replication in blood, brain and cerebrospinal fluid. *Brain Commun.* **2023**, *5*, fcac343. [CrossRef]

114. Qu, Y.; Li, J. A systematic review and meta-analysis of inflammatory biomarkers in Parkinson's disease. *npj Park. Dis.* **2023**, *9*, 18. [CrossRef]
115. Tsukita, K.; Sakamaki-Tsukita, H. High-throughput CSF proteomics and machine learning to identify proteomic signatures for parkinson disease development and progression. *Neurology* **2023**, *101*, e1434–e1447. [CrossRef]
116. Karayel, O.; Winter, S.V. Proteome profiling of cerebrospinal fluid reveals biomarker candidates for Parkinson's disease. *Cell Rep. Med.* **2022**, *3*, 100661. [CrossRef]
117. Lehnert, S.; Jesse, S. iTRAQ and multiple reaction monitoring as proteomic tools for biomarker search in cerebrospinal fluid of patients with Parkinson's disease dementia. *Exp. Neurol.* **2012**, *234*, 499–505. [CrossRef]
118. Sjödin, S.; Brinkmalm, G. Endo-lysosomal proteins and ubiquitin CSF concentrations in Alzheimer's and Parkinson's disease. *Alzheimers Res. Ther.* **2019**, *11*, 82. [CrossRef] [PubMed]
119. Xiang, C.; Cong, S. A meta-analysis of the diagnostic utility of biomarkers in cerebrospinal fluid in Parkinson's disease. *npj Park. Dis.* **2022**, *8*, 165. [CrossRef] [PubMed]
120. Hood, L. Systems biology: Integrating technology, biology, and computation. *Mech. Ageing Dev.* **2003**, *124*, 9–16. [CrossRef]
121. Hood, L. Leroy Hood expounds the principles, practice and future of systems biology. *Drug Discov. Today* **2003**, *8*, 436–438.
122. Fatima, I.; Rehman, A. Breakthroughs in AI and multi-omics for cancer drug discovery: A review. *Eur. J. Med. Chem.* **2024**, *280*, 116925. [CrossRef] [PubMed]
123. Yugi, K.; Kubota, H. Trans-Omics: How To Reconstruct Biochemical Networks Across Multiple 'Omic' Layers. *Trends Biotechnol.* **2016**, *34*, 276–290. [CrossRef] [PubMed]
124. Shin, T.H.; Manavalan, B. Silica-coated magnetic-nanoparticle-induced cytotoxicity is reduced in microglia by glutathione and citrate identified using integrated omics. *Part. Fibre Toxicol.* **2021**, *18*, 42. [CrossRef] [PubMed]
125. Shin, T.H.; Kim, S.G. Diesel-derived PM(2.5) induces impairment of cardiac movement followed by mitochondria dysfunction in cardiomyocytes. *Front. Endocrinol. (Lausanne)* **2022**, *13*, 999475. [CrossRef] [PubMed]
126. Shin, T.H.; Nithiyanandam, S. Analysis of Nanotoxicity with Integrated Omics and Mechanobiology. *Nanomaterials* **2021**, *11*, 2385. [CrossRef] [PubMed]
127. Shin, T.H.; Lee, D.Y. Integration of metabolomics and transcriptomics in nanotoxicity studies. *BMB Rep.* **2018**, *51*, 14–20. [CrossRef]
128. Shim, W.; Paik, M.J. Analysis of changes in gene expression and metabolic profiles induced by silica-coated magnetic nanoparticles. *ACS Nano* **2012**, *6*, 7665–7680. [CrossRef]
129. Shin, T.H.; Ketebo, A.A. Decrease in membrane fluidity and traction force induced by silica-coated magnetic nanoparticles. *J. Nanobiotechnol.* **2021**, *19*, 21. [CrossRef]
130. Lee, D.Y.; Lee, S.E. Strategies to Improve the Quality and Freshness of Human Bone Marrow-Derived Mesenchymal Stem Cells for Neurological Diseases. *Stem Cell. Int.* **2021**, *2021*, 8444599. [CrossRef]
131. Shin, T.H.; Lee, D.Y. Silica-Coated Magnetic Nanoparticles Decrease Human Bone Marrow-Derived Mesenchymal Stem Cell Migratory Activity by Reducing Membrane Fluidity and Impairing Focal Adhesion. *Nanomaterials* **2019**, *9*, 1475. [CrossRef]
132. Shin, T.H.; Lee, S. Quality and freshness of human bone marrow-derived mesenchymal stem cells decrease over time after trypsinization and storage in phosphate-buffered saline. *Sci. Rep.* **2017**, *7*, 1106. [CrossRef]
133. Shin, T.H.; Lee, G. Reduced lysosomal activity and increased amyloid beta accumulation in silica-coated magnetic nanoparticles-treated microglia. *Arch. Toxicol.* **2024**, *98*, 121–134. [CrossRef] [PubMed]
134. Shin, T.H.; Lee, D.Y. Silica-coated magnetic nanoparticles activate microglia and induce neurotoxic D-serine secretion. *Part. Fibre Toxicol.* **2021**, *18*, 30. [CrossRef] [PubMed]
135. Phukan, G.; Shin, T.H. Silica-coated magnetic nanoparticles impair proteasome activity and increase the formation of cytoplasmic inclusion bodies in vitro. *Sci. Rep.* **2016**, *6*, 29095. [CrossRef]
136. Evans, R.W. Complications of lumbar puncture. *Neurol. Clin.* **1998**, *16*, 83–105. [CrossRef] [PubMed]
137. Gangwal, A.; Lavecchia, A. Unleashing the power of generative AI in drug discovery. *Drug Discov. Today* **2024**, *29*, 103992. [CrossRef]
138. Basith, S.; Manavalan, B. Machine intelligence in peptide therapeutics: A next-generation tool for rapid disease screening. *Med. Res. Rev.* **2020**, *40*, 1276–1314. [CrossRef]

International Journal of
Molecular Sciences

MDPI

Review

Parkinson's Disease: Biomarkers for Diagnosis and Disease Progression

Rakesh Arya [1,†], A. K. M. Ariful Haque [2,†], Hemlata Shakya [3], Md. Masum Billah [2], Anzana Parvin [2], Md-Mafizur Rahman [2], Khan Mohammad Sakib [4], Hossain Md. Faruquee [2], Vijay Kumar [5,*] and Jong-Joo Kim [1,*]

1 Department of Biotechnology, Yeungnam University, Gyeongsan 38541, Republic of Korea; rakesharya101@yu.ac.kr
2 Department of Biotechnology and Genetic Engineering, Faculty of Biological Sciences, Islamic University, Kushtia 7003, Bangladesh; ariful.btge20.iu@gmail.com (A.K.M.A.H.); masumbillah111998@gmail.com (M.M.B.); anzana@btge.iu.ac.bd (A.P.); mmrahman@btge.iu.ac.bd (M.-M.R.); faruquee@btge.iu.ac.bd (H.M.F.)
3 Department of Biomedical Engineering, Shri G. S. Institute of Technology and Science, Indore 452003, India; hemlata.shakya19@gmail.com
4 Department of Biology, Adamjee Cantonment College, Dhaka Cantonment, Dhaka 1206, Bangladesh; khan11sakib@gmail.com
5 Department of Orthopaedic Surgery, The Johns Hopkins University School of Medicine, Baltimore, MD 21205, USA
* Correspondence: vijaykumarcbt@gmail.com (V.K.), kimjj@yu.ac.kr (J.-J.K.)
† These authors contributed equally to the work.

Abstract: Parkinson's disease (PD) is a progressive neurological disease that causes both motor and nonmotor symptoms. While our understanding of putative mechanisms has advanced significantly, it remains challenging to verify biomarkers with sufficient evidence for regular clinical use. Clinical symptoms are the primary basis for diagnosing the disease, which can be mild in the early stages and overlap with other neurological disorders. As a result, clinical testing and medical records are mostly relied upon for diagnosis, posing substantial challenges during both the initial diagnosis and the continuous disease monitoring. Recent biochemical, neuroimaging, and genetic biomarkers have helped us understand the pathophysiology of Parkinson's disease. This comprehensive study focuses on these biomarkers, which were chosen based on their relevance, methodological excellence, and contribution to the field. Biochemical biomarkers, including α-synuclein and glial fibrillary acidic protein (GFAP), can predict disease severity and progression. The dopaminergic system is widely used as a neuroimaging biomarker to diagnose PD. Numerous genes and genome wide association study (GWAS) sites have been related to the development of PD. Recent research on the SNCA gene and leucine-rich repeat protein kinase 2 (LRRK2) has shown promising results. By evaluating current studies, this review intends to uncover gaps in biomarker validation and use, while also highlighting promising improvements. It emphasizes the need for dependable and reproducible indicators in improving PD diagnosis and prognosis. These biomarkers may open up new avenues for early diagnosis, disease progression tracking, and the development of personalized treatment programs.

Keywords: Parkinson's disease; biomarker; biochemical; neuroimaging; genetic; α-synuclein

1. Introduction

The World Health Organization issued a technical brief titled "Parkinson's disease: a public health approach" in June 2022, emphasizing that the worldwide effect of Parkinson's disease (PD) is "increasing faster than for any other neurological disorder". According to the paper, the prevalence of PD has more than doubled in the last

25 years, with over 5.8 million disability-adjusted life years in 2019—an 81% rise since 2000 [1].

PD is a neurodegenerative disorder that affects the central nervous system and it affects dopamine-producing neurons in the substantia nigra region of the brain [2]. The accumulating intraneuronal aggregates of the protein α-synuclein, also known as Lewy bodies, and the gradual degradation of dopaminergic neurons in the substantia nigra are the main pathogenic alterations observed in PD patients [3]. It is a type of disorder that occurs when the brain contains an insufficient amount or quantity of dopamine. Since dopamine is linked to motor activity, dopamine depletion is the major cause of PD. Tremors, slowness and paucity of movement, stiffness or the slowing of movements, and difficulty in walking, talking, or doing other simple tasks are common symptoms that develop slowly over time [4]. The primary method for diagnosing PD is a clinical assessment. The primary symptom for diagnosis is bradykinesia, which is accompanied by at least one additional symptom such as resting tremor, rigidity, or postural instability. Clinicians must gather a patient's medical history and conduct a thorough neurological examination to exclude other medical conditions that may resemble PD [5,6]. The Movement Disorder Society (MDS) has published criteria for classifying PD as clinically established or likely based on the existence of supporting and exclusion criteria. Imaging techniques such as dopamine transporter scans and MRIs can help with the diagnosis, although they are not precise [6]. Nonmotor symptoms can occur along with other parkinsonism, including atypical parkinsonisms. Dysautonomia, a nonmotor symptom, is particularly significant in both PD and atypical parkinsonisms. Recent research emphasizes the significance of treating both motor and autonomic dysfunction to accurately differentiate between these diseases [7]. Furthermore, nonmotor symptoms such as cognitive impairment, disturbances in sleep, and mood disorders are widely established symptoms of PD [8]. PD can be treated symptomatically, typically with medication that targets both motor and nonmotor symptoms. The therapeutic landscape has developed dramatically, with a focus on symptom management to improve a patient's quality of life [9,10].

PD is incurable; however, there are numerous treatment options available, and it can be treated symptomatologically. Although some cases are hereditary and can be traced back to specific genetic mutations, the majority of cases are sporadic and are likely to be the consequence of a mix of heredity and exposure to one or more environmental factors [11,12]. The rate of PD misdiagnosis by movement disorder specialists is high, due, in part, to the lack of sensitive and precise biomarkers established for physicians to differentiate PD from other movement disorders with overlapping clinical symptoms [13]. Over the last five years, clinical diagnostic criteria aiming to improve the diagnostic accuracy of PD have been validated. However, diagnosing PD is challenging because its clinical symptoms frequently overlap with those of other neurodegenerative disorders, and current tests or biomarkers do not allow for a conclusive diagnosis in the early stages [14]. Consequently, even when the disease is fully apparent clinically, the accuracy of clinical diagnosis remains insufficient. Aside from diagnostic utility, biomarkers for PD are required to monitor disease progression and the efficacy of therapies, both of which are currently measured by assessing the severity of the motor symptoms.

In this review, we summarize what is currently known about PD biomarker research from biochemical, genetic, and neuroimaging perspectives, including the most recent advancements, their diagnostic accuracy, and potential uses. This review aims to expand our knowledge and understanding of biomarkers and Parkinson's disease broadly and discuss our suggestions for the future.

2. Methodology

The purpose of this review is to investigate the numerous biomarkers linked with PD, bringing together information gathered from biochemical, neuroimaging, and genetic studies. The research question motivating this study is: "What are the key biomarkers of Parkinson's disease that show promising results with respect to diagnosis, as identified through biochemical, neuroimaging, and genetic research?". To address this question, a complete synthesis of the available information is required, hopefully guiding future studies.

2.1. Inclusion and Exclusion Criteria

The literature search was undertaken using numerous databases, including PubMed, Scopus, and Web of Science, with an emphasis on research published within the previous decade.

Inclusion Criteria

- Peer-reviewed studies examining biochemical, neuroimaging, or genetic biomarkers associated with Parkinson's disease.
- Articles providing important and new insights into the mechanisms of these biomarkers.

Exclusion Criteria

- Non-peer-reviewed papers, research on unrelated neurological disorders, and articles with outdated and insufficient information were all excluded.
- The purpose of these selections is to provide a high-quality synthesis that captures the key research while remaining relevant to the review.

2.2. Details of Analysis and Interpretation

The analysis involved a thematic synthesis of the findings from the selected studies. Each category of biomarkers is discussed separately, summarizing recent key findings while identifying patterns and discrepancies across studies.

- Biochemical analysis: this section explores specific proteins identified as potential biomarkers, discussing their roles in pathophysiology and diagnostic utility.
- Neuroimaging insights: here, various imaging modalities are evaluated with respect to their effectiveness in detecting early changes associated with Parkinson's disease.
- Genetic factors: this part focuses on the genetic markers associated with an increased risk or familial forms of Parkinson's disease.

The interpretation highlights how these biomarkers contribute to the understanding of disease mechanisms, improving diagnostic accuracy and, potentially, guiding therapeutic strategies.

3. Biomarkers for Parkinson's Disease

A biomarker is a characteristic that can be assessed as an indicator of normal biological pathogenic processes or of responses to different exposures or interventions, including therapeutic interventions [15]. The early phases of the disease, when treatment is most effective, require the use of biomarkers. Indicators such as imaging, cerebrospinal fluid, oxidative stress, inflammation, and neuroprotection can help with the early identification of the disease [16]. The quest for biomarkers of PD is an ongoing endeavor [17]. Clinical, imaging, biochemical, and genetic biomarkers are the four types of biomarkers [18]. Individual biomarker specificity and sensitivity are not perfect, and the combination of multimodal biomarkers is likely to improve PD diagnostic accuracy. Efforts to analyze and validate putative biomarkers are uncommon, and more research is needed to identify accurate biomarkers for early disease detection and management [17] (Figure 1).

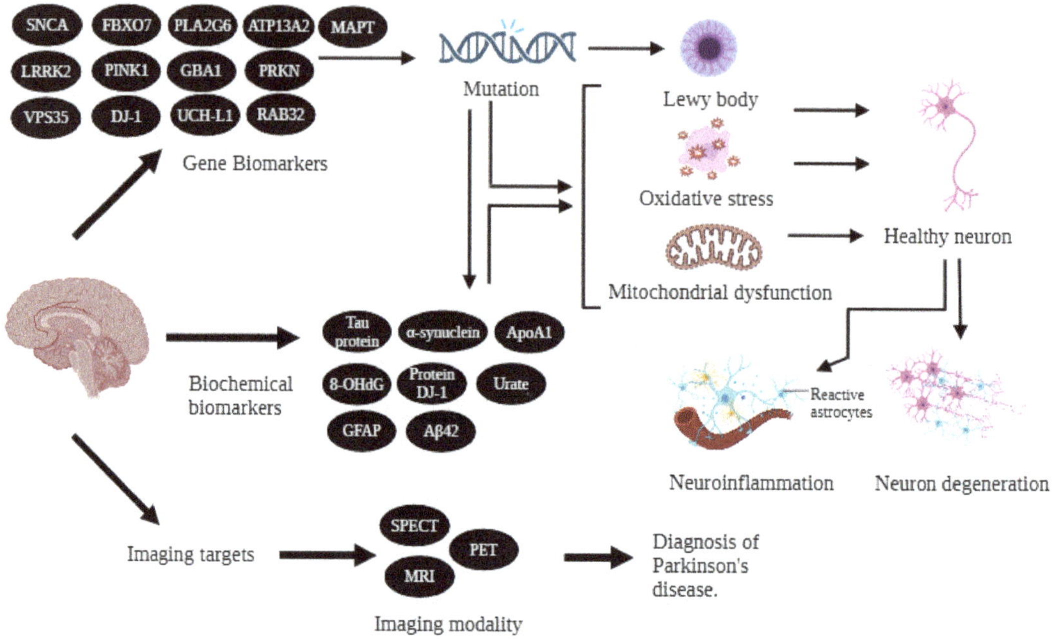

Figure 1. Schematic representation of PD biomarkers. The three primary categories of PD biomarkers are genetic, biochemical, and neuroimaging biomarkers, each leading to insights into disease mechanisms and progression. Created with www.BioRender.com.

3.1. Biochemical Biomarkers

Orexin is known to be involved in narcolepsy, but increasing evidence suggests that it may also be involved in other neurodegenerative conditions, such as Parkinson's, Alzheimer's, epilepsy, amyotrophic lateral sclerosis, and Huntington's disease [19]. A recent study has examined how orexin-A promotes dopaminergic neuron survival and function by working on orexin receptors. Orexin-A has been demonstrated to slow the death of dopaminergic neurons while also improving cognitive and motor functions, indicating that it has therapeutic promise in the context of PD [20]. Another study states that changes in orexin levels may have a role in the pathophysiology of nonmotor symptoms such as sleep disturbances and cognitive impairment. Notably, elevated plasma levels of orexin-A have been observed in individuals with early-stage PD, suggesting that orexin-A may function as a compensation mechanism against inflammatory and oxidative stress [21]. Orexin receptor antagonists may help control sleep disturbances, which are common in PD patients; however, there is concern around the fact that they may worsen both motor and nonmotor symptoms by blocking orexin signaling pathways, which are critical for maintaining dopaminergic neuron function [22]. The intermediate filament protein glial fibrillary acidic protein (GFAP) is mostly expressed in central nervous system astrocytes. When there is damage to the nervous system, astrocytes quickly release GFAP into the circulation [10]. A few studies looking at serum GFAP in PD patients have found higher levels of GFAP in the serum than in the controls [23,24]. Astrocytes produce the inflammatory marker known as glial fibrillary acidic protein, or GFAP. Research has demonstrated that plasma GFAP can indicate when dementia will start to develop in PD patients with mild cognitive impairment (PD-MCI) [25]. The type-III intermediate filament, known as glial fibrillary acidic protein (GFAP) and mostly expressed by astrocytes, can be used to identify reactive astrogliosis. The expression of PD and Alzheimer's disease (AD) is significantly heightened, both being notable neurodegenerative disorders [26,27].

The oxidation of DNA nucleosides generates 8-hydroxy-2-deoxyguanosine (8-OHdG), which serves as a valuable indicator of oxidative stress. There is evidence that the cerebrospinal fluid (CSF) of patients suffering from a range of neurodegenerative diseases has higher than normal levels of oxidative stress. A neurochemically supported diagnosis of PD may benefit from the measurement of 8-OHdG levels [28]. The possibility of polymorphism and PD altering biochemical measures has been investigated in a recent study. Parameters of oxidative stress, including 8-OHdG, ROS, and NADPH, have been shown to be strongly correlated with the risk of PD [29]. The α-synuclein (αSyn) seed amplification assay (SAA) identifies αSyn aggregates as a sign of Lewy body pathology in patient-derived biomaterials, aiding the diagnosis of dementia syndromes and PD [30]. Misfolded αSyn aggregates, or seeds, have an inherent self-replicative characteristic that allows (αSyn-SAAs) for their proliferation in vitro. In these tests, aggregate fragmentation into smaller self-propagating seeds and elongation at the expense of recombinant αSyn (rec-αSyn) are the two cyclical processes that increase the amount of αSyn seeds circulating in bodily fluids. Fluorescent dyes unique to amyloids, like thioflavin T, can detect the seeds after they have been amplified [31]. Larger analyses of the αSyn-SAA for the biochemical diagnosis of PD show that the assay classifies patients suffering from PD with high sensitivity and specificity, provides information about molecular heterogeneity, and detects prodromal individuals before diagnosis [32]. According to the results of a recent study, developing disease-modifying therapies to stop the spread of pathological proteins in PD and Alzheimer's disease (AD) may benefit from focusing on particular tau and aSyn variants with elevated post translational modification (PTM), such as pS129 aSyn and C-terminus epitopes [33]. The amount of alpha-synuclein in PD patients' saliva can vary, something which could be valuable for the early detection of this neurodegenerative disease and for its distinction from other synucleinopathies [34].

Reduced Apo A1 levels have been linked to an increased risk of PD, as well as to a more severe form of the disease and to the manifestation of PD at an earlier age [35]. A recent study using a machine learning technique has demonstrated the substantial correlation among TG, ApoA1, SNCA (rs6826785), and PD-MCI [36]. A related study has shown that, whilst PD-MCI patients have greater levels of TC, TG, and Apo A1, PD patients have considerably lower levels of lipid biomarkers. For PD-MCI patients, TC, TG, and Apo A1 might be helpful biomarkers [37].

Aβ42 plays a role in the build-up of amyloid plaques, a hallmark of AD. However, Aβ42 has a more nuanced function with respect to PD [38]. In contrast, certain research studies have indicated that PD patients experiencing cognitive decline have a higher Aβ42 burden [39]. PD with a significant probability of worsening may be identified by tau, Aβ42, and plasma EV α-synuclein combined. The results of a new study can clarify how these pathognomonic proteins interact, and they may be used in therapeutic strategies to modify disease progression or as indicators of therapy response [40].

Mutations in the DJ-1 protein generate autosomal recessive variations, and oxidized DJ-1 has been found in the brains of patients with idiopathic PD [41]. Early-onset PD with autosomal recessive inheritance is caused by mutations in PARK7 that result in nonfunctional, or in the total loss of, DJ-1 protein. In general, PARK7/DJ-1 dysfunction significantly impact the immunological responses and neuroinflammatory processes of microglia [42]. Low serum UA levels may correlate with the severity of sleep disorders in PD patients and act as a biomarker for poor sleep quality in the context of PD [43]. Low levels of serum uric acid are associated with nonmotor symptoms of PD [44]. In vitro, tau-modified α-synuclein fibrils show higher seeding activity than pure α-synuclein fibrils, resulting in mitochondrial malfunction, synaptic impairment, and neurotoxicity [45]. Tau-PET imaging enables the visualization and measurement of tau load in the brain, offering insights into the distribution and impact of tau pathology with respect to PD patients [46] (Figures 2 and 3, Table 1).

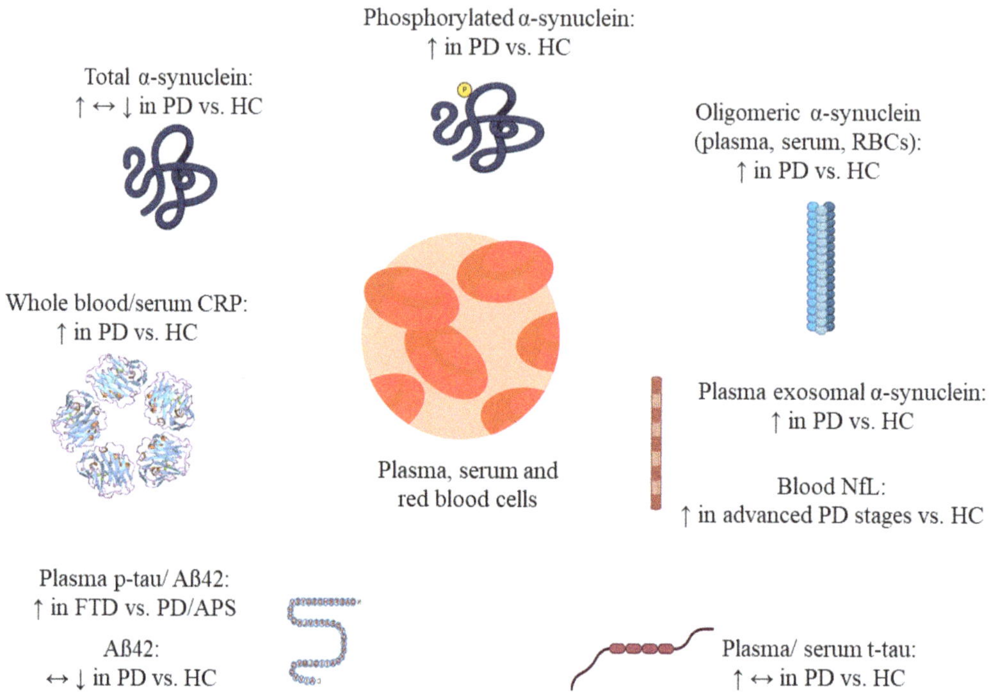

Figure 2. Possible blood-based biomarker of Parkinson's disease. This figure provides evidence supporting the potential of the blood-based biomarker for Parkinson's disease, shows significant differences in levels between PD patients and controls, good diagnostic accuracy, correlation with disease severity, dynamic changes over time, and reproducibility in an independent cohort. Total α-synuclein levels show inconsistent results, with some studies reported increase (↑), others decrease (↓), and some no significant difference (↔) between PD patients and HC. This variability may stem from differences in sample sources or detection methods. Phosphorylated α-synuclein is consistently elevated (↑) in PD patients as compared to HC. Levels of oligomeric forms of α-synuclein are increased (↑) in PD patients' plasma, serum, and red blood cells, which is more closely associated with neurotoxicity and PD pathology. Exosomal α-synuclein is also elevated (↑) in PD patients compared to HC, supporting its role as a biomarker for PD. Blood NfL levels are significantly higher (↑) in advanced stages of PD, reflecting neurodegeneration and disease progression. Tau protein levels, both in plasma and serum, are generally increased (↑) in PD, but other studies report no significant change (↔). Aβ42 levels in PD are unchanged (↔) or reduced (↓) compared to HC, reflecting study variability and potential differences in disease mechanisms. Levels of phosphorylated tau (p-tau) and Aβ42 are increased (↑) in FTD patients compared to PD or APS. CRP protein levels, both in whole blood and serum, are generally elevated (↑) in PD patients as compared to HC. PD; Parkinson's disease, HC; Healthy control, FTD; Frontotemporal dementia, APS; Atypical parkinsonian syndromes, CRP; C-Reactive protein, Aβ42; amyloid-beta 42, NfL; Neurofilament light chain, ↑; upregulated, ↓; downregulated, ↔; unchanged.

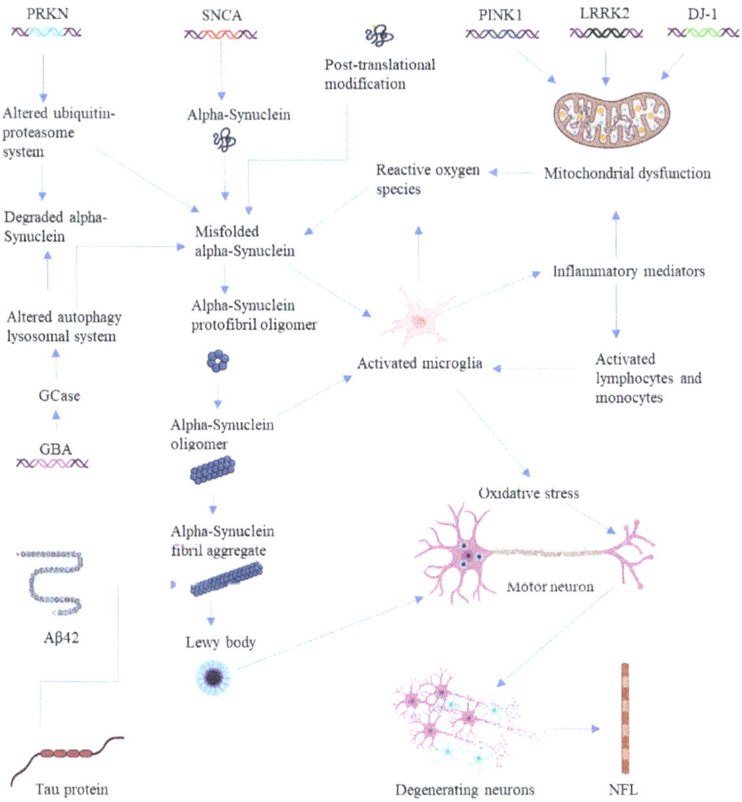

Figure 3. Cerebrospinal fluid (CSF) biomarkers of PD. This figure illustrates the complex pathways involved in PD pathology and highlights key cerebrospinal fluid (CSF) biomarkers associated with the disease. The molecular mechanisms underlying PD facilitate early diagnosis and the development of targeted therapies.

Table 1. Lists of the different biochemical biomarkers used in the diagnosis of Parkinson's disease, describing recent findings on their role with respect to PD, their prospective use, and their limitations at the onset of PD diagnosis. These biomarkers aid the understanding and diagnosis of Parkinson's disease owing to their distinct roles in the neurophysiological processes and pathological changes linked to the condition.

Biochemical Biomarkers	Findings Relative to PD	Prospective Use	Limitations	References
α-synuclein	• The α-synuclein seeding amplification assay (αSyn-SAA), is a very accurate method for identifying aberrant α-synuclein in spinal fluid.	• Early PD detection with the αSyn-SAA test has demonstrated encouraging results. In [32], it was able to determine that the disease was absent from 96% of persons without PD and that it was present in 87% of the participants with PD.	• Variations within the result. • Additional validation is needed.	[32]

Table 1. *Cont.*

Biochemical Biomarkers	Findings Relative to PD	Prospective Use	Limitations	References
Orexin	• Orexin-A promotes dopaminergic neuron survival and function by acting on orexin receptors. • Changes in orexin levels may have a role in the pathophysiology of nonmotor symptoms.	• Neuroprotective role. • Effect on sleep disturbances.	• Intricate interactions with other neurotransmitters.	[20,21]
Uric acid	• Low serum UA levels may correlate with the severity of sleep disorders in PD patients and act as a biomarker of poor sleep quality in PD patients.	• Risk factor identification. • Disease progression monitoring. • Nonmotor symptom assessment. • Therapeutic targets.	• Lack of mechanistic understanding.	[43,44,47]
Tau protein	• In vitro, tau-modified α-synuclein fibrils show higher seeding activity than pure α-synuclein fibrils, resulting in mitochondrial malfunction, synaptic impairment, and neurotoxicity. • Tau-PET imaging enables the visualization and measurement of tau load in the brain, offering insights into the distribution and impact of tau pathology across PD patients.	• Influence on its pathophysiology and contribution to disease progression. • Potential therapeutic target.	• Variability, specificity, and sensitivity.	[45,46,48]
Protein DJ1	• Mutations in PARK7 result in nonfunctional DJ-1 protein or in their total loss. In general, PARK7/DJ-1 dysfunction significantly impact the immunological responses and neuroinflammatory processes of microglia.	• Prospective therapeutic target.	• Complexity of pathogenesis. • Limited understanding of the mechanisms.	[41,42]

Table 1. *Cont.*

Biochemical Biomarkers	Findings Relative to PD	Prospective Use	Limitations	References
GFAP	• Prognostic biomarker of dementia conversion in patients with PD who have mild cognitive impairment.	• Tracking the development of the disease. • Conversion of motor subtypes in early PD.	• Lack of specificity.	[49–51]
Apo-A1	• Apo-A1 is a prominent blood biomarker associated with PD and moderate cognitive impairment (PD-MCI). • Lower levels of Apo-A1 in PD patients raise issues regarding its potential neuroprotective properties and participation in the etiology of PD.	• Diagnostic marker. • Therapeutic target. • Risk assessment and prognostic value.	• Single biomarker complexity. • Specificity, sensitivity, and variability. • Clinical translation challenges. • Complexity of lipid metabolism.	[36,37]
8-OHdG	• Specific gene mutations have been linked to higher levels of 8-OHdG in PD patients. • 8-OHdG is related to the degradation of dopaminergic neurons in PD.	• Valuable biomarker for assessing oxidative DNA damage. • 8-OHdG measurements can be used to explore the impact of genetic variants on oxidative DNA damage and susceptibility to PD.	• Interpretation of 8-OHdG levels can be complicated. • The fluctuating nature of 8-OHdG levels could complicate the assessment of disease progression. • Lack of specificity.	[29]
Aβ42	• The APOE genotype may alter amyloid-β-42 levels in CSF, indicating a more vulnerable subtype of PD characterized by significant amyloidopathy in the early stages. • Plasma extracellular vesicle (EV) tau, β-amyloid, and α-synuclein levels in PD patients are associated with clinical progression, motor and cognitive deterioration, and postural instability in affected individuals.	• Prediction of clinical outcomes. • Monitoring of disease progression. • Early-stage detection of disease.	• Lack of specificity.	[40,52]

3.2. Neuroimaging Biomarkers

Neuroimaging biomarkers are essential for PD diagnosis, tracking, and treatment. These biomarkers are obtained from many imaging modalities, including single-photon emission computed tomography (SPECT), positron emission tomography (PET), and magnetic resonance imaging (MRI). The effectiveness of a biomarker depends on the specific clinical question being investigated and on the stage of the disease. Neuroimaging biomarkers can help differentiate between several phases of PD, such as preclinical and prodromal, early, and moderate-to-late stages [53]. The major imaging targets and methods used, as well as their interventions subject to availability in clinical use, are discussed and summarized in Table 2.

Table 2. Potential applications of neuroimaging biomarkers in the context of PD.

Imaging Target	Modality	Interventions in the Context of PD	Available for Clinical Use	References
• Dopamine synthesis • Dopamine transporter	PET, SPECT	• Quantifiable loss of dopaminergic terminals. • Detecting striatal neuron loss in MSA and PSP, quantifying dopamine release, and determining receptor occupancy using D2/3-targeting anti-PD drugs.	Established and broadly available.	[54–56]
• Serotonergic system	PET	• Determining whether the raphe nuclei have lost any serotonergic nerve terminals.	Not established and not available.	[57,58]
• Cholinergic system	PET, SPECT	• VAChT loss observed in the cortex. • AChE depletion has been observed in both cortical and peripheral nerve systems. • A decrease in αβ-42 has been seen throughout the brain.	Not established and not available.	[57]
• Noradrenergic system	PET, SPECT	• Norepinephrine transporter loss detected in the midbrain and thalamus. • Identifying loss of cardiac noradrenergic nerves.	Not established and not available.	[59]
• Glucose metabolism	FDG-PET	• Identifying metabolic, blood flow, and functional connectivity characteristics associated with PD.	Not established and not available.	[60,61]
• Brain volume and atrophy	T1-weighted structural MRI	• This structural MRI method provides comprehensive brain structural information and is helpful in distinguishing PD from secondary and atypical types of parkinsonism.	Established and broadly available.	[62]

Table 2. *Cont.*

Imaging Target	Modality	Interventions in the Context of PD	Available for Clinical Use	References
• Iron in the brain	Iron-sensitive MRI (T2* relaxometry)	• The amount of iron accumulation in the substantia nigra of PD patients is correlated with the severity of their motor symptoms, and T2* relaxometry can identify elevated iron levels in this region when compared to controls.	Established and broadly available.	[62]
• Free water imaging	Diffusion MRI	• The understanding of structural alterations in PD patients can be aided by MRI methods that evaluate tissue integrity and microstructures, especially in areas like the substantia nigra.	Not established and not available.	[63]
• Neuromelanin	MRI	• This MRI method provides information about changes in the substantia nigra, a region affected by PD, by visualizing the neuromelanin content in this area.	Not established and not available.	[64]
• Cerebral blood flow	PET, SPECT, fMRI	• It is helpful to distinguish PD from secondary and atypical types of parkinsonism.	Not established and not available.	[65]
• 18 Kd translocator protein	TPSO-PET	• It provides information about microglial activation and neuroinflammation in PD patients.	Not established and not available.	[66]
• 123I-metaiodobenzy-lguanidine (123I-MIBG)	SPECT/scintigraphy	• It has a correlation with the clinical profile and severity of PD. It can identify cardiac sympathetic denervation with greater sensitivity.	Established and available.	[67]
• Beta amyloid	PET	• Patients with PD dementia (PDD) frequently have significant deposits of amyloid-β (Aβ) which contribute to memory loss and accelerate the rate of cognitive impairment.	Established and available.	[68]
• Tau	PET	• Tau-PET imaging studies have identified aberrant tau deposition in the brain of PD patients with cognitive impairment (PDCI), including PD dementia (PDD) and PD mild cognitive impairment (PD-MCI).	Established and available.	[46]

T2*: observed or effective relaxation time.

3.2.1. Imaging Strategies for PD Research and Diagnosis

Single-Photon Emission Computed Tomography (SPECT)

Single-photon emission computed tomography (SPECT) is a nuclear imaging technology that involves injecting a radioactive tracer into the patient's body, such a tracer emitting single gamma ray photons. These gamma rays are detected by a rotational gamma camera and converted into light photons. The detected photons are utilized to create a three-dimensional picture of the radiotracer distribution in the body using mathematical algorithms. SPECT imaging gives useful information about many physiological processes and can be used to diagnose diseases, stage them, and evaluate therapy responses. SPECT tracers can be used for scintigraphy, which records a single projection picture of the tracer's distribution inside the body [69].

Positron Emission Tomography (PET)

In Positron Emission Tomography (PET), a patient is given a radiotracer injection. The radiotracer releases positrons, which interact with electrons to produce two 511-KeV photons that are released in opposing directions. External detectors pick up these photons, which reveal details about the body's radioactivity distribution. The annihilation photons are captured by rings of detectors in the PET scanners, enabling the localization of radioactive decay events. Based on the identification of these annihilation events, PET imaging helps with disease diagnosis and treatment by offering important insights into physiological processes [70].

Magnetic Resonance Imaging (MRI) and Functional Magnetic Resonance Imaging (fMRI)

Magnetic resonance imaging (MRI) relies on the interaction of hydrogen nuclei in the body with intense magnetic fields and radio waves. When a patient is placed in an MRI machine, their hydrogen atoms align with the magnetic field. Radiofrequency pulses are then used to cause the atoms to produce energy, which is measured by the MRI scanner. This radiated radiation is utilized to generate detailed photographs of the body's internal structure using the distribution of the hydrogen atoms [71].

The idea behind functional magnetic resonance imaging (fMRI) is to estimate brain activity by measuring changes in blood oxygenation and flow. When neuronal activity increases in a brain area, neurovascular coupling causes localized changes in blood flow, volume, and oxygenation. fMRI detects the blood oxygenation level dependent (BOLD) signal, which is sensitive to variations in the proportion of oxygenated to deoxygenated hemoglobin in the blood. The magnetic characteristics of oxygenated and deoxygenated hemoglobin differ, allowing BOLD signal alterations to be directed at areas of increased cerebral activity, allowing for the mapping of brain function [72].

3.2.2. Imaging Targets Used in PD

Dopaminergic System

Dopaminergic system imaging is an effective biomarker in the early stages of the medical condition and provides a means of evaluating neurodegeneration even in the preclinical and prodromal phases of PD.

Dopamine Synthesis

Certain neurotransmitters, which are substances that neurons utilize for communication, require the enzyme L-amino acid decarboxylase (AADC) in order to be produced. Dopamine synthesis specifically requires AADC, and a number of neurological diseases are marked by the aberrant function or loss of dopamine-producing neurons. Thus, monitoring AADC activity may provide information on how the disease is progressing [73]. FMT-PET (6-fluoro-meta-tyrosine positron emission tomography) is a high-resolution imaging technology used to directly measure the activity of the AADC enzyme in the brain. FMT-PET has been found to identify variations in AADC activity over time, a fact which may aid the prediction of PD in individuals suffering from rapid eye movement (REM) sleep behavior disorder (RBD). Longitudinal research has discovered that RBD patients who later developed PD or dementia with Lewy bodies exhibited substantially reduced AADC activity levels on follow-up FMT-PET scans compared to baseline [74]. F-DOPA PET detects the uptake of the F-DOPA tracer, which is decarboxylated to dopamine by the

AADC enzyme. The reduced absorption of F-DOPA in PD patients shows altered AADC activity, which is a significant part of the disease process [75].

Dopamine Transporter

DAT imaging evaluates presynaptic dopamine deficits by targeting the DAT using positron emission tomography (PET) or single-photon emission computed tomography (SPECT) methods [76]. Both DAT-SPECT and DAT-PET imaging are useful methods for diagnosing dopamine insufficiency and for tracking the course of PD. While PET imaging offers a superior resolution and the potential for multimodal target evaluation, SPECT imaging is more widely used in clinical practice. In nonadvanced PD patients, the PET scan employing [18F] FE-PE2I has been shown to enable accurate assessment of the DAT deficit in several nigrostriatal areas, offering important insights into the course and severity of the disease. The putamen's and sensorimotor striatum's DAT availability has been found to be significantly correlated with scores on the Movement Disorder Society unified PD rating scale part III (MDS-UPDRS-III), especially when it comes to bradykinesia and rigidity symptoms. These findings validate the use of PET radioligand [18F]FE-PE2I as a marker of the severity of PD [77]. A recent study has shown that the diagnosis and clinical treatment of parkinsonism patients are significantly impacted by DAT-SPECT imaging. Following the findings of the DAT scan, 37% of patients experienced changes in diagnosis and 42% experienced changes in clinical treatment [78].

Vesicular Monoamine Transporter Type 2

VMAT2 is found in synaptic vesicle membranes and stores dopamine. VMAT2 is thought to be involved in monoamine absorption into synaptic vesicles. Two VMAT2 radioligands, [11C]DTBZ and [18F]FP-DTBZ, have been utilized to determine the degree of nigrostriatal deficit in PD patients using PET. Studies have revealed that VMAT2 radioligands such as [18F]FE-DTBZ-d4 have better stability and exhibit a better brain uptake capacity than other tracers, making them attractive for the assessment of VMAT2 in the brain [79].

Dopamine Receptors

Dopamine receptor imaging can reveal information about neurons on the postsynaptic side of nerve terminals, as most receptors are found on these membranes. Dopamine receptor imaging using SPECT and PET is critical for distinguishing multiple system atrophy (MSA) and progressive supranuclear palsy (PSP) from PD. These imaging methods can help assess the development rate of PD and aid the selection of suitable therapies, particularly with respect to individuals receiving neuroleptics [80].

Serotonergic System

The monoamine neurotransmitter serotonin is created from the amino acid tryptophan. It is then released into the synaptic cleft, stored in synaptic vesicles, and then re-captured by the corresponding transporter protein [81]. Compared to healthy controls, PD patients have fewer available 5-HT1B receptors in their brain. Creative ability and receptor availability are also associated, pointing toward a potential role of 5-HT1B receptors in the development of PD cognitive symptoms [82,83]. There is a correlation between the degree of apathy in prodromal PD and the integrity of the serotonin transporter signal in the dorsal raphe nucleus [84]. In both striatal and extra-striatal brain areas, the loss of serotonergic innervation is thought to be an early sign of PD [57].

Cholinergic System

LRRK2 mutation carriers, both healthy and suffering from PD, have greater cortical AChE activity compared to healthy controls and patients with idiopathic PD [85]. With PET and SPECT tracers, it is possible to scan this transporter. According to different studies, the above tracers show a considerably reduced availability of VAChT in the cortical areas of

PD patients compared to healthy controls [86]. PD is characterized by substantial posterior cortical cholinergic degradation in nondemented individuals [87,88], with indications of progression from posterior to anterior over time [89]. GBA-mutated individuals with PD are more likely to experience early cognitive decline, as well as a higher reduction in cortical cholinergic denervation, even with normal cognition [90].

Noradrenergic System (NA)

The non-motor symptoms of PD may be caused by noradrenergic transmission deficits. In PD patients, noradrenergic activity decreases and is linked to cognitive impairment and RBD [91]. Prodromal PD does not show a significant decrease in NA transporters. However, there are some indications of substantially decreased availability in the sensorimotor cortex and thalamus [92]. Recent research suggests that NA receptors are reduced in both the motor insular cortex and the thalamus and putamen [93].

Cerebral Blood Flow and Glucose Metabolism

PD alters the normal functioning of the brain, including blood flow, oxygen and energy consumption, and interregional communication. The PET tracer [18F]-FDG can detect brain energy use [60].

Metabolic imaging with [18F]-FDG is able to identify PD-related metabolic patterns (PDRP) in PD patients. PDRP expression is assessed scan by scan, enabling an objective assessment of disease activity in individual patients, something which can help with the clinical evaluation and monitoring of PD development [61]. Recent research indicates that fMRI brain responses to DBS stimulation in PD patients might be an objective indicator of clinical responses [94].

Other Imaging Targets

A recent study has used structural brain scans to create and assess a robust explainable deep learning model for PD categorization in order to distinguish between PD patients and healthy controls. This study focuses on training a 3D convolutional neural network (CNN) utilizing one of the biggest collections of T1-weighted magnetic resonance imaging datasets [62]. Standard diffusion measures are not as accurate as free water (FW)-corrected diffusion measures. A recent study has assessed whole brain white matter and corrected diffusion metrics for these structures in patients suffering from PD, multiple system atrophy (MSA), and progressive supranuclear palsy (PSP). The study reports distinct differences in the patterns of FW alterations between PD and atypical parkinsonian disorders, indicating the potential use of whole brain FW maps as markers during the diagnosis of these disorders [95].

The transverse relaxation time (T2) of tissues is measured using T2 relaxometry in imaging to provide information on the composition and properties of the tissue. Iron sensitivity in imaging enables the identification and description of iron accumulation in tissues, especially in the substantia nigra (SN), using methods such as transverse MR relaxometry and quantitative susceptibility mapping (QSM) [96]. The advent of quantitative susceptibility mapping (QSM) has enabled the noninvasive quantification of brain iron levels in vivo via MRI scans, contributing to the understanding of iron-associated pathogenesis, and has the potential to contribute to the development of iron-based biomarkers of PD [97]. The distribution of neuromelanin and iron in the SN has been evaluated in a study using T2 relaxometry alongside iron-sensitive imaging. Such an approach allows for the visualization of the dorsal part of the substantia nigra pars compacta, which contains a considerable quantity of neuromelanin. Since increased iron deposition in PD patients can affect some imaging characteristics, such as the tail sign in nigrosome1, i.e., a cluster of dopaminergic neurons in the SN that contain neuromelanin, iron sensitivity affects how neuromelanin distribution is seen in imaging investigations [64]. Strong evidence suggests that neuroinflammation contributes significantly to the onset and development of PD and other neurodegenerative diseases. The activation of microglia plays a vital role in driving the inflammatory response of the brain [65]. PET ligands that target the 18 kDa translocator protein (TSPO) can be used to reveal neuroinflammation in vivo. When microglia are activated, there is a significant increase in the expression of TSPO, which is mostly expressed in the outer membrane of the mitochondria.

Numerous studies have documented a significant increase in microglial activation in PD patients [66]. There are recognized procedures for measuring the sympathetic denervation of the heart, such as SPECT or scintigraphy using iodine radio-labelled MIBG. The approach has been utilized to divide PD into two-to-three subcategories despite significant flaws [67,98]. The uptake of 18F-florzolotau has been found to be considerably greater in the cortical areas of patients suffering from PD-related dementia compared to PD patients with normal cognition and healthy controls, particularly in the temporal lobes. Notably, 18F-florzolotau uptake in the occipital lobe of individuals suffering from PD with dementia has been found to be significantly associated with cognitive impairment, as measured by mini-mental state examination (MMSE) scores [99]. According to logistic regression analysis, tau PET has a greater correlation with change management than amyloid PET across all participants and dementia types [100]. The α-synuclein PET tracer allows for the identification of prodromal and early-stage synucleinopathies and can provide information on the progression of the disease. It can also help differentiate parkinsonisms caused by synucleinopathies, tauopathies, and other diseases. However, no radiotracers have been approved for clinical use in humans [101]. The original magnetic resonance parkinsonism index (MRPI) is a useful technique for distinguishing between progressive supranuclear palsy-parkinsonism (PSP-P) and PD [102]. However, the MRPI 2.0 has shown improved diagnostic accuracy, with an area under the curve (AUC) of 0.91 with respect to the identification of these diseases in their early stages, something which is critical for a swift and accurate diagnosis. The MRPI 2.0 includes additional measures, such as the breadth of the third ventricle, which improve its sensitivity and specificity compared to the original MRPI. In recent investigations, it has demonstrated a sensitivity of 100% and a specificity of roughly 94.3% when distinguishing PSP-P from PD, making it a reliable tool for clinicians [103,104].

3.3. Genetic Biomarkers

Genetic biomarkers of PD are genetic variants related to the disease that can be found using genome-wide association studies (GWAS) and machine learning algorithms. These biomarkers can aid the diagnosis and progression monitoring of PD [105]. Recent advancements in genetic research methodologies have offered new perspectives on the role of the genetic background in the onset of PD. Therefore, PD is a multifaceted genetic condition triggered by a combination of aging, environmental factors, and inherited elements. There are two main approaches to identify genetic variables in PD patients. One approach consists of looking into unusual Mendelian variants of PD and finding the genes that cause them. The other entails the detection of risk variations via genetic statistical analyses of huge groups of subjects [106]. Research into Mendelian PD genetics has identified 24 genes or loci listed in the online Mendelian inheritance in man (OMIM) database that play a role in PD progression. The PRKN gene is responsible for autosomal recessive juvenile PD, while the leucine-rich repeat kinase 2 (LRRK2) variant causes autosomal dominant PD [106]. LRRK2 architecture impacts kinase activation, and increased LRRK2 substrate phosphorylation may contribute to PD etiology. Evidence suggests that oxidative stress and/or endolysosomal stress may increase LRRK2 kinase activity in idiopathic PD patients. Exposure to environmental toxins linked to idiopathic PD in epidemiological studies has also been shown to trigger LRRK2 kinase activity [107]. Through whole-exome sequencing, the vacuolar protein sorter-35 (VPS35) gene has been identified as the initial causal gene for autosomal dominant PD [108]. In a Japanese study, coiled-coil-helix domain containing 2 (CHCHD2) and prosaposin (PSAP) have been identified as novel genes associated with familial PD, albeit with relatively low prevalence in Mendelian PD [109,110]. DJ-1, FBXO7, SYNJ1, EIF4G1, GIGYF2, HTRA2, PLA2G6, DNAJC6, ATP13A2, and DNAJC13 are other families of PD-causing genes. However, the functions of these gene products, such as the management of oxidative stress, intracellular trafficking, phospholipid membranes, mitochondrial processes, and the ubiquitin–proteasome system, have all been associated with the development of PD. GBA1 variants associated with PD, such as p.E326K, p.T369M, p.N370S, and p.L444P, reduce the activity of glucocerebrosidase, the enzyme encoded by GBA1, thereby impairing the lysosomal breakdown of α-synuclein [111].

Despite the progress in genetic analysis technology and high-throughput methods, making new discoveries remains challenging. For instance, whole-exome sequencing of small PD families often reveals hundreds of candidate causative variants; however, typically, each family has only one causal variant, regardless of its size. Consequently, researchers must employ various strategies to reduce the number of potential variants from hundreds to a single one. It is estimated that at least 80% of the families with familial PD have a single candidate mutation, leaving the underlying gene unidentified (Table 3).

Table 3. Genetic biomarkers and their potential roles in the context of PD.

Biomarker	Chromosome Location	Characteristics	Mutations	Pathological Features	References
LRRK2	12q12	Contains GTPase and autosomal dominant.	R1441G, R1441C, R1441H, Y1699C, G2019S, R1628P, G2385R and I2020T	It can cause dopaminergic neuronal cell death, reduced dopamine neurotransmission, and protein synthesis and degradation problems.	[112]
PINK1	1p36.12	Autosomal recessive, with eight exons encoding a 581 amino acid serine/threonine protein kinase.	p.Gln456Stop, p.Leu347Pro	Hypereflexia, dystonia, early l-dopa-induced dyskineasis, and mental and behavioral difficulties can occur.	[113,114]
PRKN	6q25.2-27	G403C mutations cause a glycine to arginine amino acid change.	G403C, deletion mutation of exon 6	Exon 6 deletion mutations cause EOPD.	[115,116]
UCH-L1	4p13	Autosomal dominant.	p.Arg178Gln, p.Ala216Asp	Visual impairment, impaired light perception, optic atrophy, and more.	[117–119]
SNCA	4q21.3-q22	SNCA is a sarcopterygian-specific gene. Five exons encode this autosomal dominant gene.	A30P, E46K, H50Q, G51D, A53T	Lewy body dementia is characterized by cognitive fluctuations, visual hallucinations, REM sleep behavior, and other symptoms.	[120–122]
VPS35	16q12	Autosomal dominant.	D620N	The mutation impairs LAMP2a retrieval from the endosome to the Golgi, resulting in a drop in LAMP2a levels within neuronal cells. This induces an increase in α-synuclein buildup within these cells.	[123]
GBA	1q21	Autosomal recessive.	N370S and L444P	GBA mutations cause mitochondrial and lysosomal dysfunction.	[124]

4. MicroRNAs and Peptides Show Potential to Be Next-Generation Biomarkers

MicroRNAs (miRNAs) and peptides are increasingly recognized as promising biomarkers of PD, with significant potential to improve early detection and diagnostic accuracy. A recent study has demonstrated the potential of circulating miRNAs as biomarkers for the detection of PD and associated disorders. A review by Giri et al. highlights that α-synuclein and miRNA can be detected in the urine of PD patients and can be used as early biomarkers in the diagnosis and prognosis of neurodegenerative diseases such as PD [125]. A meta-analysis has found that biofluid miRNAs may successfully differentiate between PD patients and healthy individuals [126,127]. Several studies have frequently associated specific miRNAs, including miR-24-3p, miR-146a-5p, and miR-331-5p, with PD, suggesting their potential as valuable diagnostic markers [127,128]. Previous research has demonstrated that the functional transfer of miRNAs via extracellular vesicles is being explored for its potential to detect not only PD, but also related disorders, such as random eye movement sleep behavior disorder, thereby enhancing its clinical relevance [129]. A study has found overexpression of miR-132 in plasma samples from male individuals suffering from PD compared to female patients, and it has also established the role of miR-132 in the regulation of the protein Nurr1, which affects dopamine regulation [130]. There are several other miRNAs, such as miR-7-1-5p, miR-223-3p (serum) [131], hsa-miR-144-3p (serum) [132], miR-24, miR-195, miR-19b (serum-EVs) [133], miR-22-3p, miR-10b-5p, and miR-151a-3p (CSF) [134], that are deregulated in different biofluids and that can be used as potential biomarkers for the early diagnosis of PD and for the monitoring of disease progression.

It has been reported that a panel of peptides can predict PD before the symptoms appear with an accuracy of 79%. This panel contains biomarkers such as the granulin precursor and the complement C3, both of which are associated with disease severity, indicating their potential in early detection and therapeutic use [135]. Another study has discovered cerebral fluid peptides that might successfully distinguish PD patients from healthy controls [136]. Innovative peptide-based methods are under development to precisely target risky α-synuclein oligomers. A plasmonic biosensor based on α-helical peptides has the potential to detect these oligomers early and improve diagnosis and therapy outcomes [137]. A variation of the small humanin-like peptide 2 (SHLP2) protein has been found to act as a protective factor against PD, lowering the risk by almost 50% in some groups. This finding highlights the significance of mitochondrial-derived microproteins in determining PD risk and paves the way for new treatment strategies [138].

5. Conclusions

Although Parkinson's disease is caused by a combination of environmental variables and genetic predispositions, its diagnosis requires a multidisciplinary approach. Biomarkers are critical for improving our understanding of PD and its treatment. Identification and confirmation of biomarkers can lead to improved patient care, more accurate diagnoses, and better treatment outcomes.

In terms of early PD diagnosis, even before the clinical symptoms manifest, the α-synuclein seeding amplification assay (αSyn-SAA) represents a significant advancement. GFAP may be a valuable biomarker for predicting cognitive decline and dementia conversion in PD patients. In prospective cohort studies, combining blood biomarkers for the prognosis and diagnosis of PD can increase diagnostic accuracy. The diagnosis of PD can be rendered more sensitive and specific by choosing the right signs for combination detection. The dopaminergic system is now the most widely accessible, well-established, and validated biomarker in clinical practice. Understanding the genetic variables affecting PD risk, onset, and development is critical to develop therapies that can reduce disease progression. Numerous genes and GWAS sites have been linked to the development of PD. We need to continue the investigation into possible genetic risk factors and work together to grasp their implications at the molecular and biological level. Understanding genetic and functional information is crucial for developing proper treatments for PD. Correlating imaging data with biochemical markers can increase PD diagnosis accuracy. It can also help assess therapy success, particularly during the early stages when neuroprotective medications are most beneficial. This literature review highlights the continuous efforts to identify new indicators, improve the efficacy of existing biomarkers, and close the diagnostic gaps. With further research, the application of accurate biomarkers in clinical practice has the potential to transform the way PD is managed, improve the accuracy of diagnosis, and, ultimately, benefit patients. In addition to enriching our present knowledge of PD, this review highlights interesting directions for further research and advancement in the hunt for efficient preventative and treatment methods.

6. Future Prospects

Future research should combine biochemical, neuroimaging, and genetic markers to provide a complete biomarker profile of PD. This multimodal method may increase diagnostic accuracy and assist early diagnosis by capturing the disease's multidimensional character. Longitudinal studies that investigate biomarker changes over time are needed among individuals at risk of developing PD or in the early stages of the disease. Investigating gene–environment interactions might provide crucial insights into PD etiology. Despite these advancements, the establishment of a comprehensive biomarker panel specifically for PD remains a focus for future research [127]. Large-scale proteomics techniques have revealed possible diagnostic markers in serum samples, including ALCAM, contactin 1, CD36, DUS3, NEGR1, Notch1, TrkB, and BTK, all of which correlate with clinical scores and disease progression [139]. Additionally, employing a neurology 4-plex-A biomarker panel alongside plasma α-synuclein has demonstrated potential with respect to the differentiation of PD patients from healthy individuals and with respect to the prediction of clinical disease progression [140]. When combined with clinical factors, blood biomarkers such as serum neurofilament light (NfL) and genetic status (GBA and APOE) have been shown to be beneficial in PD prognostic modeling, offering a stronger prediction of poor outcomes [141]. Furthermore, extracellular vesicles (EVs) produced from PD patients have shown promise as a source of novel biomarkers for diagnosis, prognosis, and therapeutic monitoring [142].

Author Contributions: Conceptualization, R.A., A.K.M.A.H. and H.M.F.; writing—original draft preparation, R.A., A.K.M.A.H., H.S. and H.M.F.; writing—review and editing, R.A., A.K.M.A.H., H.S., M.M.B., A.P., M.-M.R., K.M.S., H.M.F. and V.K.; visualization, R.A., M.M.B., A.P., M.-M.R., K.M.S., H.M.F. and V.K.; supervision, J.-J.K. All authors have read and agreed to the published version of the manuscript.

Funding: This research received no external funding.

Institutional Review Board Statement: Not applicable.

Informed Consent Statement: Not applicable.

Data Availability Statement: No new data were created or analyzed in this study.

Conflicts of Interest: The authors declare no conflicts of interest.

References

1. World Health Organization. *Parkinson Disease: A Public Health Approach: Technical Brief*; WHO: Geneva, Switzerland, 2022; ISBN 9789240050990.
2. Samii, A.; Nutt, J.G.; Ransom, B.R. Parkinson's Disease. *Lancet* **2004**, *363*, 1783–1793. [CrossRef] [PubMed]
3. Jellinger, K.A. Neuropathology of Sporadic Parkinson's Disease: Evaluation and Changes of Concepts. *Mov. Disord.* **2012**, *27*, 8–30. [CrossRef] [PubMed]
4. Gökçal, E.; Gur, V.E.; Selvitop, R.; Babacan Yildiz, G.; Asil, T. Motor and Non-Motor Symptoms in Parkinson's Disease: Effects on Quality of Life. *Noro Psikyatr. Ars.* **2017**, *54*, 143–148. [CrossRef] [PubMed]
5. Munhoz, R.P.; Tumas, V.; Pedroso, J.L.; Silveira-Moriyama, L. The Clinical Diagnosis of Parkinson's Disease. *Arq. Neuropsiquiatr.* **2024**, *82*, s00431777775. [CrossRef] [PubMed]
6. de Klerk, G.W.; van Laar, T.; Meles, S.K. A Retrospective Study of the MDS Criteria for Prodromal Parkinson's Disease in the General Population. *NPJ Park. Dis.* **2024**, *10*, 125. [CrossRef]

7. Alster, P.; Madetko-Alster, N. Significance of Dysautonomia in Parkinson's Disease and Atypical Parkinsonisms. *Neurol. Neurochir. Pol.* **2024**, *58*, 147–149. [CrossRef]
8. Schapira, A.H.V.; Chaudhuri, K.R.; Jenner, P. Non-Motor Features of Parkinson Disease. *Nat. Rev. Neurosci.* **2017**, *18*, 509. [CrossRef]
9. Nandanwar, D.; Truong, D.D. Treatment of Parkinson's Disease. In *Reference Module in Neuroscience and Biobehavioral Psychology*; Elsevier: Amsterdam, The Netherlands, 2024.
10. Liddelow, S.A.; Barres, B.A. Reactive Astrocytes: Production, Function, and Therapeutic Potential. *Immunity* **2017**, *46*, 957–967. [CrossRef]
11. Adam, H.; Gopinath, S.C.B.; Md Arshad, M.K.; Adam, T.; Parmin, N.A.; Husein, I.; Hashim, U. An Update on Pathogenesis and Clinical Scenario for Parkinson's Disease: Diagnosis and Treatment. *3 Biotech* **2023**, *13*, 142. [CrossRef]
12. Vázquez-Vélez, G.E.; Zoghbi, H.Y. Parkinson's Disease Genetics and Pathophysiology. *Annu. Rev. Neurosci.* **2021**, *44*, 87–108. [CrossRef]
13. Tolosa, E.; Garrido, A.; Scholz, S.W.; Poewe, W. Challenges in the Diagnosis of Parkinson's Disease. *Lancet Neurol.* **2021**, *20*, 385–397. [CrossRef] [PubMed]
14. Rajput, A.H.; Rajput, A. Accuracy of Parkinson Disease Diagnosis Unchanged in 2 Decades. *Neurology* **2014**, *83*, 386–387. [CrossRef] [PubMed]
15. Biomarkers Definitions Working Group. Biomarkers and Surrogate Endpoints: Preferred Definitions and Conceptual Framework. *Clin. Pharmacol. Ther.* **2001**, *69*, 89–95. [CrossRef] [PubMed]
16. Lotankar, S.; Prabhavalkar, K.S.; Bhatt, L.K. Biomarkers for Parkinson's Disease: Recent Advancement. *Neurosci. Bull.* **2017**, *33*, 585–597. [CrossRef] [PubMed]
17. Li, T.; Le, W. Biomarkers for Parkinson's Disease: How Good Are They? *Neurosci. Bull.* **2020**, *36*, 183–194. [CrossRef]
18. Emamzadeh, F.N.; Surguchov, A. Parkinson's Disease: Biomarkers, Treatment, and Risk Factors. *Front. Neurosci.* **2018**, *12*, 612. [CrossRef]
19. Berhe, D.F.; Gebre, A.K.; Assefa, B.T. Orexins Role in Neurodegenerative Diseases: From Pathogenesis to Treatment. *Pharmacol. Biochem. Behav.* **2020**, *194*, 172929. [CrossRef]
20. Alrouji, M.; Al-Kuraishy, H.M.; Al-Gareeb, A.I.; Zaafar, D.; Batiha, G.E.-S. Orexin Pathway in Parkinson's Disease: A Review. *Mol. Biol. Rep.* **2023**, *50*, 6107–6120. [CrossRef]
21. Alster, P.; Madetko-Alster, N.; Otto-Ślusarczyk, D.; Migda, A.; Migda, B.; Struga, M.; Friedman, A. Role of Orexin in Pathogenesis of Neurodegenerative Parkinsonisms. *Neurol. Neurochir. Pol.* **2023**, *57*, 335–343. [CrossRef]
22. Braun, A.; Manavis, J.; Yamanaka, A.; Ootsuka, Y.; Blumbergs, P.; Bobrovskaya, L. The Role of Orexin in Parkinson's Disease. *J. Neurosci. Res.* **2024**, *102*, e25322. [CrossRef]
23. Su, W.; Chen, H.B.; Li, S.H.; Wu, D.Y. Correlational Study of the Serum Levels of the Glial Fibrillary Acidic Protein and Neurofilament Proteins in Parkinson's Disease Patients. *Clin. Neurol. Neurosurg.* **2012**, *114*, 372–375. [CrossRef] [PubMed]
24. Oeckl, P.; Halbgebauer, S.; Anderl-Straub, S.; Steinacker, P.; Huss, A.M.; Neugebauer, H.; von Arnim, C.A.F.; Diehl-Schmid, J.; Grimmer, T.; Kornhuber, J.; et al. Glial Fibrillary Acidic Protein in Serum Is Increased in Alzheimer's Disease and Correlates with Cognitive Impairment. *J. Alzheimers Dis.* **2019**, *67*, 481–488. [CrossRef] [PubMed]
25. Liu, T.; Zuo, H.; Ma, D.; Song, D.; Zhao, Y.; Cheng, O. Cerebrospinal Fluid GFAP Is a Predictive Biomarker for Conversion to Dementia and Alzheimer's Disease-Associated Biomarkers Alterations among de Novo Parkinson's Disease Patients: A Prospective Cohort Study. *J. Neuroinflamm.* **2023**, *20*, 167. [CrossRef]
26. Abdelhak, A.; Foschi, M.; Abu-Rumeileh, S.; Yue, J.K.; D'Anna, L.; Huss, A.; Oeckl, P.; Ludolph, A.C.; Kuhle, J.; Petzold, A.; et al. Blood GFAP as an Emerging Biomarker in Brain and Spinal Cord Disorders. *Nat. Rev. Neurol.* **2022**, *18*, 158–172. [CrossRef] [PubMed]
27. Li, D.; Liu, X.; Liu, T.; Liu, H.; Tong, L.; Jia, S.; Wang, Y. Neurochemical Regulation of the Expression and Function of Glial Fibrillary Acidic Protein in Astrocytes. *Glia* **2020**, *68*, 878–897. [CrossRef]
28. Gmitterová, K.; Heinemann, U.; Gawinecka, J.; Varges, D.; Ciesielczyk, B.; Valkovic, P.; Benetin, J.; Zerr, I. 8-OHdG in Cerebrospinal Fluid as a Marker of Oxidative Stress in Various Neurodegenerative Diseases. *Neurodegener. Dis.* **2009**, *6*, 263–269. [CrossRef]
29. Karahalil, B.; Miser Salihoğlu, E.; Elkama, A.; Orhan, G.; Saygın, E.; Yardim Akaydin, S. Individual Susceptibility Has a Major Impact on Strong Association between Oxidative Stress, Defence Systems and Parkinson's Disease. *Basic Clin. Pharmacol. Toxicol.* **2022**, *130*, 158–170. [CrossRef]
30. Bräuer, S.; Rossi, M.; Sajapin, J.; Henle, T.; Gasser, T.; Parchi, P.; Brockmann, K.; Falkenburger, B.H. Kinetic Parameters of Alpha-Synuclein Seed Amplification Assay Correlate with Cognitive Impairment in Patients with Lewy Body Disorders. *Acta Neuropathol. Commun.* **2023**, *11*, 162. [CrossRef]
31. Concha-Marambio, L.; Pritzkow, S.; Shahnawaz, M.; Farris, C.M.; Soto, C. Seed Amplification Assay for the Detection of Pathologic Alpha-Synuclein Aggregates in Cerebrospinal Fluid. *Nat. Protoc.* **2023**, *18*, 1179–1196. [CrossRef]
32. Siderowf, A.; Concha-Marambio, L.; Lafontant, D.-E.; Farris, C.M.; Ma, Y.; Urenia, P.A.; Nguyen, H.; Alcalay, R.N.; Chahine, L.M.; Foroud, T.; et al. Assessment of Heterogeneity among Participants in the Parkinson's Progression Markers Initiative Cohort Using α-Synuclein Seed Amplification: A Cross-Sectional Study. *Lancet Neurol.* **2023**, *22*, 407–417. [CrossRef]

33. van der Gaag, B.L.; Deshayes, N.A.C.; Breve, J.J.P.; Bol, J.G.J.M.; Jonker, A.J.; Hoozemans, J.J.M.; Courade, J.-P.; van de Berg, W.D.J. Distinct Tau and Alpha-Synuclein Molecular Signatures in Alzheimer's Disease with and without Lewy Bodies and Parkinson's Disease with Dementia. *Acta Neuropathol.* **2024**, *147*, 14. [CrossRef] [PubMed]
34. Nijakowski, K.; Owecki, W.; Jankowski, J.; Surdacka, A. Salivary Biomarkers for Parkinson's Disease: A Systematic Review with Meta-Analysis. *Cells* **2024**, *13*, 340. [CrossRef] [PubMed]
35. Qiang, J.K.; Wong, Y.C.; Siderowf, A.; Hurtig, H.I.; Xie, S.X.; Lee, V.M.-Y.; Trojanowski, J.Q.; Yearout, D.; Leverenz, J.B.; Montine, T.J.; et al. Plasma Apolipoprotein A1 as a Biomarker for Parkinson Disease. *Ann. Neurol.* **2013**, *74*, 119–127. [CrossRef] [PubMed]
36. Deng, X.; Ning, Y.; Saffari, S.E.; Xiao, B.; Niu, C.; Ng, S.Y.E.; Chia, N.; Choi, X.; Heng, D.L.; Tan, Y.J.; et al. Identifying Clinical Features and Blood Biomarkers Associated with Mild Cognitive Impairment in Parkinson Disease Using Machine Learning. *Eur. J. Neurol.* **2023**, *30*, 1658–1666. [CrossRef] [PubMed]
37. Deng, X.; Saffari, S.E.; Ng, S.Y.E.; Chia, N.; Tan, J.Y.; Choi, X.; Heng, D.L.; Xu, Z.; Tay, K.-Y.; Au, W.-L.; et al. Blood Lipid Biomarkers in Early Parkinson's Disease and Parkinson's Disease with Mild Cognitive Impairment. *J. Park. Dis.* **2022**, *12*, 1937–1943. [CrossRef]
38. Gu, L.; Guo, Z. Alzheimer's Aβ42 and Aβ40 Peptides Form Interlaced Amyloid Fibrils. *J. Neurochem.* **2013**, *126*, 305–311. [CrossRef]
39. Mihaescu, A.S.; Valli, M.; Uribe, C.; Diez-Cirarda, M.; Masellis, M.; Graff-Guerrero, A.; Strafella, A.P. Beta Amyloid Deposition and Cognitive Decline in Parkinson's Disease: A Study of the PPMI Cohort. *Mol. Brain* **2022**, *15*, 79. [CrossRef]
40. Chan, L.; Chung, C.-C.; Hsieh, Y.-C.; Wu, R.-M.; Hong, C.-T. Plasma Extracellular Vesicle Tau, β-Amyloid, and α-Synuclein and the Progression of Parkinson's Disease: A Follow-up Study. *Ther. Adv. Neurol. Disord.* **2023**, *16*, 1–14. [CrossRef]
41. Repici, M.; Giorgini, F. DJ-1 in Parkinson's Disease: Clinical Insights and Therapeutic Perspectives. *J. Clin. Med.* **2019**, *8*, 1377. [CrossRef]
42. Lind-Holm Mogensen, F.; Scafidi, A.; Poli, A.; Michelucci, A. PARK7/DJ-1 in Microglia: Implications in Parkinson's Disease and Relevance as a Therapeutic Target. *J. Neuroinflamm.* **2023**, *20*, 95. [CrossRef]
43. Zhou, C.; Gu, M.; Yin, L.; Yin, W.; Liu, J.; Zhu, Y.; Yin, K.; Lei, C.; Xu, Z.; Yang, X. Low Serum Uric Acid Levels May Be a Potential Biomarker of Poor Sleep Quality in Patients with Parkinson's Disease. *Sleep Med.* **2023**, *105*, 9–13. [CrossRef] [PubMed]
44. Shi, X.; Zheng, J.; Ma, J.; Wang, Z.; Sun, W.; Li, M.; Huang, S.; Hu, S. Low Serum Uric Acid Levels Are Associated with the Nonmotor Symptoms and Brain Gray Matter Volume in Parkinson's Disease. *Neurol. Sci.* **2022**, *43*, 1747–1754. [CrossRef] [PubMed]
45. Pan, L.; Li, C.; Meng, L.; Tian, Y.; He, M.; Yuan, X.; Zhang, G.; Zhang, Z.; Xiong, J.; Chen, G.; et al. Tau Accelerates α-Synuclein Aggregation and Spreading in Parkinson's Disease. *Brain* **2022**, *145*, 3454–3471. [CrossRef] [PubMed]
46. Zhang, J.; Jin, J.; Su, D.; Feng, T.; Zhao, H. Tau-PET Imaging in Parkinson's Disease: A Systematic Review and Meta-Analysis. *Front. Neurol.* **2023**, *14*, 1145939. [CrossRef] [PubMed]
47. Dănău, A.; Dumitrescu, L.; Lefter, A.; Popescu, B.O. Serum Uric Acid Levels in Parkinson's Disease: A Cross-Sectional Electronic Medical Record Database Study from a Tertiary Referral Centre in Romania. *Medicina* **2022**, *58*, 245. [CrossRef]
48. Jadhav, S.; Avila, J.; Schöll, M.; Kovacs, G.G.; Kövari, E.; Skrabana, R.; Evans, L.D.; Kontsekova, E.; Malawska, B.; de Silva, R.; et al. A Walk through Tau Therapeutic Strategies. *Acta Neuropathol. Commun.* **2019**, *7*, 22. [CrossRef]
49. Lin, J.; Ou, R.; Li, C.; Hou, Y.; Zhang, L.; Wei, Q.; Pang, D.; Liu, K.; Jiang, Q.; Yang, T.; et al. Plasma Glial Fibrillary Acidic Protein as a Biomarker of Disease Progression in Parkinson's Disease: A Prospective Cohort Study. *BMC Med.* **2023**, *21*, 420. [CrossRef]
50. Tang, Y.; Han, L.; Li, S.; Hu, T.; Xu, Z.; Fan, Y.; Liang, X.; Yu, H.; Wu, J.; Wang, J. Plasma GFAP in Parkinson's Disease with Cognitive Impairment and Its Potential to Predict Conversion to Dementia. *NPJ Park. Dis.* **2023**, *9*, 23. [CrossRef]
51. Che, N.; Ou, R.; Li, C.; Zhang, L.; Wei, Q.; Wang, S.; Jiang, Q.; Yang, T.; Xiao, Y.; Lin, J.; et al. Plasma GFAP as a Prognostic Biomarker of Motor Subtype in Early Parkinson's Disease. *NPJ Park. Dis.* **2024**, *10*, 48. [CrossRef]
52. Zenuni, H.; Bovenzi, R.; Bissacco, J.; Grillo, P.; Simonetta, C.; Mascioli, D.; Pieri, M.; Bernardini, S.; Sancesario, G.M.; Stefani, A.; et al. Clinical and Neurochemical Correlates of the APOE Genotype in Early-Stage Parkinson's Disease. *Neurobiol. Aging* **2023**, *131*, 24–28. [CrossRef]
53. Mitchell, T.; Lehéricy, S.; Chiu, S.Y.; Strafella, A.P.; Stoessl, A.J.; Vaillancourt, D.E. Emerging Neuroimaging Biomarkers Across Disease Stage in Parkinson Disease. *JAMA Neurol.* **2021**, *78*, 1262. [CrossRef] [PubMed]
54. Prange, S.; Theis, H.; Banwinkler, M.; van Eimeren, T. Molecular Imaging in Parkinsonian Disorders—What's New and Hot? *Brain Sci.* **2022**, *12*, 1146. [CrossRef] [PubMed]
55. af Bjerkén, S.; Axelsson, J.; Larsson, A.; Flygare, C.; Remes, J.; Strandberg, S.; Eriksson, L.; Bäckström, D.; Jakobson Mo, S. Reliability and Validity of Visual Analysis of [18F]FE-PE2I PET/CT in Early Parkinsonian Disease. *Nucl. Med. Commun.* **2023**, *44*, 397–406. [CrossRef] [PubMed]
56. Pineda-Pardo, J.A.; Sánchez-Ferro, Á.; Monje, M.H.G.; Pavese, N.; Obeso, J.A. Onset Pattern of Nigrostriatal Denervation in Early Parkinson's Disease. *Brain* **2022**, *145*, 1018–1028. [CrossRef] [PubMed]
57. de Natale, E.R.; Wilson, H.; Politis, M. Serotonergic Imaging in Parkinson's Disease. *Prog. Brain Res.* **2021**, *261*, 303–338.
58. Maillet, A.; Météreau, E.; Tremblay, L.; Favre, E.; Klinger, H.; Lhommée, E.; Le Bars, D.; Castrioto, A.; Prange, S.; Sgambato, V.; et al. Serotonergic and Dopaminergic Lesions Underlying Parkinsonian Neuropsychiatric Signs. *Mov. Disord.* **2021**, *36*, 2888–2900. [CrossRef]

59. Lancini, E.; Bartl, F.; Haag, L.; Rühling, M.; Ashton, N.J.; Zetterberg, H.; Düzel, E.; Hämmerer, D.; Betts, M. CSF and PET Biomarkers for Noradrenergic Dysfunction in Neurodegenerative Disease: A Systematic Review and Meta-analysis. *Alzheimers Dement.* **2021**, *17*, e055942. [CrossRef]
60. Meles, S.K.; Teune, L.K.; de Jong, B.M.; Dierckx, R.A.; Leenders, K.L. Metabolic Imaging in Parkinson Disease. *J. Nucl. Med.* **2017**, *58*, 23–28. [CrossRef]
61. Meles, S.K.; Renken, R.J.; Pagani, M.; Teune, L.K.; Arnaldi, D.; Morbelli, S.; Nobili, F.; van Laar, T.; Obeso, J.A.; Rodríguez-Oroz, M.C.; et al. Abnormal Pattern of Brain Glucose Metabolism in Parkinson's Disease: Replication in Three European Cohorts. *Eur. J. Nucl. Med. Mol. Imaging* **2020**, *47*, 437–450. [CrossRef]
62. Camacho, M.; Wilms, M.; Mouches, P.; Almgren, H.; Souza, R.; Camicioli, R.; Ismail, Z.; Monchi, O.; Forkert, N.D. Explainable Classification of Parkinson's Disease Using Deep Learning Trained on a Large Multi-Center Database of T1-Weighted MRI Datasets. *Neuroimage Clin.* **2023**, *38*, 103405. [CrossRef]
63. Alushaj, E.; Handfield-Jones, N.; Kuurstra, A.; Morava, A.; Menon, R.S.; Owen, A.M.; Sharma, M.; Khan, A.R.; MacDonald, P.A. Increased Iron in the Substantia Nigra Pars Compacta Identifies Patients with Early Parkinson's Disease: A 3T and 7T MRI Study. *Neuroimage Clin.* **2024**, *41*, 103577. [CrossRef] [PubMed]
64. Lee, H.; Baek, S.-Y.; Kim, E.-J.; Huh, G.Y.; Lee, J.-H.; Cho, H. MRI T_2 and T_2* Relaxometry to Visualize Neuromelanin in the Dorsal Substantia Nigra Pars Compacta. *Neuroimage* **2020**, *211*, 116625. [CrossRef] [PubMed]
65. Muzio, L.; Viotti, A.; Martino, G. Microglia in Neuroinflammation and Neurodegeneration: From Understanding to Therapy. *Front. Neurosci.* **2021**, *15*, 742065. [CrossRef] [PubMed]
66. Best, L.; Ghadery, C.; Pavese, N.; Tai, Y.F.; Strafella, A.P. New and Old TSPO PET Radioligands for Imaging Brain Microglial Activation in Neurodegenerative Disease. *Curr. Neurol. Neurosci. Rep.* **2019**, *19*, 24. [CrossRef] [PubMed]
67. Horsager, J.; Andersen, K.B.; Knudsen, K.; Skjærbæk, C.; Fedorova, T.D.; Okkels, N.; Schaeffer, E.; Bonkat, S.K.; Geday, J.; Otto, M.; et al. Brain-First versus Body-First Parkinson's Disease: A Multimodal Imaging Case-Control Study. *Brain* **2020**, *143*, 3077–3088. [CrossRef]
68. Kantarci, K.; Lowe, V.J.; Chen, Q.; Przybelski, S.A.; Lesnick, T.G.; Schwarz, C.G.; Senjem, M.L.; Gunter, J.L.; Jack, C.R.; Graff-Radford, J.; et al. β-Amyloid PET and Neuropathology in Dementia with Lewy Bodies. *Neurology* **2020**, *94*, e282–e291. [CrossRef]
69. Accorsi, R. Brain Single-Photon Emission CT Physics Principles. *Am. J. Neuroradiol.* **2008**, *29*, 1247–1256. [CrossRef]
70. Livieratos, L. Basic Principles of SPECT and PET Imaging. In *Radionuclide and Hybrid Bone Imaging*; Springer: Berlin/Heidelberg, Germany, 2012; pp. 345–359.
71. van Geuns, R.-J.M.; Wielopolski, P.A.; de Bruin, H.G.; Rensing, B.J.; van Ooijen, P.M.A.; Hulshoff, M.; Oudkerk, M.; de Feyter, P.J. Basic Principles of Magnetic Resonance Imaging. *Prog. Cardiovasc. Dis.* **1999**, *42*, 149–156. [CrossRef]
72. Kim, S.-G.; Bandettini, P.A. Principles of Functional MRI. In *BOLD fMRI*; Springer: New York, NY, USA, 2010; pp. 3–22.
73. Kaasinen, V.; Vahlberg, T. Striatal Dopamine in Parkinson Disease: A Meta-analysis of Imaging Studies. *Ann. Neurol.* **2017**, *82*, 873–882. [CrossRef]
74. Wassenberg, T.; Molero-Luis, M.; Jeltsch, K.; Hoffmann, G.F.; Assmann, B.; Blau, N.; Garcia-Cazorla, A.; Artuch, R.; Pons, R.; Pearson, T.S.; et al. Consensus Guideline for the Diagnosis and Treatment of Aromatic L-Amino Acid Decarboxylase (AADC) Deficiency. *Orphanet J. Rare Dis.* **2017**, *12*, 12. [CrossRef]
75. Stormezand, G.N.; Schreuder, R.S.B.H.; Brouwers, A.H.; Slart, R.H.J.A.; Elsinga, P.H.; Walenkamp, A.M.E.; Dierckx, R.A.J.O.; Glaudemans, A.W.J.M.; Luurtsema, G. The Effects of Molar Activity on [18F]FDOPA Uptake in Patients with Neuroendocrine Tumors. *EJNMMI Res.* **2021**, *11*, 88. [CrossRef] [PubMed]
76. Ikeda, K.; Ebina, J.; Kawabe, K.; Iwasaki, Y. Dopamine Transporter Imaging in Parkinson Disease: Progressive Changes and Therapeutic Modification after Anti-Parkinsonian Medications. *Intern. Med.* **2019**, *58*, 1665–1672. [CrossRef]
77. Kerstens, V.S.; Fazio, P.; Sundgren, M.; Halldin, C.; Svenningsson, P.; Varrone, A. [18F]FE-PE2I DAT Correlates with Parkinson's Disease Duration, Stage, and Rigidity/Bradykinesia Scores: A PET Radioligand Validation Study. *EJNMMI Res.* **2023**, *13*, 29. [CrossRef] [PubMed]
78. Tsang, K.; Walker, R. Dopamine Transporter Single Photon Emission Computed Tomography (DaT-SPECT) Use in the Diagnosis and Clinical Management of Parkinsonism: An 8-Year Retrospective Study. *J. Neurol.* **2023**, *270*, 2550–2558. [CrossRef] [PubMed]
79. Nag, S.; Jahan, M.; Tóth, M.; Nakao, R.; Varrone, A.; Halldin, C. PET Imaging of VMAT2 with the Novel Radioligand [^{18}F]FE-DTBZ-D4 in Nonhuman Primates: Comparison with [^{11}C]DTBZ and [^{18}F]FE-DTBZ. *ACS Chem. Neurosci.* **2021**, *12*, 4580–4586. [CrossRef] [PubMed]
80. Kaasinen, V.; Vahlberg, T.; Stoessl, A.J.; Strafella, A.P.; Antonini, A. Dopamine Receptors in Parkinson's Disease: A Meta-Analysis of Imaging Studies. *Mov. Disord.* **2021**, *36*, 1781–1791. [CrossRef]
81. Visser, A.K.D.; van Waarde, A.; Willemsen, A.T.M.; Bosker, F.J.; Luiten, P.G.M.; den Boer, J.A.; Kema, I.P.; Dierckx, R.A.J.O. Measuring Serotonin Synthesis: From Conventional Methods to PET Tracers and Their (Pre)Clinical Implications. *Eur. J. Nucl. Med. Mol. Imaging* **2011**, *38*, 576–591. [CrossRef]
82. Varrone, A.; Svenningsson, P.; Forsberg, A.; Varnäs, K.; Tiger, M.; Nakao, R.; Halldin, C.; Nilsson, L.-G.; Farde, L. Positron Emission Tomography Imaging of 5-Hydroxytryptamine1B Receptors in Parkinson's Disease. *Neurobiol. Aging* **2014**, *35*, 867–875. [CrossRef]

83. Varrone, A.; Svenningsson, P.; Marklund, P.; Fatouros-Bergman, H.; Forsberg, A.; Halldin, C.; Nilsson, L.; Farde, L. 5-HT_{1B} Receptor Imaging and Cognition: A Positron Emission Tomography Study in Control Subjects and Parkinson's Disease Patients. *Synapse* **2015**, *69*, 365–374. [CrossRef]
84. Barber, T.R.; Griffanti, L.; Muhammed, K.; Drew, D.S.; Bradley, K.M.; McGowan, D.R.; Crabbe, M.; Lo, C.; Mackay, C.E.; Husain, M.; et al. Apathy in Rapid Eye Movement Sleep Behaviour Disorder Is Associated with Serotonin Depletion in the Dorsal Raphe Nucleus. *Brain* **2018**, *141*, 2848–2854. [CrossRef]
85. Liu, S.-Y.; Wile, D.J.; Fu, J.F.; Valerio, J.; Shahinfard, E.; McCormick, S.; Mabrouk, R.; Vafai, N.; McKenzie, J.; Neilson, N.; et al. The Effect of LRRK2 Mutations on the Cholinergic System in Manifest and Premanifest Stages of Parkinson's Disease: A Cross-Sectional PET Study. *Lancet Neurol.* **2018**, *17*, 309–316. [CrossRef] [PubMed]
86. van der Zee, S.; Vállez García, D.; Elsinga, P.H.; Willemsen, A.T.M.; Boersma, H.H.; Gerritsen, M.J.J.; Spikman, J.M.; van Laar, T. [^{18}F]Fluoroethoxybenzovesamicol in Parkinson's Disease Patients: Quantification of a Novel Cholinergic Positron Emission Tomography Tracer. *Mov. Disord.* **2019**, *34*, 924–926. [CrossRef] [PubMed]
87. Horsager, J.; Okkels, N.; Hansen, A.K.; Damholdt, M.F.; Andersen, K.H.; Fedorova, T.D.; Munk, O.L.; Danielsen, E.H.; Pavese, N.; Brooks, D.J.; et al. Mapping Cholinergic Synaptic Loss in Parkinson's Disease: An [18F]FEOBV PET Case-Control Study. *J. Park. Dis* **2022**, *12*, 2493–2506. [CrossRef] [PubMed]
88. van der Zee, S.; Kanel, P.; Gerritsen, M.J.J.; Boertien, J.M.; Slomp, A.C.; Müller, M.L.T.M.; Bohnen, N.I.; Spikman, J.M.; van Laar, T. Altered Cholinergic Innervation in De Novo Parkinson's Disease with and Without Cognitive Impairment. *Mov. Disord.* **2022**, *37*, 713–723. [CrossRef] [PubMed]
89. Bohnen, N.I.; Roytman, S.; Kanel, P.; Müller, M.L.T.M.; Scott, P.J.H.; Frey, K.A.; Albin, R.L.; Koeppe, R.A. Progression of Regional Cortical Cholinergic Denervation in Parkinson's Disease. *Brain Commun.* **2022**, *4*, fcac320. [CrossRef]
90. Slingerland, S.; van der Zee, S.; Carli, G.; Slomp, A.C.; Boertien, J.M.; d'Angremont, E.; Bohnen, N.I.; Albin, R.L.; van Laar, T. Cholinergic Innervation Topography in GBA-Associated De Novo Parkinson's Disease Patients. *Brain* **2024**, *147*, 900–910. [CrossRef]
91. Sommerauer, M.; Fedorova, T.D.; Hansen, A.K.; Knudsen, K.; Otto, M.; Jeppesen, J.; Frederiksen, Y.; Blicher, J.U.; Geday, J.; Nahimi, A.; et al. Evaluation of the Noradrenergic System in Parkinson's Disease: An 11C-MeNER PET and Neuromelanin MRI Study. *Brain* **2018**, *141*, 496–504. [CrossRef]
92. Knudsen, K.; Fedorova, T.D.; Hansen, A.K.; Sommerauer, M.; Otto, M.; Svendsen, K.B.; Nahimi, A.; Stokholm, M.G.; Pavese, N.; Beier, C.P.; et al. In-Vivo Staging of Pathology in REM Sleep Behaviour Disorder: A Multimodality Imaging Case-Control Study. *Lancet Neurol.* **2018**, *17*, 618–628. [CrossRef]
93. Laurencin, C.; Lancelot, S.; Brosse, S.; Mérida, I.; Redouté, J.; Greusard, E.; Lamberet, L.; Liotier, V.; Le Bars, D.; Costes, N.; et al. Noradrenergic Alterations in Parkinson's Disease: A Combined 11C-Yohimbine PET/Neuromelanin MRI Study. *Brain* **2024**, *147*, 1377–1388. [CrossRef]
94. Boutet, A.; Madhavan, R.; Elias, G.J.B.; Joel, S.E.; Gramer, R.; Ranjan, M.; Paramanandam, V.; Xu, D.; Germann, J.; Loh, A.; et al. Predicting Optimal Deep Brain Stimulation Parameters for Parkinson's Disease Using Functional MRI and Machine Learning. *Nat. Commun.* **2021**, *12*, 3043. [CrossRef]
95. Shah, A.; Prasad, S.; Indoria, A.; Pal, P.K.; Saini, J.; Ingalhalikar, M. Free Water Imaging in Parkinson's Disease and Atypical Parkinsonian Disorders. *J. Neurol.* **2024**, *271*, 2521–2528. [CrossRef] [PubMed]
96. Sun, H.; Walsh, A.J.; Lebel, R.M.; Blevins, G.; Catz, I.; Lu, J.-Q.; Johnson, E.S.; Emery, D.J.; Warren, K.G.; Wilman, A.H. Validation of Quantitative Susceptibility Mapping with Perls' Iron Staining for Subcortical Gray Matter. *Neuroimage* **2015**, *105*, 486–492. [CrossRef] [PubMed]
97. Guan, X.; Lancione, M.; Ayton, S.; Dusek, P.; Langkammer, C.; Zhang, M. Neuroimaging of Parkinson's Disease by Quantitative Susceptibility Mapping. *Neuroimage* **2024**, *289*, 120547. [CrossRef] [PubMed]
98. Totsune, T.; Baba, T.; Sugimura, Y.; Oizumi, H.; Tanaka, H.; Takahashi, T.; Yoshioka, M.; Nagamatsu, K.; Takeda, A. Nuclear Imaging Data-Driven Classification of Parkinson's Disease. *Mov. Disord.* **2023**, *38*, 2053–2063. [CrossRef] [PubMed]
99. Tang, Y.; Li, L.; Hu, T.; Jiao, F.; Han, L.; Li, S.; Xu, Z.; Fan, Y.; Sun, Y.; Liu, F.; et al. In Vivo ^{18}F-Florzolotau Tau Positron Emission Tomography Imaging in Parkinson's Disease Dementia. *Mov. Disord.* **2023**, *38*, 147–152. [CrossRef]
100. Shimohama, S.; Tezuka, T.; Takahata, K.; Bun, S.; Tabuchi, H.; Seki, M.; Momota, Y.; Suzuki, N.; Morimoto, A.; Iwabuchi, Y.; et al. Impact of Amyloid and Tau PET on Changes in Diagnosis and Patient Management. *Neurology* **2023**, *100*, E264–E274. [CrossRef]
101. Raval, N.R.; Madsen, C.A.; Shalgunov, V.; Nasser, A.; Battisti, U.M.; Beaman, E.E.; Juhl, M.; Jørgensen, L.M.; Herth, M.M.; Hansen, H.D.; et al. Evaluation of the α-Synuclein PET Radiotracer (D3)-[11C]MODAG-001 in Pigs. *Nucl. Med. Biol.* **2022**, *114–115*, 42–48. [CrossRef]
102. Muleiro Alvarez, M.; Cano-Herrera, G.; Osorio Martínez, M.F.; Vega Gonzales-Portillo, J.; Monroy, G.R.; Murguiondo Pérez, R.; Torres-Ríos, J.A.; van Tienhoven, X.A.; Garibaldi Bernot, E.M.; Esparza Salazar, F.; et al. A Comprehensive Approach to Parkinson's Disease: Addressing Its Molecular, Clinical, and Therapeutic Aspects. *Int. J. Mol. Sci.* **2024**, *25*, 7183. [CrossRef]
103. Quattrone, A.; Sarica, A.; Buonocore, J.; Morelli, M.; Bianco, M.G.; Calomino, C.; Aracri, F.; De Maria, M.; Vescio, B.; Vaccaro, M.G.; et al. Differentiating between Common PSP Phenotypes Using Structural MRI: A Machine Learning Study. *J. Neurol.* **2023**, *270*, 5502–5515. [CrossRef]

104. Quattrone, A.; Bianco, M.G.; Antonini, A.; Vaillancourt, D.E.; Seppi, K.; Ceravolo, R.; Strafella, A.P.; Tedeschi, G.; Tessitore, A.; Cilia, R.; et al. Development and Validation of Automated Magnetic Resonance Parkinsonism Index 2.0 to Distinguish Progressive Supranuclear Palsy-Parkinsonism From Parkinson's Disease. *Mov. Disord.* **2022**, *37*, 1272–1281. [CrossRef]
105. Bao, Y.; Wang, L.; Yu, F.; Yang, J.; Huang, D. Parkinson's Disease Gene Biomarkers Screened by the LASSO and SVM Algorithms. *Brain Sci.* **2023**, *13*, 175. [CrossRef] [PubMed]
106. Buniello, A.; MacArthur, J.A.L.; Cerezo, M.; Harris, L.W.; Hayhurst, J.; Malangone, C.; McMahon, A.; Morales, J.; Mountjoy, E.; Sollis, E.; et al. The NHGRI-EBI GWAS Catalog of Published Genome-Wide Association Studies, Targeted Arrays and Summary Statistics 2019. *Nucleic Acids Res.* **2019**, *47*, D1005–D1012. [CrossRef] [PubMed]
107. Rocha, E.M.; Keeney, M.T.; Di Maio, R.; De Miranda, B.R.; Greenamyre, J.T. LRRK2 and Idiopathic Parkinson's Disease. *Trends Neurosci.* **2022**, *45*, 224–236. [CrossRef] [PubMed]
108. Vilariño-Güell, C.; Wider, C.; Ross, O.A.; Dachsel, J.C.; Kachergus, J.M.; Lincoln, S.J.; Soto-Ortolaza, A.I.; Cobb, S.A.; Wilhoite, G.J.; Bacon, J.A.; et al. VPS35 Mutations in Parkinson Disease. *Am. J. Hum. Genet.* **2011**, *89*, 162–167. [CrossRef] [PubMed]
109. Funayama, M.; Ohe, K.; Amo, T.; Furuya, N.; Yamaguchi, J.; Saiki, S.; Li, Y.; Ogaki, K.; Ando, M.; Yoshino, H.; et al. CHCHD2 Mutations in Autosomal Dominant Late-Onset Parkinson's Disease: A Genome-Wide Linkage and Sequencing Study. *Lancet Neurol.* **2015**, *14*, 274–282. [CrossRef]
110. Oji, Y.; Hatano, T.; Ueno, S.-I.; Funayama, M.; Ishikawa, K.; Okuzumi, A.; Noda, S.; Sato, S.; Satake, W.; Toda, T.; et al. Variants in Saposin D Domain of Prosaposin Gene Linked to Parkinson's Disease. *Brain* **2020**, *143*, 1190–1205. [CrossRef]
111. Blauwendraat, C.; Reed, X.; Krohn, L.; Heilbron, K.; Bandres-Ciga, S.; Tan, M.; Gibbs, J.R.; Hernandez, D.G.; Kumaran, R.; Langston, R.; et al. Genetic Modifiers of Risk and Age at Onset in GBA Associated Parkinson's Disease and Lewy Body Dementia. *Brain* **2020**, *143*, 234–248. [CrossRef]
112. Rui, Q.; Ni, H.; Li, D.; Gao, R.; Chen, G. The Role of LRRK2 in Neurodegeneration of Parkinson Disease. *Curr. Neuropharmacol.* **2018**, *16*, 1348–1357. [CrossRef]
113. Vizziello, M.; Borellini, L.; Franco, G.; Ardolino, G. Disruption of Mitochondrial Homeostasis: The Role of PINK1 in Parkinson's Disease. *Cells* **2021**, *10*, 3022. [CrossRef]
114. Quinn, P.M.J.; Moreira, P.I.; Ambrósio, A.F.; Alves, C.H. PINK1/PARKIN Signalling in Neurodegeneration and Neuroinflammation. *Acta Neuropathol. Commun.* **2020**, *8*, 189. [CrossRef]
115. Konovalova, E.V.; Lopacheva, O.M.; Grivennikov, I.A.; Lebedeva, O.S.; Dashinimaev, E.B.; Khaspekov, L.G.; Fedotova, E.Y.; Illarioshkin, S.N. Mutations in the Parkinson's Disease-Associated PARK2 Gene Are Accompanied by Imbalance in Programmed Cell Death Systems. *Acta Nat.* **2015**, *7*, 146–149. [CrossRef]
116. Fang, Y.-Q.; Mao, F.; Zhu, M.-J.; Li, X.-H. Compound Heterozygous Mutations in PARK2 Causing Early-Onset Parkinson Disease. *Medicine* **2019**, *98*, e14228. [CrossRef] [PubMed]
117. Ham, S.J.; Lee, D.; Xu, W.J.; Cho, E.; Choi, S.; Min, S.; Park, S.; Chung, J. Loss of UCHL1 Rescues the Defects Related to Parkinson's Disease by Suppressing Glycolysis. *Sci. Adv.* **2021**, *7*, eabg4574. [CrossRef] [PubMed]
118. Shibata, N.; Motoi, Y.; Tomiyama, H.; Ohnuma, T.; Kuerban, B.; Tomson, K.; Komatsu, M.; Hattori, N.; Arai, H. Lack of Genetic Association of the UCHL1 Gene with Alzheimer's Disease and Parkinson's Disease with Dementia. *Dement. Geriatr. Cogn. Disord.* **2012**, *33*, 250–254. [CrossRef]
119. Rydning, S.L.; Backe, P.H.; Sousa, M.M.L.; Iqbal, Z.; Øye, A.-M.; Sheng, Y.; Yang, M.; Lin, X.; Slupphaug, G.; Nordenmark, T.H.; et al. Novel UCHL1 Mutations Reveal New Insights into Ubiquitin Processing. *Hum. Mol. Genet.* **2017**, *26*, 1031–1040. [CrossRef]
120. Olgiati, S.; Thomas, A.; Quadri, M.; Breedveld, G.J.; Graafland, J.; Eussen, H.; Douben, H.; de Klein, A.; Onofrj, M.; Bonifati, V. Early-Onset Parkinsonism Caused by Alpha-Synuclein Gene Triplication: Clinical and Genetic Findings in a Novel Family. *Park. Relat. Disord.* **2015**, *21*, 981–986. [CrossRef]
121. Siddiqui, I.J.; Pervaiz, N.; Abbasi, A.A. The Parkinson Disease Gene SNCA: Evolutionary and Structural Insights with Pathological Implication. *Sci. Rep.* **2016**, *6*, 24475. [CrossRef]
122. Magistrelli, L.; Contaldi, E.; Comi, C. The Impact of SNCA Variations and Its Product Alpha-Synuclein on Non-Motor Features of Parkinson's Disease. *Life* **2021**, *11*, 804. [CrossRef]
123. Mohan, M.; Mellick, G.D. Role of the VPS35 D620N Mutation in Parkinson's Disease. *Park. Relat. Disord.* **2017**, *36*, 10–18. [CrossRef]
124. Milenkovic, I.; Blumenreich, S.; Futerman, A.H. GBA Mutations, Glucosylceramide and Parkinson's Disease. *Curr. Opin. Neurobiol.* **2022**, *72*, 148–154. [CrossRef]
125. Giri, B.; Seamon, M.; Banerjee, A.; Chauhan, S.; Purohit, S.; Morgan, J.; Baban, B.; Wakade, C. Emerging Urinary Alpha-Synuclein and MiRNA Biomarkers in Parkinson's Disease. *Metab. Brain Dis.* **2022**, *37*, 1687–1696. [CrossRef] [PubMed]
126. Zhang, W.T.; Wang, Y.J.; Yao, Y.F.; Zhang, G.X.; Zhang, Y.N.; Gao, S.S. Circulating MicroRNAs as Potential Biomarkers for the Diagnosis of Parkinson's Disease: A Meta-Analysis. *Neurología* **2024**, *39*, 573–583. [CrossRef] [PubMed]
127. Guévremont, D.; Roy, J.; Cutfield, N.J.; Williams, J.M. MicroRNAs in Parkinson's Disease: A Systematic Review and Diagnostic Accuracy Meta-Analysis. *Sci. Rep.* **2023**, *13*, 16272. [CrossRef] [PubMed]
128. Shaheen, N.; Shaheen, A.; Osama, M.; Nashwan, A.J.; Bharmauria, V.; Flouty, O. MicroRNAs Regulation in Parkinson's Disease, and Their Potential Role as Diagnostic and Therapeutic Targets. *NPJ Park. Dis.* **2024**, *10*, 186. [CrossRef] [PubMed]
129. Li, Y.; Cao, Y.; Liu, W.; Chen, F.; Zhang, H.; Zhou, H.; Zhao, A.; Luo, N.; Liu, J.; Wu, L. Candidate Biomarkers of EV-MicroRNA in Detecting REM Sleep Behavior Disorder and Parkinson's Disease. *NPJ Park. Dis.* **2024**, *10*, 18. [CrossRef]

130. Yang, Z.; Li, T.; Li, S.; Wei, M.; Qi, H.; Shen, B.; Chang, R.C.-C.; Le, W.; Piao, F. Altered Expression Levels of MicroRNA-132 and Nurr1 in Peripheral Blood of Parkinson's Disease: Potential Disease Biomarkers. *ACS Chem. Neurosci.* **2019**, *10*, 2243–2249. [CrossRef]
131. Citterio, L.A.; Mancuso, R.; Agostini, S.; Meloni, M.; Clerici, M. Serum and Exosomal MiR-7-1-5p and MiR-223-3p as Possible Biomarkers for Parkinson's Disease. *Biomolecules* **2023**, *13*, 865. [CrossRef]
132. Zago, E.; Dal Molin, A.; Dimitri, G.M.; Xumerle, L.; Pirazzini, C.; Bacalini, M.G.; Maturo, M.G.; Azevedo, T.; Spasov, S.; Gómez-Garre, P.; et al. Early Downregulation of Hsa-MiR-144-3p in Serum from Drug-Naïve Parkinson's Disease Patients. *Sci. Rep.* **2022**, *12*, 1330. [CrossRef]
133. Cao, X.-Y.; Lu, J.-M.; Zhao, Z.-Q.; Li, M.-C.; Lu, T.; An, X.-S.; Xue, L.-J. MicroRNA Biomarkers of Parkinson's Disease in Serum Exosome-like Microvesicles. *Neurosci. Lett.* **2017**, *644*, 94–99. [CrossRef]
134. dos Santos, M.C.T.; Barreto-Sanz, M.A.; Correia, B.R.S.; Bell, R.; Widnall, C.; Perez, L.T.; Berteau, C.; Schulte, C.; Scheller, D.; Berg, D.; et al. MiRNA-Based Signatures in Cerebrospinal Fluid as Potential Diagnostic Tools for Early Stage Parkinson's Disease. *Oncotarget* **2018**, *9*, 17455–17465. [CrossRef]
135. Hällqvist, J.; Bartl, M.; Dakna, M.; Schade, S.; Garagnani, P.; Bacalini, M.-G.; Pirazzini, C.; Bhatia, K.; Schreglmann, S.; Xylaki, M.; et al. Plasma Proteomics Identify Biomarkers Predicting Parkinson's Disease up to 7 Years before Symptom Onset. *Nat. Commun.* **2024**, *15*, 4759. [CrossRef] [PubMed]
136. Latosinska, A.; Frantzi, M.; Siwy, J. Peptides as "Better Biomarkers"? Value, Challenges, and Potential Solutions to Facilitate Implementation. *Mass Spectrom. Rev.* **2024**, *43*, 1195–1236. [CrossRef] [PubMed]
137. Giarola, J.F.; Santos, J.; Estevez, M.-C.; Ventura, S.; Pallarès, I.; Lechuga, L.M. An α-Helical Peptide-Based Plasmonic Biosensor for Highly Specific Detection of α-Synuclein Toxic Oligomers. *Anal. Chim. Acta* **2024**, *1304*, 342559. [CrossRef] [PubMed]
138. Kim, S.-J.; Miller, B.; Hartel, N.G.; Ramirez, R.; Braniff, R.G.; Leelaprachakul, N.; Huang, A.; Wang, Y.; Arpawong, T.E.; Crimmins, E.M.; et al. A Naturally Occurring Variant of SHLP2 Is a Protective Factor in Parkinson's Disease. *Mol. Psychiatry* **2024**, *29*, 505–517. [CrossRef]
139. Abdi, I.Y.; Bartl, M.; Dakna, M.; Abdesselem, H.; Majbour, N.; Trenkwalder, C.; El-Agnaf, O.; Mollenhauer, B. Cross-Sectional Proteomic Expression in Parkinson's Disease-Related Proteins in Drug-Naïve Patients vs. Healthy Controls with Longitudinal Clinical Follow-Up. *Neurobiol. Dis.* **2023**, *177*, 105997. [CrossRef]
140. Youssef, P.; Hughes, L.; Kim, W.S.; Halliday, G.M.; Lewis, S.J.G.; Cooper, A.; Dzamko, N. Evaluation of Plasma Levels of NFL, GFAP, UCHL1 and Tau as Parkinson's Disease Biomarkers Using Multiplexed Single Molecule Counting. *Sci. Rep.* **2023**, *13*, 5217. [CrossRef]
141. Vijiaratnam, N.; Lawton, M.; Heslegrave, A.J.; Guo, T.; Tan, M.; Jabbari, E.; Real, R.; Woodside, J.; Grosset, K.; Chelban, V.; et al. Combining Biomarkers for Prognostic Modelling of Parkinson's Disease. *J. Neurol. Neurosurg. Psychiatry* **2022**, *93*, 707–715. [CrossRef]
142. Leggio, L.; Paternò, G.; Vivarelli, S.; Falzone, G.G.; Giachino, C.; Marchetti, B.; Iraci, N. Extracellular Vesicles as Novel Diagnostic and Prognostic Biomarkers for Parkinson's Disease. *Aging Dis.* **2021**, *12*, 1494. [CrossRef]

International Journal of
Molecular Sciences

MDPI

Review

An Overview of Epigenetic Changes in the Parkinson's Disease Brain

Anthony Klokkaris and Anna Migdalska-Richards *

Department of Clinical and Biomedical Sciences, Faculty of Health and Life Sciences, University of Exeter Medical School, University of Exeter, Exeter EX2 5DW, UK; ak735@exeter.ac.uk
* Correspondence: a.migdalska-richards@exeter.ac.uk

Abstract: Parkinson's disease is a progressive neurodegenerative disorder, predominantly of the motor system. Although some genetic components and cellular mechanisms of Parkinson's have been identified, much is still unknown. In recent years, emerging evidence has indicated that non-DNA-sequence variation (in particular epigenetic mechanisms) is likely to play a crucial role in the development and progression of the disease. Here, we present an up-to-date overview of epigenetic processes including DNA methylation, DNA hydroxymethylation, histone modifications and non-coding RNAs implicated in the brain of those with Parkinson's disease. We will also discuss the limitations of current epigenetic research in Parkinson's disease, the advantages of simultaneously studying genetics and epigenetics, and putative novel epigenetic therapies.

Keywords: epigenetics; Parkinson's disease; DNA methylation; DNA hydroxymethylation; histone modifications; non-coding RNAs; microRNAs; epigenetic therapy

1. Introduction

Parkinson's disease (PD) is a debilitating movement disorder and the second-leading cause of neurodegeneration, affecting more than ten million individuals worldwide [1]. The prevalence of the disease increases sharply with age, with 1.7% affected by the age of 80–84 [2]. With a rapidly aging population, the number of people affected is only expected to increase. It is estimated that the number of individuals with PD will double between 2015 and 2040 [3]. PD already significantly contributes to the global burden of disease, with an economic burden of USD 51.9 billion per year in the United States alone [4].

Parkinson's is associated with a number of motor symptoms, including resting tremor, bradykinesia, rigidity and postural instability [5]. In addition, the vast majority of individuals with PD also experience a range of non-motor complications, such as sleep disturbances, olfactory deficits, autonomic dysfunction, cognitive impairment and depression [5,6]. While some non motor symptoms such as hyposmia, depression and REM sleep behavior disorder typically precede the onset of motor symptoms, other symptoms such as cognitive impairment, dementia, anhedonia and psychosis more commonly appear later [6,7]. Parkinson's is also linked with substantial morbidity and mortality, with pneumonia and cardiovascular complications accounting for the majority of deaths [8].

Pathologically, PD is characterized by the progressive loss of dopaminergic neurons in the substantia nigra pars compacta (Figure 1). These neurons project to the striatum, and consequently, striatal dopamine levels are typically depleted in diseased individuals, accounting for many of the motor defects [9]. Dopaminergic neuron loss is normally also accompanied by the formation of α-synuclein aggregates in intracellular Lewy body inclusions, the pathological hallmark of PD [9]. Increased levels of α-synuclein may be toxic to human neurons and are believed to lead to further dopaminergic neuron loss [10]. It has also been suggested that misfolded α-synuclein may propagate along anatomically connected axons in a prion-like manner, spreading from the substantia nigra to other regions of the brain [11,12].

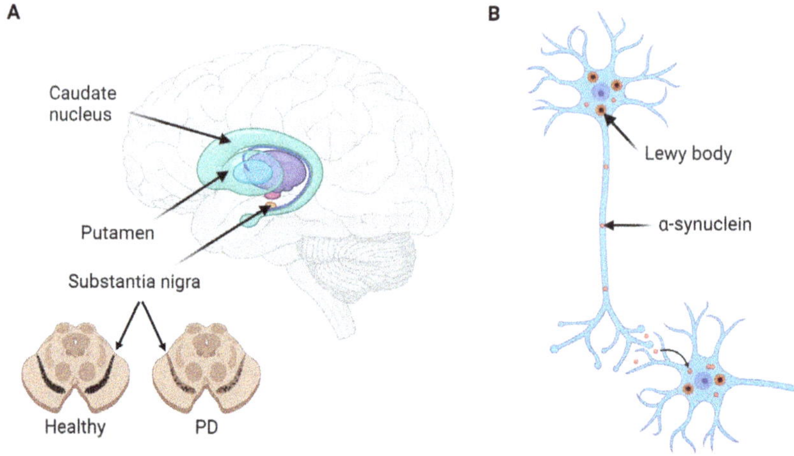

Figure 1. Overview of PD pathology. Abbreviations: Parkinson's disease (PD). (**A**) Schematic of the brain highlighting the substantia nigra, putamen and caudate nucleus (the latter two are both part of the striatum). Dopaminergic neurons are depleted in the PD substantia nigra, as shown below. (**B**) Schematic of two neurons affected by α-synuclein and Lewy body pathology, which have spread from one neuron to the next. Created with BioRender.com.

Several mechanisms are thought to play a role in the pathogenesis of PD, such as mitochondrial dysfunction, oxidative stress, neuroinflammation, protein aggregation, defective autophagy and environmental toxin-induced effects [13–18], with many of these mechanisms interlinked. For example, protein homeostasis can be disrupted by both increased protein aggregation and reduced protein clearance, while the generation of reactive oxygen species can be exacerbated by concurrent elevation of oxidative stress and impairment of mitophagy [15,18,19].

Rare mutations in a handful of genes are known to directly cause PD (Table 1), including α-synuclein (*SNCA*), leucine-rich repeat kinase 2 (*LRRK2*), PTEN-induced kinase 1 (*PINK1*), parkin (*PRKN*), parkinsonism-associated deglycase (*DJ-1*) and ATPase cation transporting 13A2 (*ATP13A2*) [20]. These genes were initially identified as monogenic causes of the disease primarily through linkage analysis of large families affected by PD. In addition, certain genetic risk factors are associated with the disease, including particular mutations in the glucocerebrosidase 1 (*GBA1*) and *LRRK2* genes [21–23]. For example, *GBA1* mutations are estimated to result in PD development in up to 30% of heterozygous carriers by the age of 80 [24–27].

Table 1. Key rare mutations that cause PD.

Gene Name	Abbreviation	Mode of Inheritance
α-synuclein	*SNCA*	Autosomal dominant
leucine-rich repeat kinase 2	*LRRK2*	Autosomal dominant
PTEN-induced kinase 1	*PINK1*	Autosomal recessive
parkin	*PRKN*	Autosomal recessive
Parkinsonism-associated deglycase	*DJ-1*	Autosomal recessive
ATPase cation transporting 13A2	*ATP13A2*	Autosomal recessive

Identification of additional genetic loci associated with PD has come from genome-wide association studies (GWASs), which test for allele frequency differences of hundreds of thousands of genetic variants across large numbers of individuals. A recent large multi-ancestry meta-analysis of PD GWASs identified 78 independent genome-wide significant loci (including in *SNCA* and *LRRK2*), 12 of which were not previously reported [28]. A PD

GWAS meta-analysis was also carried out in 2019 [22], with the risk variants explaining 16–36% of PD heritable risk, highlighting that the etiology of the disease remains considerably unknown. It is likely that yet-to-be-identified PD genetic risk may be facilitated by larger-scale GWASs. However, given that the majority of GWAS variants lie in non-coding, regulatory regions of the genome [29] and PD concordance amongst monozygotic twins is only 17% [30], additional factors such as epigenetic variation are likely to play a pivotal role.

Epigenetics is the study of molecular processes that influence the regulation of gene expression without altering the underlying DNA sequence [31]. Key epigenetic mechanisms include DNA methylation, DNA hydroxymethylation, histone modifications, and actions mediated by non-coding RNAs such as microRNAs (miRNAs) [32]. Epigenetic processes often affect the accessibility of DNA to the transcription machinery, thus regulating gene expression levels [32]. Classic examples of epigenetic regulation include cell fate determination and specialization in embryonic development, X-chromosome inactivation and genomic imprinting [33]. Another example relevant to PD is neuronal function, where epigenetics is implicated in the synaptic plasticity of learning and memory formation [34,35]. Epigenetic dysregulation can lead to disease and has been particularly well studied in cancer [36]. In recent years, epigenetics has also been implicated in other neurodegenerative disorders, such as Alzheimer's disease [37], and there is increasing evidence that epigenetics may play an important role in Parkinson's.

Epigenetic changes might be influenced by genetic variation, and a few studies have started to integrate genetic and epigenetic information in PD [22,38–41]. On the other hand, epigenetic changes could be influenced by the environment. Mounting evidence suggests that long-lasting phenotypic changes can result from alterations in epigenetic make-up that are caused by environmental cues [33,42,43]. One such cue is stress, and several studies have demonstrated that stress, via epigenetic alterations, may impact the development, age of onset and progression of neurodegenerative and neuropsychiatric disorders, including PD, Huntington's disease and depression [44–46]. In addition to stress, other factors such as prior head injuries and environmental contaminants (including toxins, heavy metals, pesticides and herbicides) have been associated with PD [47], and it is plausible that these factors also act via epigenetics to confer PD susceptibility. Finally, further evidence suggesting a link between the environment and epigenetic make-up comes from studies of monozygotic twins, which show that older twins have more epigenetic differences (and differential gene expression) compared to younger twins [48–50]. It is thought that these differences arise from twins spending an increasing amount of time in different environments over their lives. Therefore, the low concordance rate between twins for PD might be attributed to different environments, which, in turn, may cause distinct epigenetic changes that differentially influence disease susceptibility.

Importantly, epigenetic studies in Parkinson's brains may not only unravel new mechanisms underlying disease development and progression but also uncover new drug targets and so advance novel therapeutic approaches. The development of new PD treatments would be particularly impactful as there are currently no treatments that can cure or modify the disease. The few treatments that do exist, including levodopa (L-DOPA), dopamine agonists and monoamine oxidase-B inhibitors, are symptomatic and only alleviate symptoms temporarily [51]. Moreover, these treatments become substantially less effective after only a few years and typically lead to problematic side-effects such as dyskinesia [52]. Therefore, new disease-modifying therapies that can slow, prevent or even reverse PD progression are urgently required.

In this review, we first summarize key findings related to epigenetics within the brain in Parkinson's, covering DNA methylation, DNA hydroxymethylation, histone modifications and non-coding RNAs, focusing on the utility of epigenetics in identifying novel disease mechanisms. We then discuss the limitations of current epigenetic research in PD, the advantages of simultaneously studying genetics and epigenetics, the potential for epigenetic-based therapeutics and possible future directions.

2. DNA Methylation

DNA methylation is the most well studied epigenetic modification and consists of the addition of a methyl group to the 5-carbon position of a cytosine in a CpG dinucleotide (cytosine followed by guanine in the 5′-to-3′ direction) (Figure 2) [53,54]. This process is catalyzed by DNA methyltransferases (DNMTs), which are involved in either de novo (DNMT3a and 3b) or maintenance (DNMT1) methylation [54].

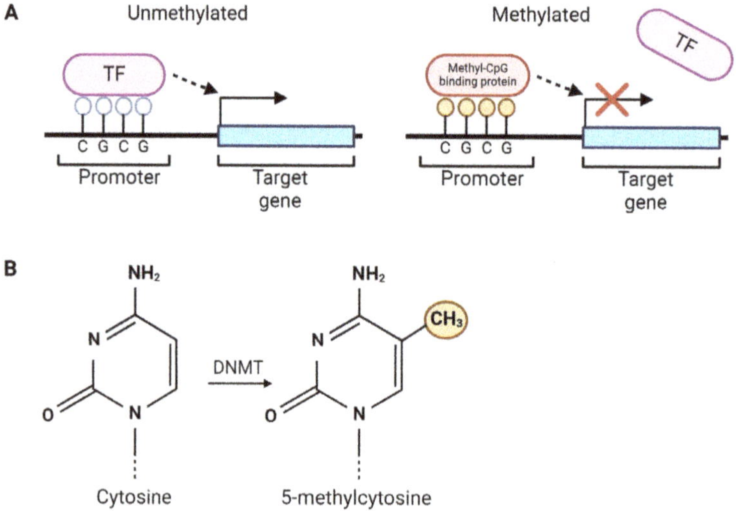

Figure 2. DNA methylation overview. Abbreviations: transcription factor (TF), DNA methyltransferase (DNMT). (**A**) Schematic of an unmethylated (left) and methylated (right) CpG island in a gene promotor and target gene. When unmethylated, transcription factors can typically bind, allowing the gene to be expressed. When methylated, transcription factor binding is typically prevented, and a methyl-CpG binding protein is recruited. Both of these events can lead to the gene being silenced. (**B**) Structural formulas of cytosine and 5-methylcytosine. DNA methylation is catalyzed by DNMT enzymes. Created with BioRender.com.

CpG sites (CpGs) are usually concentrated in CpG islands, which are commonly located in gene promoters [54]. Methylation of CpG islands within promoters has typically been associated with transcriptional silencing, either through blocking transcription factor binding or recruiting methyl-CpG binding proteins that themselves mediate transcriptional repression [53,55–57]. However, this is not always the case; for example, gene body DNA methylation has been associated with an increased level of gene expression and the regulation of alternative splicing [58–60].

2.1. Single-Gene Analyses

In Parkinson's, initial DNA methylation studies mainly focused on genes already hypothesized to be associated with the disease (Table 2). Following bisulfite conversion of DNA, these studies mostly used either Sanger sequencing or pyrosequencing to measure DNA methylation. Pyrosequencing is a method of DNA sequencing that can be used to measure the degree of DNA methylation at individual CpG sites situated in close proximity [61]. The most extensively studied candidate gene in PD is *SNCA*, which encodes the α-synuclein protein. Point mutations in *SNCA* (including the Ala53Thr missense mutation) were the first identified genetic causes of PD [62–64], and *SNCA* locus amplifications (such as duplications and triplications) are also known to cause familial PD [65,66].

It is well established that expression can be regulated by DNA methylation levels within regulatory regions [53], and a number of studies have investigated whether DNA methylation within intron 1 of the *SNCA* gene is altered in PD. The majority of studies reported *SNCA* intron 1 DNA hypomethylation in PD brains [67–70]; however, others found no significant differences [38,71].

Jowaed et al. went further and examined the effect of inhibiting DNA methylation in SK-N-SH cells (a neuroblastoma cell line), which resulted in an increase in *SNCA* mRNA expression [67]. Thus, a plausible mechanism for the high levels of α-synuclein found in PD brains is DNA hypomethylation of the promoter region of *SNCA*.

A recent study highlighted that *SNCA* expression may be regulated by the binding of methyl-CpG binding protein 2 (MeCP2) to methylated CpG sites in *SNCA* intron 1 [70]. Using chromatin immunoprecipitation and qPCR in SK-N-SH cells, they showed that MeCP2 binds to *SNCA* intron 1 and that its binding is dependent on the degree of DNA methylation in intron 1. In the human cortex, *SNCA* intron 1 was hypomethylated in PD compared to controls, and the binding of MeCP2 also appeared to be lower in PD. In addition, CRISPR/Cas9 editing of SK-N-SH cells suggested that increased or decreased binding of MeCP2 to *SNCA* intron 1 is associated with a reduction or gain in α-synuclein expression, respectively [70].

Furthermore, based on a suggestion that α-synuclein may itself interact with the epigenetic machinery, Desplats et al. conducted a series of immunoprecipitation experiments using human brain homogenates that showed that α-synuclein does indeed interact with DNMT1 [69]. Next, they showed increased DNMT1 staining in the cytoplasm (compared to the nucleus) in the cells of PD brains compared to controls. Supported by further experiments in rat cells, they then concluded that DNMT1 may be sequestered to the cytoplasm by α-synuclein, which causes global DNA hypomethylation in PD brains [69].

A more recent study looked at *SNCA* intron 1 methylation separately in bulk frontal cortex, neuronal nuclei and glial nuclei (isolated using fluorescence-activated nuclear sorting) [72]. They observed borderline significant hypomethylation in neuronal nuclei but no significant differences in glial nuclei or bulk cortex. This may explain the mixed results of earlier studies where bulk samples were analyzed and highlights a potential advantage of profiling DNA methylation separately in individual cell types.

Another recent study started to take into account genetic variation by separating PD individuals with *GBA1* mutations from idiopathic PD individuals [73]. Significant hypomethylation of several CpGs in *SNCA* intron 1 in the frontal cortex was observed in PD-*GBA1* but not in idiopathic PD. This demonstrates that segregating PD cases according to genetic subtype might help uncover DNA methylation changes.

In addition to the *SNCA* gene, DNA methylation of other PD-related genes has also been investigated. These include *PRKN*, *PINK1* (both involved in mitochondrial function and mitophagy), *DJ-1* (involved in oxidative stress protection) and *MAPT* (which encodes the microtubule-associated protein tau that forms neurofibrillary tangles most notably in Alzheimer's disease) [74–77]. For example, *MAPT* DNA methylation changes in PD varied according to brain region (hypermethylation in the cerebellum and hypomethylation in the putamen). The same study also suggested that the higher incidence of PD seen in males can possibly be explained by the higher *MAPT* DNA methylation levels observed in females and indicated that increased *MAPT* methylation was associated with later disease onset [77].

Some studies have investigated DNA methylation of genes that interact with key PD genes. For example, Su et al. examined peroxisome proliferator-activated receptor gamma coactivator 1α (*PGC-1α*) [78]. They found increased *PGC-1α* promoter methylation in the substantia nigra of PD individuals, along with a decrease in *PGC-1α* gene and protein expression. Previously, *PGC-1α* was shown to interact with parkin to regulate mitochondrial biogenesis and protect dopaminergic neurons [79]. It has also been associated with increased sensitivity to neurodegeneration [80].

It is worth noting that there are some key limitations of candidate gene studies. In general, small sample sizes are used, limiting their statistical power to detect significant methylation changes. Secondly, by their nature, they are based on a priori hypotheses, limiting the identification of novel genes epigenetically dysregulated in PD that have not been previously associated with neurodegenerative diseases. Thirdly, they are only testing one gene at a time rather than looking within the context of the whole genome (where it is now standard practice to correct for multiple testing) and are therefore subject to less statistical rigor. Therefore, rather than focusing on single genes, the field has shifted toward genome-wide approaches, which are discussed in the next section.

Table 2. Summary of DNA methylation studies in single genes in PD brains.

Gene(s)	Samples	Methylation Sites Studied	Key Findings	Reference/Year
SNCA	Substantia nigra and cortex (6 PD, 6 controls), putamen (6 PD, 8 controls)	23 CpGs in intron 1	Hypomethylation in all brain areas	[67]/2010
SNCA	Anterior cingulate cortex (11 PD, 8 controls), putamen (6 PD, 4 controls), substantia nigra (3 PD, 3 controls)	13 CpGs in intron 1	Hypomethylation in substantia nigra only	[68]/2010
SNCA	Brain homogenates (4 PD, 4 controls)	Gross methylation of intron 1	Hypomethylation	[69]/2011
SNCA	Cerebral cortex (12 PD, 12 controls)	Gross methylation of intron 1	No overall differences but methylation associated with rs3756063 SNP	[38]/2015
SNCA	Substantia nigra (8 PD, 8 controls)	23 CpGs in intron 1	No differences	[71]/2017
SNCA	Bulk frontal cortex (20 PD, 20 controls), sorted neuronal and glial nuclei (12 PD, 12 controls)	23 CpGs in intron 1	Bulk tissue and glial nuclei: No differences across all CpGs Neuronal nuclei: Hypomethylation in PD across all CpGs (borderline significant)	[72]/2021
SNCA	Frontal cortex (9 PD-*GBA1*, 11 idiopathic PD, 6 controls), putamen (6 PD-*GBA1*, 9 idiopathic PD, 6 controls), substantia nigra (8 PD-*GBA1*, 13 idiopathic PD, 3 controls)	17 CpGs (8 sites in intron 1 located further from TSS, 6 CpGs in intron 1 located close to TSS and 3 CpGs within promotor)	Frontal cortex: Hypomethylation in 5 CpGs located further from TSS in intron 1 in PD-*GBA1* compared to controls Hypomethylation in 2 CpGs in intron 1 in PD-*GBA1* compared to idiopathic PD No significant differences in any CpGs located close to TSS in intron 1 or in the promoter Putamen: No significant differences in intron 1 or the promoter Substantia nigra: 1 CpG in the promoter hypomethylated in idiopathic PD compared to both controls and PD-*GBA1* No significant differences in intron 1	[73]/2023
SNCA	Cortex (2PD, 2 controls)	23 CpGs in intron 1	Hypomethylation	[70]/2024
PAD2	Cortex (white matter) (2 PD, 0 controls)	Gross methylation of portion of intron 1, exon 1, and distal extension	No differences	[81]/2007

Table 2. *Cont.*

Gene(s)	Samples	Methylation Sites Studied	Key Findings	Reference/Year
TNF-α	Substantia nigra and cortex (7 PD, 8 controls), striatum (3 PD, 2 controls)	10 CpGs in promoter	No differences	[82]/2008
MAPT, *PSEN1*, *APP* and *UCHL1*	Frontal cortex (8 PD, 17 controls)	*MAPT*: 20 CpGs near exon 0.41 + 31 CpGs in intronic regions *PESN1*: 26 CpGs at TSS *APP*: 18 + 16 CpGs in promoter regions *UCHL1*: 37 CpGs around TSS	No differences	[76]/2009
PRKN	Substantia nigra, cerebellum, occipital cortex (5 PD, 2 controls)	Gross methylation of promoter	No differences	[74]/2013
MAPT	Cerebellum, putamen, anterior cingulate cortex (28 PD, 12 controls)	6 CpGs within promoter/intron 1	Hypermethylation in cerebellum, hypomethylation in putamen, no significant differences in anterior cingulate cortex	[77]/2014
ADORA2A	Putamen (25 PD, 26 controls)	108 CpGs in 5′ untranslated region	Hypomethylation at 2 CpGs	[83]/2014
PGC-1α	Substantia nigra (10 PD, 10 controls)	Gross methylation of promoter	Hypermethylation (of mainly non-CpG dinucleotides) and decreased expression	[78]/2015
mtDNA D-loop	Substantia nigra (10 PD, 10 controls)	CpGSs and non-CpGs in D-loop	Hypomethylation in nearly all CpGs and non-CpGs	[84]/2016
SNCA, *LRRK2*, *PRKN*, *PINK1* and *DJ-1*	Substantia nigra, occipital cortex, parietal cortex (5 PD, 5 controls)	*SNCA*: 6–8 CpGs in promoter, 5 CpGs in exon 1 *LRRK2:* 9 CpGs in promoter, 9 CpGs in exon 1 *PRKN:* 4 CpGs in promoter, 9 CpGs in promoter overlapping with intron 1 *PINK1*: 5 CpGs in promoter/exon 1 *DJ-1*: 6 CpGs in promoter, 8 CpGs in promoter overlapping with exon 1	No significant differences at whole CpG islands, although differential methylation observed at some specific CpGs in *SNCA*, *PRKN* and *PINK1*	[75]/2018
CYP2E1, *TP73*, *C21ORF56* and *CDH13*	Cortex (14 PD, 10 controls)	*CYP2E1*: 10 CpGs (including promoter region) *TP73*: 5 CpGs *C21ORF56*: 2 CpGs *CDH13*: 1 CpG	Hypomethylation of *CYP2E1*, *TP73* and *C21ORF56* No difference in *CDH13* methylation	[85]/2022

Abbreviations: CpG site (CpG), transcription start site (TSS), single nucleotide polymorphism (SNP).

2.2. Epigenome-Wide Association Studies

Several epigenome-wide association studies (EWASs) have identified DNA methylation changes at both previously implicated and novel genes in PD brains (Table 3). Most of these studies used methylation arrays, which measure DNA methylation changes at many CpG sites throughout the genome. The first PD brain EWAS was performed by Kaut et al.

using the Illumina HumanMethylation27 BeadChip, which covers 27,500 CpGs [86]. A few genes were found to be differentially methylated in either the putamen or cortex, but only *CYP2E1* (cytochrome P450 2E1) was significantly altered (hypomethylated) in both. This gene, which interestingly had already been implicated in PD pathology, encodes a protein that provides protection against xenobiotic exposure. The study also showed that decreased methylation of *CYP2E1* resulted in increased *CYP2E1* mRNA expression levels in the cortex.

Several subsequent studies used the Illumina HumanMethylation450 BeadChip, a substantially larger array that covers over 450,000 methylation sites. For example, Masliah et al. identified many differentially methylated CpGs between healthy and PD individuals [87]. Some of the top changing loci in this study were associated with four previously identified PD risk genes, including *MAPT*. It is worth noting that similar DNA methylation patterns were observed in both the brain and blood, suggesting that blood could potentially act as a surrogate for studying brain DNA methylation changes. This could have profound implications for developing easy-to-perform lab-based diagnostic tests for PD. Another study using the 450 BeadChip looked at DNA methylation changes in PD in relation to environmental exposures, namely plantation work and exposure to organochlorines [88]. Exposure to such pesticides has been associated with PD risk [89]. Since epigenetics may mediate environmental risk factors, the authors profiled DNA methylation changes in the brain and identified a few differentially methylated loci in PD individuals who were exposed to plantation work for 10+ years compared to non-exposed PD individuals [88]. Two of the loci differentially methylated between the organochlorine exposure groups were annotated to the *DNAJC15* gene, which is involved in protein translocation into the mitochondria and the regulation of the Hsp70 class of chaperones [90].

Over the last few years, there has been an increasing focus on looking at epigenetics in different brain cell types, and there are now a few PD studies that have isolated neuronal nuclei from the brain for DNA methylation profiling. Kochmanski et al. used the Infinium MethylationEPIC BeadChip (which contains probes for over 850,000 methylation sites) to profile DNA methylation changes in neuronal nuclei isolated from the parietal cortex [91]. They focused on sex-specific DNA methylation changes, given that there are known differences in disease risk, progression and severity between men and women, and sex-specific effects have not been investigated in previous EWASs. They found only three differentially methylated CpGs in males compared to 87 differentially methylated CpGs in females. The most significantly differentially methylated CpG in males was in the *PARK7* locus (hypomethylated) encoding the DJ-1 protein, while the most significantly differentially methylated CpG in females was within the *ATXN1* gene (hypermethylated). *DJ-1* hypomethylation may result in increased DJ-1 protein levels; however, *DJ-1* mutations in PD individuals have been linked with an apparent loss of function of DJ-1, leading to the accumulation of reactive oxygen species and decreased protection against oxidative stress [92].

Not all studies of genome-wide DNA methylation used arrays. For example, Marshall et al. used bisulfite padlock probe sequencing (a targeted bisulfite deep-sequencing method) to investigate genome-wide enhancer DNA methylation in neurons isolated from the prefrontal cortex [93]. A total of 1799 differentially methylated cytosines in enhancers were observed, which were predicted to target 2885 genes, including various PD risk genes (such as *DJ-1* and *PRKN*) as well as *TET2* (which plays a key role in DNA hydroxymethylation—see next section). Furthermore, *TET2* was also upregulated at the mRNA level in both neurons and bulk cytosol. This group also used the same technique to look at enhancer and promoter DNA methylation in neuronal nuclei isolated from each brain hemisphere [94]. They found that hemispheric asymmetry in DNA methylation was greater in PD individuals compared to controls and that this may be associated with the lateralization of disease symptoms.

Additionally, Gordevicius et al. used bisulfite padlock probe sequencing to profile DNA methylation of autophagy-lysosome pathway genes in the olfactory bulb and neuronal nuclei isolated from the prefrontal cortex [95]. In the olfactory bulb, there were

1142 differentially methylated sites, including *SNCA*, which was hypomethylated. In prefrontal cortex neurons, there were 70 differentially methylated sites; however, when only including PD brains with Braak stages 3–4, there were 110 differentially methylated sites, which significantly overlapped with the changes in the whole PD cohort. The authors suggest that alterations to DNA methylation might, therefore, begin earlier on in the disease before the prefrontal cortex is affected by Lewy body pathology.

Interestingly, some EWASs have identified DNA methylation changes of genes that interact with proteins encoded by key PD genes. For example, Dashtipour et al. observed robust hypermethylation of the *SNCAIP* gene, which encodes synphilin-1, in the cortex of PD individuals [96]. Synphilin-1 has been shown to interact with α-synuclein and regulate its degradation [97]. In another study, Young et al. identified hypermethylation of the *ARFGAP1* gene (which encodes a GTPase activating protein) in the dorsal motor nucleus of the vagus of PD individuals [98]. ARFGAP1 has been shown, both in vitro and in vivo, to interact with and regulate the activity of LRRK2 [99,100]. In particular, downregulation of *ARFGAP1* was shown to ameliorate the toxicity caused by *LRRK2* mutations [99,100], perhaps suggesting that *ARFGAP1* hypermethylation may have a protective role in PD brains.

Several of the aforementioned studies used Gene Ontology analysis to examine the functions of genes with altered methylation. Some of the key processes affected were neuronal development and differentiation, synaptic transmission, neurotransmitter transport, dopaminergic synapse, metabolism, immunity, bile acid secretion, cell cycle and cell signaling (including the Wnt pathway), all of which have been implicated in PD pathogenesis [87,91,93,94,98,101].

Some of the studies above report quite large numbers of differentially methylated CpGs; however, it is worth noting that not all studies accounted for multiple-testing correction. While several studies did report false discovery rate (FDR) corrected significance values, it may be more suitable to use a more stringent significance threshold of $p < 9 \times 10^{-8}$, which has been proposed to control for the false-positive rate in EPIC array studies [102]. This threshold was used in the Kochmanski et al. study [91].

Table 3. Summary of epigenome-wide association studies in PD brains.

Samples	Method	Key Findings	Reference/Year
Frontal cortex, cerebellum (399 healthy individuals)	Illumina HumanMethylation27 BeadChip	Differential methylation of 1 or more CpGs correlated with SNPs at PARK16/1q32, GPNMB/7p15, and STX1B/16p11 loci	[103]/2011
Cortex, putamen (6 PD, 6 controls)	Illumina HumanMethylation27 BeadChip	Cortex: Hypomethylation of *CYP2E1* and *PPP4R2* Putamen: Hypomethylation of *CYP2E1* and *LOC84245* Hypermethylation of *DEFA1* and *CHFR*	[86]/2012
Frontal cortex (5 PD, 6 controls)	Illumina HumanMethylation450 BeadChip	317 hypermethylated and 2591 hypomethylated CpGs	[87]/2013
Substantia nigra (39 PD, 13 controls)	Illumina HumanMethylation450 BeadChip	Hypermethylation of 1 CpG (cg10917602) associated with PD susceptibility	[104]/2016
Frontal cortex (12 PD, 12 controls)	Illumina HumanMethylation450 BeadChip	2794 differentially methylated CpGs in the frontal cortex of PD cases and 328 differentially methylated CpGs, majority hypomethylated. Clear pattern of *SNCAIP* hypermethylation	[96]/2017

Table 3. *Cont.*

Samples	Method	Key Findings	Reference/Year
Dorsal motor nucleus of the vagus, substantia nigra, cingulate gyrus (38 PD, 41 controls)	Illumina HumanMethylation450 BeadChip and Infinium MethylationEPIC BeadChip	234 differentially methylated regions in the dorsal motor nucleus of the vagus (including *ARFGAP1* hypermethylation), 44 in the substantia nigra and 141 in the cingulate gyrus	[98]/2019
Temporal lobe (13 PD with 0 years of plantation work, 4 PD with 10+ years of plantation work) (12 PD with 0–2 organochlorines, 4 PD with 4+ organochlorines)	Illumina HumanMethylation450 BeadChip	7 differentially methylated loci between PD individuals with 10+ vs. 0 years of plantation work exposure 8 differentially methylated loci between PD individuals with 4+ vs. 0–2 organochlorines in the brain 2 different loci annotated to *DNAJC15* which were differentially methylated (in both brain and blood) between the organochlorine exposure groups	[88]/2020
Prefrontal cortex neuronal nuclei (discovery cohort: 57 PD, 48 controls) (replication cohort: 26 PD, 31 controls)	Bisulfite padlock probe sequencing of enhancers and promoters (633,803 modified cytosines)	6207 CpGs in PD showing hemispheric asymmetry in DNA methylation (3894 CpGs showed a greater hemispheric asymmetry in PD compared to controls). These targeted 4691 genes, including PD risk genes. More DNA methylation and transcriptomic differences seen in hemisphere matched to symptom-dominant side Above findings validated in replication cohort 37 PD risk genes showing more hemispheric asymmetry in PD and/or greater differences in symptom-dominant hemisphere, including *SNCA*, *ITPKB*, *SATB1*, *ANK2* and *CAMK2D*	[94]/2020
Prefrontal cortex neuronal nuclei (57 PD, 48 controls) (22 PD Braak 3–4, 48 controls)	Bisulfite padlock probe sequencing (31,590 enhancers)	1799 differentially methylated cytosines in enhancers (mainly hypermethylated) 2172 differentially methylated cytosines in enhancers when comparing PD Braak stage 3–4 (prior to Lewy body pathology reaching the cortex) and controls Differentially methylated enhancers targeted 2885 genes, including 15 different PD risk genes and *TET2* 76 of the genes with dysregulated enhancers were also transcriptionally altered	[93]/2020

Table 3. *Cont.*

Samples	Method	Key Findings	Reference/Year
Olfactory bulb (9 PD, 14 controls), prefrontal cortex neuronal nuclei—discovery cohort (52 PD including 20 Braak stage 3–4, 42 controls), prefrontal cortex neuronal nuclei—replication cohort (13 PD, 15 controls)	Bisulfite padlock probe sequencing of autophagy-lysosome pathway genes (143,553 CpGs in olfactory bulb) (130,733 CpGs and 696,665 non-CpGs in prefrontal cortex discovery cohort) (110,397 CpGs in prefrontal cortex replication cohort)	Olfactory bulb: 1142 differentially methylated CpGs affecting 353 genes (mostly hypermethylated) (*SNCA* hypomethylation) Prefrontal cortex neuronal nuclei: 70 differentially methylated CpGs affecting 58 genes (mostly hypermethylated) in discovery cohort 110 differentially methylated CpGs affecting 87 genes in PD Braak stage 3–4 1131 differentially methylated CpGs affecting 341 genes in replication cohort	[95]/2021
Prefrontal cortex (27 PD, 26 controls)	Whole-genome bisulfite sequencing	No association between mitochondrial DNA methylation and disease status	[105]/2022
Cortex (14 PD, 10 controls)	Illumina HumanMethylation450 BeadChip	35 hypomethylated and 22 hypermethylated genes (not significant after *p*-value adjustment). Included 5 CpGs hypomethylated in *CYP2E1* and 6 CpGs hypomethylated in *C21ORF56*	[85]/2022
Sorted neuronal nuclei from parietal cortex (50 PD, 50 controls)	Infinium MethylationEPIC BeadChip	3 and 87 differentially methylated CpGs in males and females, respectively, including *PARK7* hypomethylation (males), *ATXN1* hypermethylation (females) and *SLC17A6* hypomethylation (females) 258 and 214 differentially methylated regions in males and females, respectively, including *NR4A2* (males) and *SLC17A6* (females) 1 differentially methylated region completely overlaps between sexes (annotated to *PTPRN2*)—hypermethylated in males and hypomethylated in females	[91]/2022
Primary motor cortex (40 PD, 38 controls)	Infinium MethylationEPIC BeadChip	3062 hypomethylated and 1251 hypermethylated CpGs 2.07 years of accelerated epigenetic age in PD compared to controls	[101]/2022
Prefrontal cortex neurons	Targeted bisulfite sequencing	667 differentially methylated genes, including 107 associated with stool butyrate levels	[106]/2022
Prefrontal cortex (19 PDD, 18 controls)	Infinium MethylationEPIC BeadChip	1151 differentially methylated CpGs (82% hypomethylated) 1 differentially methylated region in *OTX2* gene	[107]/2023

Abbreviations: epigenome-wide association study (EWAS), Parkinson's disease dementia (PDD). Note: Of all the studies included in this table, only Kochmanski et al. [91] used the significance threshold of $p < 9 \times 10^{-8}$, which has been proposed to control for the false-positive rate in EPIC array studies.

3. DNA Hydroxymethylation

DNA hydroxymethylation is another, although less studied, DNA modification in which the ten-eleven translocation (TET) class of enzymes catalyze the oxidation of 5-

methylcytosine to 5-hydroxymethylcytosine (by replacing the hydrogen atom at the carbon 5 position with a hydroxymethyl group) (Figure 3), which serves as an intermediate step in the DNA demethylation pathway [54,108,109]. DNA hydroxymethylation has recently gained interest as a potentially important epigenetic modification and is particularly abundant in neurons in the brain [110,111]. To date, only a handful of studies have looked at DNA hydroxymethylation in PD.

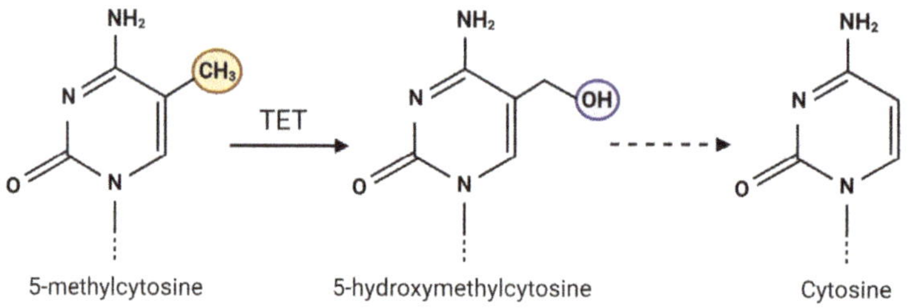

Figure 3. DNA hydroxymethylation overview. Abbreviations: ten-eleven translocation (TET). Structural formulas depicting the conversion of 5-methylcytosine to 5-hydroxymethylcytosine, catalyzed by TET. 5-hydroxymethylcytosine can also serve as an intermediate in the DNA demethylation pathway and be converted back to cytosine. Created with BioRender.com.

Using ELISA, Stöger et al. analyzed both DNA methylation and DNA hydroxymethylation in the cerebellum of 36 PD and 27 control individuals [112]. Although no significant differences were found in DNA methylation levels, PD individuals had significantly higher overall 5-hydroxymethylcytosine levels compared to the control group.

Kaut et al. examined DNA hydroxymethylation in the substantia nigra (eight PD and eight controls), cerebellum (eight PD and eight controls) and temporal gyrus (ten PD and ten controls) using an immunohistochemical approach [113]. They observed a significantly higher percentage of 5-hydroxymethylcytosine-immunoreactive cells in the PD cerebellum, supporting the Stöger study. However, no differences were seen in the substantia nigra or the neocortex. The lack of differences in these brain regions could be related to the small sample sizes used here.

Marshall et al. investigated whether *TET2* enhancer dysregulation affected hydroxymethylation levels in prefrontal cortex neurons [93]. Using hydroxymethylated DNA immunoprecipitation sequencing in 20 PD and 23 control individuals, they found that hydroxymethylation was considerably increased at gene bodies, promotors and enhancers in PD neurons and overlapped with the epigenetically modified enhancers. Experiments in mice showed that *TET2* silencing was neuroprotective. Therefore, TET2 could be a potential therapeutic target in PD [93].

Finally, Min et al. looked at genome-wide DNA hydroxymethylation in the substantia nigra of twelve PD and nine control individuals using hMe-Seal, which uses chemical labeling and affinity purification coupled with sequencing [114]. They identified 1800 hyperhydroxymethylated and 2319 hypohydroxymethylated regions (FDR < 0.05). In contrast, and perhaps surprisingly, when using methylated DNA immunoprecipitation sequencing to profile genome-wide DNA methylation, no differentially methylated regions were identified. Gene Ontology and Kyoto Encyclopedia of Genes and Genomes pathway analyses were also carried out, which highlighted that genes annotated to differentially hydroxymethylated regions were enriched for processes including neurogenesis and neuronal differentiation and were involved in multiple signaling pathways.

4. Histone Modifications

The third type of epigenetic mechanism that will be considered here is histone modifications. These are post-translational modifications to the N-terminal tails of histone octamers and include methylation, acetylation, phosphorylation, ubiquitination, SUMOylation and ADP-ribosylation [115,116]. These marks typically regulate chromatin structure and so alter the accessibility of DNA to the transcriptional machinery, with highly compacted chromatin (heterochromatin) associated with transcriptional repression and open chromatin (euchromatin) associated with transcriptional activation (Figure 4). Whether a given modification is repressive or active usually depends on the exact residue that is marked, although acetylation (which is the most well studied type of histone modification in PD) is normally associated with transcriptional activation [115,116].

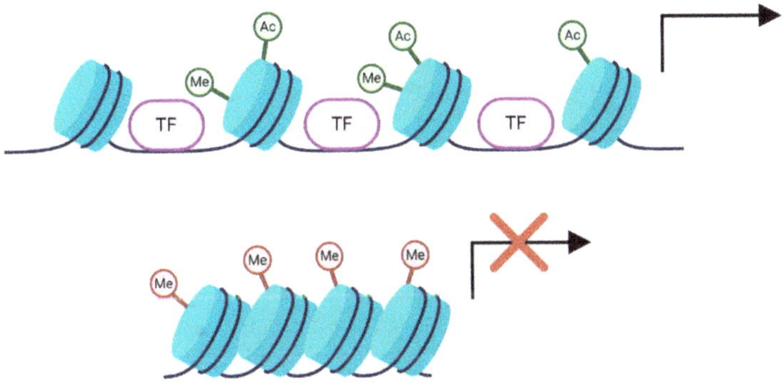

Figure 4. Histone modifications overview. Abbreviations: acetylation (Ac), methylation (Me). The top picture depicts an open chromatin region where the histone octamers contain active acetylation and active methylation marks, resulting in transcription factor binding and gene expression. The bottom picture depicts a compacted chromatin region where the histone octamers contain repressive methylation marks. This prevents transcription factors from binding, resulting in gene silencing. Created with BioRender.com.

So far, only a few studies have investigated histone modifications in PD brains (Table 4). For example, Gebremedhin and Rademacher used Western blotting to examine global H3 acetylation in the primary motor cortex and found a significant increase in H3K14 and H3K18 acetylation and a significant decrease in H3K9 acetylation, which correlated with both Lewy body stage and substantia nigra pigmentation scores [117]. In contrast, Harrison et al. found that H3K9 acetylation was increased in the substantia nigra of PD brains, with the size of the increase correlated with the Braak stage [118]. Further work is therefore needed to establish whether the direction of change in H3K9 acetylation is dependent on, for example, different brain regions.

A genome-wide study of histone acetylation was performed in the prefrontal cortex of PD brains, with a focus on H3K27 acetylation [119]. They identified 2877 H3K27-hyperacetylated regions and 14 H3K27-hypoacetylated regions, indicating a genome-wide dysregulation of H3K27 acetylation in the disease. Some of these regions were linked to known PD or neurodegenerative disease risk genes, including *SNCA*, *MAPT*, *PRKN*, *DJ-1* and amyloid precursor protein (*APP*). Perhaps interestingly, there was little correlation between H3K27 acetylation and gene transcription in the PD group (while a positive correlation was observed in the control group), suggesting that H3K27 acetylation may become decoupled from gene expression in PD.

Finally, Guhathakurta et al. studied a different histone mark—methylation—in *SNCA* regulatory regions and found that H3K4me3 was increased in PD bulk substantia nigra as well as neuronal nuclei isolated from this region [120]. Next, using CRISPR-Cas9 technology, they demonstrated that the reduction in H3K4me3 at the *SNCA* promoter leads to a decrease in α-synuclein levels both in SH-SY5Y cells and in idiopathic PD-iPSC-derived dopaminergic neurons.

Table 4. Summary of histone modification studies in PD brains.

Samples	Method	Key Findings	Reference/Year
Midbrain, cerebral cortex, cerebellar cortex (5 PD, 5 controls)	Western blotting and immunostaining	Midbrain: Increased acetylation of H2AK5, H2BK15, H3K9 and H4K5 (in 2–3 PD individuals) Downregulation of HDAC1, HDAC2, HDAC4, HDAC6 and SirT1 Higher proportion of acetylated midbrain dopaminergic neurons Cerebral cortex: Increased acetylation (H2AK5, H2BK15, H3K9 and H4K5) in 1 PD individual only Cerebellar cortex: Increased acetylation of H2BK15	[121]/2016
Primary motor cortex (9 PD, 8 controls)	Western botting	Increased acetylated H3–total H3 ratio Increased acetylated H3K14–total H3 ratio Increased acetylated H3K18–total H3 ratio Decreased acetylated H3K9–total H3 ratio	[117]/2016
Substantia nigra (8 early PD, 12 late PD, 10 controls)	Western blotting	Increased acetylation at H3K9 in late PD Correlation between level of histone acetylation and Braak stage	[118]/2018
Prefrontal cortex (global acetylation: 13 PD, 13 controls), Prefrontal cortex, striatum and cerebellar cortex (7 PD, 7 controls), Prefrontal cortex (discovery cohort: 17 PD, 11 controls), Prefrontal cortex (replication cohort: 10 PD, 11 controls)	Western blotting ChIP-seq (genome-wide H3K27 acetylation)	Prefrontal cortex: Increased global histone acetylation Increased acetylation at H3K27, H2BK15, H3K9/14, H3K56 and H4K12 No significant changes at H2AK5, H4K5 and H4K16 Striatum and cerebellar cortex: Increased acetylation at H3K27 Discovery study: 2877 H3K27-hyperacetylated regions (corresponding to 1434 genes) and 14 hypoacetylated regions (corresponding to 9 genes) Replication study: 2486 H3K27-hyperacetylated regions (corresponding to 946 genes) and 227 hypoacetylated regions (corresponding to 253 genes) 275 hyperacetylated genes (*DLG2* and *TNRC6B* most significant) and 2 hypoacetylated genes (*PTPRH*, *JUP*) replicated across both cohorts	[119]/2021
Substantia nigra (18 PD, 9 controls), substantia nigra neuronal nuclei (7 PD, 6 controls)	ChIP (H3K4me3, H3K27ac and H3K27me3)	Increased H3K4me3 at *SNCA* regulatory region No significant difference in H3K27ac Increased H3K4me3 at *SNCA* promoter/intron 1 Positive correlation between H3K4me3 and α-synuclein expression	[120]/2021
Substantia nigra (9 PD, 9 controls)	ChIP-seq (H3K27ac)	Identification of 2770 downregulated and 2910 upregulated cis-regulatory elements	[39]/2023

Abbreviations: chromatin immunoprecipitation (ChIP), histone deacetylase (HDAC), lysine (K).

5. Non-Coding RNAs

As with DNA methylation and histone modifications, non-coding RNAs are another epigenetic mechanism involved in the regulation of gene expression. These RNAs include miRNAs, long non-coding RNAs (lncRNAs), small nucleolar RNAs (snoRNAs), small inhibitory RNAs (siRNAs) and piwi-interacting RNAs (piRNAs) [122,123]. In PD, the most well studied of these are miRNAs, and this section will focus on miRNA and lncRNA expression studies.

miRNAs are small non-coding RNA molecules consisting of 20–22 nucleotides that bind to the 3′ untranslated region of target mRNAs, resulting in gene silencing by either translational repression or mRNA degradation (Figure 5) [124,125].

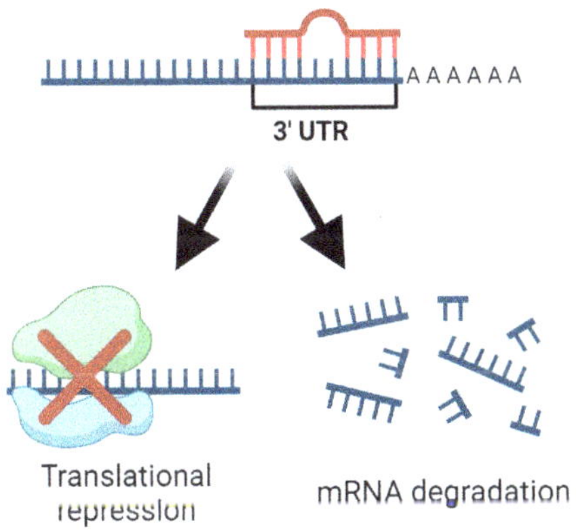

Figure 5. microRNAs overview. Abbreviations: microRNA (miRNA), untranslated region (UTR). Schematic illustrating how a miRNA binds to the 3′UTR of a target mRNA, which results in gene silencing either by translational repression or mRNA degradation. Created with BioRender.com.

A number of studies have observed miRNA expression changes in the brains of PD individuals (Table 5). Several studies focused both on the identification of putative miRNAs dysregulated in PD and the subsequent functional analyses of those miRNAs in vitro. For example, suppression of miR-133b (the expression of which was significantly decreased in PD midbrain tissue) affected the expression of dopaminergic neuron markers in primary embryonic rat midbrain cultures. This suggested that miR-133b is involved in regulating the maturation and function of midbrain dopaminergic neurons [126]. In another example, overexpression of miR-126 (the expression of which was significantly increased in PD brains) impaired IGF-1/PI3K signaling in human SH-SY5Y cells, increased vulnerability to the neurotoxin 6-OHDA and reduced trophic support [127]. In yet another example, increased expression of a number of miRNAs (including miR-224, miR-379 and miR-26b in SH-SY5Y cells) resulted in a loss of expression of LAMP-2A and hsc70, key proteins involved in chaperone-mediated autophagy, as well as increased accumulation of α-synuclein, a substrate normally degraded by chaperone mediated autophagy [128].

Multiple studies identified differentially expressed miRNAs that are directly or indirectly involved in the regulation of key PD genes, including miR-34b and miR-34c (depletion of which led to *DJ-1* and *PRKN* downregulation and *SNCA* upregulation), miR-205 (its downregulation resulted in *LRRK2* upregulation), miR-127-5p and miR-16-5p (both previously shown to regulate *GBA1* expression) [129–133]. The resulting altered expression of key PD genes mimics the upregulation or downregulation caused by genetic mutations in those genes, providing further evidence for their involvement in the pathogenesis and progression of the disease.

Interestingly, some miRNAs that showed altered expression in the brain were also found to be dysregulated in peripheral tissues. Some more commonly implicated miRNAs in both the brain and periphery include miR-30, miR-29, let-7, miR-485, miR-132 and miR-133b [126,132,134–148]. One study found 56 miRNAs dysregulated in both the brain and leukocytes [136]. Together, this highlights that peripheral samples may be suitable to use as a surrogate to identify brain miRNA changes, which can consequently help to identify novel pathways and mechanisms that are dysregulated in PD. Furthermore, the identification of PD-specific miRNA changes in easily accessible peripheral samples might lead to the development of new tests that would aid PD diagnosis.

Pathways involving some of the identified miRNAs include cell cycle regulation, apoptosis, axon guidance, neurogenesis, synaptic function, inflammation, oxidative stress, metabolism, ubiquitin/proteasome pathway, endocytosis and cell signaling (including insulin, NF-κB, p53 and mTOR) [127,134,136,138,149,150]. Interestingly, Ravanidis et al. found that certain processes (including ubiquitin-mediated proteolysis, circadian rhythm and axon guidance) are implicated in idiopathic but not genetic (*SNCA* or *GBA1*) PD cases [151].

Another important class of non-coding RNAs is lncRNAs, which are over 200 nucleotides long [152]. Altered lncRNA expression has been observed in PD brain regions, including the substantia nigra, cerebellum and cortex (Table 6). As with miRNAs, some studies provided functional insights into the roles that dysregulated lncRNAs might play in the pathogenesis of PD. For example, the expression of AL049437 was significantly upregulated in PD substantia nigra, and the reduction in its expression in SH-SY5Y cells led to increases in cell viability, tyrosine hydroxylase secretion and mitochondrial transmembrane potential and mass [153]. Another study found that the downregulation of six lncRNAs in PD substantia nigra and three lncRNAs in PD cerebellum was accompanied by a significant increase in *SNCA* mRNA levels in the substantia nigra and a significant decrease in *LRRK2* and *PINK1* mRNA levels in the substantia nigra and cerebellum. This highlights a putative correlation between the expression levels of lncRNAs and nearby PD genes [154].

Table 5. Summary of microRNA studies in PD brains.

Samples	Method	Key Findings	PD-Associated Pathophysiology	Reference/Year
Midbrain, cerebellum and cortex (3 PD, 5 controls)	RT-qPCR (panel of 224 miRNA precursors), RNase protection assay, qPCR and Northern blotting (including mature miR-133b)	Downregulation of miR-133b (precursor and mature)	Involved in the regulation of dopaminergic neuron maturation and function	[126]/2007
Amygdala (11 PD, 6 controls) Amygdala (13 PD, 12 controls), frontal cortex (14 PD, 21 controls), cerebellum (11 PD, 17 controls) and substantia nigra (7 PD, 6 controls)	miRCURY LNA™ *miRNA* microarray (17 miRNAs) RT-qPCR (miR-637, miR-34b, miR-34c)	Downregulation of miR-637 and miR-34c-5p in amygdala Downregulation of miR-34b and miR-34c validated in amygdala, frontal cortex, substantia nigra and cerebellum (only miR-34c significant in cerebellum) Could not confirm change in miR-637 expression (measured in amygdala)	Depletion of miR-34b and miR-34c led to *DJ-1* and *PRKN* downregulation and *SNCA* upregulation	[129]/2011
Substantia nigra and amygdala (6 PD, 5 controls)	RT-qPCR (8 miRNAs)	6 miRNAs upregulated in substantia nigra 2 miRNAs upregulated in amygdala	Target and associated with a reduction in *lamp-2a* and *hsc70* levels, key proteins involved in chaperone-mediated autophagy	[128]/2013
Frontal cortex (15 PD, 11 controls), striatum (5 PD and 4 controls)	RT-qPCR (miR-205)	Downregulation of miR-205	Downregulation of miR-205 resulted in *LRRK2* upregulation	[131]/2013

Table 5. *Cont.*

Samples	Method	Key Findings	PD-Associated Pathophysiology	Reference/Year
Substantia nigra (8 PD, 4 controls)	TaqMan low-density arrays (733 miRNAs) and TaqMan assays (miR-198, miR-548d, miR-385-5p and miR-135b)	10 miRNAs downregulated (including miR-135b) 1 miRNA upregulated (miR-548d)	Predicted target genes included *SNCA*, *PRKN*, *LRRK2*, *ATXN1*, *SNCAIP* and *GBA*	[137]/2014
Laser capture microdissected dopaminergic neurons (8 PD, 8 controls)	Human MicroRNA TaqMan Arrays (379 miRNAs)	Dysregulation of miRNA expression profile Upregulation of miR-126	miR-126 overexpression impaired IGF-1/PI3K/AKT signaling	[127]/2014
Midbrain tissue, laser capture microdissected dopaminergic neurons (5 PD, 8 controls)	RT-qPCR (miR-133b)	Downregulation of miR-133b in midbrain tissue No differences in miR-133b levels in dopaminergic neurons		[146]/2014
Putamen (25 PD, 26 controls)	RT-qPCR (miR-34b and c)	Downregulation of miR-34b, particularly in early stages	Adenosine A2A receptor ($A_{2A}R$) identified as potential target of miR-34b	[83]/2014
Laser capture microdissected dopaminergic neurons (8 PD, 8 controls)	Human MicroRNA TaqMan Arrays	109 miRNAs upregulated (miR-132 significantly upregulated) 50 miRNAs downregulated 14 significantly differentially expressed miRNAs associated with target genes Trend toward upregulation in males and downregulation in females	Targets associated with several aspects of PD pathogenesis, including cellular function and dopaminergic neuron identity	[134]/2015
Prefrontal cortex (29 PD, 33 controls)	Small RNA sequencing—Illumina HiSeq 2000 (911 miRNAs)	125 differentially expressed miRNAs (downregulation of miR-10b-5p)	Including miR-127-5p and miR-16-5p, both previously shown to regulate *GBA1* expression	[132]/2016
Putamen (12 PD—mostly L-DOPA treated, 12 controls)	Human v2 miRNA expression assay kit (800 miRNAs) and RT-qPCR (4 miRNAs)	6 miRNAs upregulated 7 miRNAs downregulated Upregulation of miR-3195 and miR-204-5p Downregulation of miR-155-5p and miR-219-2-3p	miRNAs associated with inflammatory response and oxidative stress	[138]/2016
Amygdala (14 PD, 7 controls)	RNA-seq	42 differentially expressed miRNAs in premotor-stage PD compared to controls 103 differentially expressed miRNAs in motor-stage PD compared to controls		[155]/2016

Table 5. *Cont.*

Samples	Method	Key Findings	PD-Associated Pathophysiology	Reference/Year
Anterior cingulate gyrus (22 PD, 10 controls)	TaqMan miR array (744 miRNAs)	43 miRNAs upregulated	13 of these each predicted to regulate at least one of *DJ-1*, *PRKN*, *PINK1*, *LRRK2*, *SNCA* or *HTRA2*	[135]/2016
	RT-qPCR (13 miRNAs)	5 miRNAs upregulated	Predicted to each regulate at least one of *SNCA*, *PRKN* or *LRRK2* and additional genes involved in normal cellular function	
Prefrontal cortex (29 PD, 36 controls)	RNA-seq (99 novel miRNAs)	Upregulation of miR-46 and miR-236 Downregulation of miR-225		[156]/2016
Prefrontal cortex (29 PD, 33 controls)	RNA-seq	321 differentially expressed miRNAs		[136]/2017
Substantia nigra (6 PD, 5 controls)	RT-qPCR (miR-7)	Downregulation of miR-7	Depletion of miR-7 results in increased α-synuclein expression, dopaminergic neuron loss and reduced striatal dopamine content	[157]/2017
Cingulate gyrus (8 PD, 8 controls)	RNA-seq	44 miRNAs upregulated 55 miRNAs downregulated		[158]/2018
Substantia nigra (4 PD, 4 controls)	In situ hybridization (miR-425)	Downregulation of miR-425	miR-425 deficiency triggers necroptosis of dopaminergic neurons	[159]/2019
Prefrontal cortex (15 PD, 10 controls)	RT-qPCR (10 miRNAs)	3 miRNAs downregulated (miR-124, miR-144 and miR-218)	Target *KPNB1/A3/A4*, which were all upregulated in PD brains. Inhibition of these miRNAs activates NF-κB signaling	[149]/2020
Midbrain (19 PD, 12 controls)	Small and total RNA-seq, RT-qPCR (4 miRNAs)	4 miRNAs upregulated (miR-539-3p, miR-376a-5p, miR-218-5p, miR-369-3p)	Targets of miR-369-3p (*GTF2H3*) and miR-218-5p (*RAB6C*) downregulated	[160]/2022
Superior temporal gyrus (214 PD, 47 controls)	TaqMan Advanced miRNA Assays (10 miRNAs)	3 miRNAs downregulated (miR-132-3p, miR-132-5p and miR-129-5p)	miR-132-3p/-5p significantly associated with α-synuclein Braak stage and may interact with *SNCA* mRNA	[161]/2022
Midbrain (5 PD, 5 controls)	RT-qPCR (miR-132-3p)	Upregulation of miR-132-3p	*GLRX* identified as potential miR-132-3p target, *GLRX* mRNA and protein expression decreased in PD	[162]/2022

Table 5. *Cont.*

Samples	Method	Key Findings	PD-Associated Pathophysiology	Reference/Year
Middle frontal gyrus (16 PD Braak stage 4, 9 PD Braak stage 5–6, 19 PDD (Braak stage 5–6), 19 controls)	RNA-seq RT-qPCR (let-7e-3p, miR-424-3p and miR-543)	9 miRNAs downregulated 3 miRNAs upregulated (combined PD groups) Upregulation of let-7e-3p in PD with Braak 5–6 compared to PDD in both gray and white matter Upregulation of miR-424-3p in PD with Braak 5–6 in both gray and white matter compared to controls, and in PD in gray matter compared to PDD Upregulation of miR-543 in PD compared to controls in white matter only	*SIRT1* identified as potential miR-543 target	[150]/2022

Abbreviation: reverse transcription polymerase chain reaction (RT-qPCR)

Table 6. Summary of long non-coding RNA studies in PD brains.

Samples	Method	Key Findings	PD-Associated Pathophysiology	Reference/Year
Substantia nigra, amygdala (5 PD, 5 controls)	RT-qPCR (3 lncRNAs)	3 lncRNAs upregulated in amygdala (RP11-462G22.1, RP11-79P5.3 and U1) RP11-462G22.1 and RP11-79P5.3 upregulated in substantia nigra		[163]/2014
Anterior cingulate gyrus neurons (20 PD, 10 controls)	RT-qPCR (90 lncRNAs)	4 lncRNAs upregulated (lincRNA-p21, Malat1, SNHG1 and TncRNA) Downregulation of H19 lncRNA		[164]/2017
Substantia nigra (11 PD, 14 controls)	Affymetrix Human Genome U133A Array (698 lncRNAs)	42 lncRNAs upregulated (AL049437 most significantly upregulated) 45 lncRNAs downregulated (AK021630 most significantly downregulated)	Reduction in AL049437 expression led to increases in cell viability, tyrosine hydroxylase secretion and mitochondrial transmembrane potential and mass	[153]/2017
Substantia nigra and cerebellum (9 PD, 8 controls)	RT-qPCR (6 lncRNAs)	6 lncRNAs downregulated in substantia nigra 3 lncRNAs downregulated in cerebellum (AK127687, UCHL1-AS1, MAPT-AS1)	Accompanied by increased *SNCA* mRNA levels in the substantia nigra and decreased *LRRK2* and *PINK1* mRNA levels in both brain regions	[154]/2019
Substantia nigra (29 PD, 24 controls)	RT-qPCR (NEAT1 lncRNA)	Upregulation of NEAT1	Neuroprotective agents induce NEAT1 upregulation	[165]/2019
Superior frontal gyrus (23 Braak Lewy body stage 0 controls, 61 PD/PDD/incidental Lewy body disease—subdivided into 19 Braak Lewy body stage 1–4, 19 Braak Lewy body stage 5, 23 Braak Lewy body stage 6)	RNA-seq	Differential expression of 34 lncRNAs between groups		[166]/2023
Substantia nigra (57 PD, 43 controls)	Bioinformatics analysis of microarray data	37 lncRNAs upregulated 68 lncRNAs downregulated		[167]/2023

Abbreviation: long non-coding RNA (lncRNA).

6. Discussion

Figure 6 provides a summary of the potential roles of DNA methylation, histone modifications and miRNAs in PD from the studies discussed, including some of the key genes and biological processes affected.

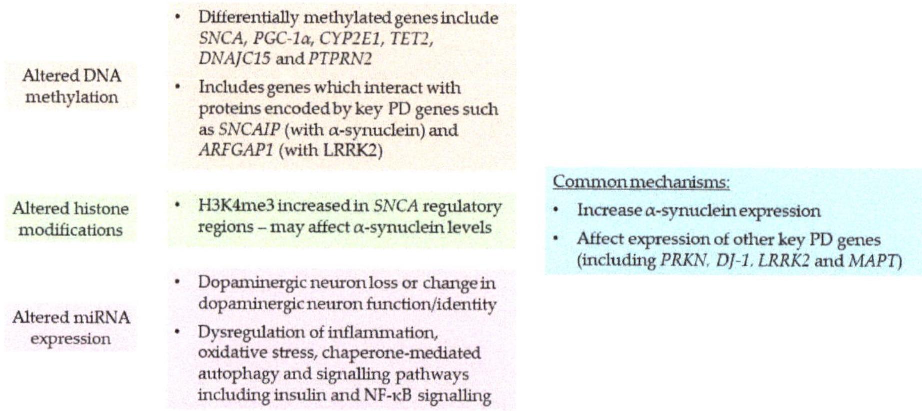

Figure 6. Summary of how epigenetics may affect PD pathogenesis. This figure highlights the potential roles of DNA methylation, histone modifications and miRNA expression in PD. Studies have related all three epigenetic mechanisms to an increase in α-synuclein expression and changes in the expression of multiple PD risk genes.

6.1. Limitations of Current Studies

Firstly, several of the candidate gene DNA methylation and miRNA studies used small sample sizes and might, therefore, lack the power to identify significant epigenetic changes. This could explain some of the discrepancies between the findings reported. For example, while several studies observed *SNCA* intron 1 hypomethylation [67–70], not all studies have replicated this [38,71]. Even when the same brain regions were analyzed, for example, substantia nigra, cortex or putamen, there were discrepant findings between studies [38,67,68,71]. There are also some inconsistent findings between different miRNA studies. For example, some of the key dysregulated miRNAs identified by earlier studies include miR-34b/c, miR-133b and miR-205 [126,129,131]. However, several subsequent studies failed to observe a difference for some or all of these miRNAs [132,134,135,137,155]. Therefore, performing studies using greater sample sizes may help provide more conclusive answers where there are conflicting findings.

Furthermore, not all miRNA studies used the same techniques; the earlier studies (up to 2014) almost exclusively used RT-qPCR, whereas some of the later studies used microarrays and RNA-seq. Therefore, differences in findings between studies could be influenced by the use of different laboratory approaches as well as small sample sizes. A few of the miRNA brain studies used more than one technique to analyze the same miRNAs, for example, using arrays followed by RT-qPCR [129,135,138]. This type of approach should be encouraged to validate initial findings and help prioritize miRNAs for further investigation in future studies. Another factor to consider is the disease stage of the samples; for example, Miñones-Moyano et al. observed miR-34b/c downregulation in the early stage of the disease [129], while Cardo et al. failed to replicate this finding using advanced-stage brains [137]. Pantano et al. compared miRNA expression in both early and late-stage PD to controls and suggested that miRNA profiles vary according to disease stage [155]. Therefore, the variation in findings between studies could also be influenced by differences in the disease stages of the brains profiled.

Furthermore, it is well known that PD pathology differs between cell types. In particular, in addition to neurons (whose role in PD via α-synuclein accumulation and dopaminergic neuron loss is well established) [9], multiple studies have shown that oligodendrocytes and other types of glial cells are also involved [168–176]. However, despite this, almost all PD epigenetic studies performed to date have only analyzed bulk brain tissue. This is a substantial drawback since results then reflect changes in a mixture of both neuronal and different non-neuronal cell types, and it is known that epigenetic profiles are highly cell type-specific [177,178]. Furthermore, results may be confounded by the proportions of the different cell populations present, which may vary from sample to sample. Therefore, bulk brain tissue studies mask the epigenetic disease signatures of distinct cell types, hampering both mechanistic understanding and the search for novel drug targets.

Novel protocols for fluorescence-activated nuclear sorting are now emerging, allowing the study of epigenetics in individual cell types [179–181]. In the last few years, there have been initial studies in PD that isolated neuronal nuclei from human brain tissue and detected genome-wide DNA methylation changes in these nuclei [93–95,106]. Since Gu et al. observed *SNCA* hypomethylation in neuronal nuclei but not in glial nuclei or bulk frontal cortex, perhaps some of the discrepancies in the earlier candidate gene studies could be attributed to differences in cell proportions [72]. Similarly, one study compared miRNA expression changes in isolated neurons to bulk brain tissue. While miR-133b was downregulated in PD midbrain tissue, no change in the level of this miRNA was observed in laser capture microdissected dopaminergic neurons [146], perhaps indicating that the difference found in bulk brain tissue is driven by glial cells. Separating cell populations in epigenetic studies could also make the signals less noisy due to the enrichment of a certain cell type, so studies will likely require smaller sample sizes to achieve the same statistical power compared to bulk brain studies. Cell type-specific approaches are crucial going forward to provide a more detailed picture of the role of epigenetics in PD.

In addition to this, studies so far have generally not considered different genetic subtypes of PD. Smith et al. reported *SNCA* intron 1 hypomethylation in PD individuals with *GBA1* mutations but *SNCA* promotor hypomethylation in idiopathic PD [73], highlighting the existence of potentially distinct epigenetic profiles between PD individuals with *GBA1* mutations and idiopathic PD cases. These findings could, perhaps, explain why previous studies (which do not separate PD-*GBA1* and idiopathic PD) have led to mixed results regarding *SNCA* methylation. This demonstrates the putative importance of separating the two groups in future studies as well as further stratifying by other known PD genes to gain a better understanding of DNA methylation changes in the disease.

6.2. Utility of Combining Genetics and Epigenetics Studies

Integrating genetic and epigenetic information is essential to further understand the etiology of and mechanisms involved in diseases. The majority of GWAS variants are found in regulatory regions, and it is known that epigenetic changes can be genetically influenced [182]. Genetic variants that are associated with the levels of DNA methylation at specific positions in the genome are called methylation quantitative trait loci (mQTLs).

One study examined the effects of PD risk variants in the *SNCA* locus on intron 1 methylation levels and identified rs3756063 as an mQTL associated with *SNCA* intron 1 methylation [38]. This SNP was found to be localized 30 bp upstream of a GATA transcription factor binding motif within *SNCA* intron 1 and additionally alters a CpG dinucleotide itself, suggesting it may affect the interaction between GATA and the *SNCA* gene. GATA transcription factors have previously been shown to regulate *SNCA* expression [183].

Other studies have examined the overlap between SNPs in the *SNCA* gene and epigenetic marks. Sharma et al. found that several SNPs distributed throughout the *SNCA* gene (which were previously associated with PD) were located within or near peaks of certain histone modification marks and DNA binding motif sequences [40]. In another example, certain SNPs associated with PD were present in enhancer (H3K27-acetylated) regions in the *SNCA* and *PARK16* loci [41].

In addition, summary-data-based Mendelian randomization (SMR) is an approach that combines summary-level GWAS data with DNA methylation (mQTL) or gene expression (eQTL) data to identify genes with methylation or expression levels that are either pleiotropically or causally associated with a particular trait [184]. SMR works by triangulating SNP association statistics with the phenotypic trait on SNP association statistics with methylation/expression to provide an estimate of the effect of a genetically determined increase in DNA methylation or gene expression on the disease phenotype [184,185]. The 2019 PD GWAS meta-analysis by Nalls et al. used SMR and found a significant association between the methylation or expression of 151 genes (around PD risk variants) and a potential causal change in PD risk [22].

Finally, multiomic approaches can be used to aid in the identification of genes involved in PD and provide more insights into the roles of specific cell types. A recent study combining chromatin accessibility (snATAC-seq), transcriptomic (snRNA-seq), histone modification (bulk H3K27ac ChIP-seq) and GWAS data investigated cell type-specific disruptions in the PD substantia nigra [39]. A total of 2770 downregulated and 2910 upregulated cis-regulatory elements (cREs) were identified in this brain region in PD. Integration of the dysregulated cREs and PD GWAS SNPs led to the identification of 656 target genes (potential PD candidate genes) in specific cell types (including dopaminergic neurons, GABAergic neurons, oligodendrocytes, oligodendrocyte precursor cells, astrocytes, microglia, endothelial cells and pericytes). Over 50% of these target genes were assigned to only one or two cell types, indicating a high degree of cell type specificity. Oligodendrocyte and microglial cREs were strongly enriched in more than one PD GWAS, and the authors suggest that these are the two key cell types associated with the disease (also supported by RNA-seq data). Active regulatory associations involving known PD genes were annotated to different cell types. For example, certain PD GWAS SNPs were specific to dopaminergic neurons, *SNCA* was involved in most cell types, and *MAPT* was mainly associated with oligodendrocytes, OPCs and astrocytes. In addition, it was found that transcription factor binding is disrupted by PD GWAS variants in cis-regulatory regions. The authors conclude that common risk variants exhibit cell type-specific regulatory functions and that the combination of both genetic predisposition and epigenetic alterations contributes to PD-related cellular processes and pathogenesis.

This study highlights the potential advantages of integrating genetic, epigenetic and transcriptomic information to understand more about the roles of PD risk variants in particular cell types. Future multiomic studies should be performed, perhaps in different brain regions, in addition to incorporating DNA (hydroxy)methylation and proteomic data to build a more comprehensive understanding of PD risk and mechanisms, including cell type-specific changes.

6.3. *Epigenetic Therapies*

An understanding of epigenetics in Parkinson's could lead to new drug targets and potential novel therapeutic approaches. Such therapies could take advantage of the fact that epigenetic changes are reversible, and thus, it might be plausible to correct the altered epigenetic modifications in PD.

Research in cancer shows that therapies based on correcting DNA methylation or histone modification changes are possible. Here, several drugs are in clinical trials, and a few are FDA-approved. This includes DNMT inhibitors such as azacitidine, approved for the treatment of myelodysplastic syndrome and acute myeloid leukemia, as well as HDAC inhibitors such as panobinostat, a third-line treatment for multiple myeloma [186].

A number of studies in cell and animal models of PD have found that HDAC inhibitors such as sodium butyrate and valproic acid (both increasing acetylation) can be neuroprotective, reduce α-synuclein levels, increase autophagy, restore striatal dopamine levels and attenuate motor and non-motor changes [187–198]. However, at present, it is unclear whether further increasing histone acetylation would be beneficial in individuals with PD, as the studies to date have predominantly reported the presence of widespread hyperacetylation in PD brains.

An alternative and more specific approach may be to module the expression of genes that are epigenetically dysregulated in PD. For example, short hairpin RNA mini-circles targeting α-synuclein were delivered using exosomes into a PD mouse model, which reduced α-synuclein aggregation and dopaminergic neuron loss and improved disease symptoms [199]. CRISPR/Cas9-based epigenome editing technologies can also be used here. A recent study trialed a lentiviral vector-based therapy containing a deactivated form of CRISPR/Cas9 which targeted *SNCA* intron 1 specifically in dopaminergic and cholinergic neurons [200]. Introducing the therapy into iPSC-derived dopaminergic and cholinergic neurons from a *SNCA* triplication patient led to the downregulation of *SNCA* mRNA and protein in these specific cell types as well as increasing neuronal viability and mitochondrial function. In another example, the observed reduction in α-synuclein levels following the demethylation of H3K4 at the *SNCA* promoter using CRISPR/Cas9 [120] suggests that a possible future PD treatment avenue might involve region-specific methylation or demethylation of DNA or certain histones to reverse epigenetic changes in the disease.

Furthermore, the expression of miRNAs can be modulated by either miRNA mimics or antagomirs. For example, Hu et al. observed a decrease in miR-425 expression in the human substantia nigra, while injection of miR-425 mimics into the substantia nigra of MPTP-treated mice attenuated necroptosis activation, protected against dopaminergic neuron loss, increased striatal dopamine levels and reduced deficits in motor activity [159]. In another example, Zhou et al. found that miR-103a-3p was predicted to bind parkin and was upregulated in both MPP+-induced SH-SY5Y cells and MPTP-induced mouse models [201]. This miRNA was shown in another study to be upregulated in the blood of L-DOPA-treated PD patients [144]. Zhou et al. observed that transfection of a miR-103a-3p antagomir into both the cell model and striatum of the mouse model led to the upregulation of parkin and improved mitophagy [201]. Additionally, it resulted in neuroprotection in the mouse model, preventing the loss of TH-positive cells in the substantia nigra. This suggests that modulating the levels of these miRNAs described may be beneficial in PD. However, finding an appropriate miRNA to target may be difficult as a single miRNA can have multiple mRNA targets; therefore, downstream and off-target effects need to be considered [202].

To sum up, epigenetic-based therapies are an exciting prospect for PD and other neurodegenerative disorders. However, this is a rather long way off, with further advances in mechanistic understanding required and some challenges needing to be overcome.

6.4. Future Directions

There are a number of avenues that could be explored in the near future to further our understanding of PD epigenetics. Firstly, larger-scale EWASs should be performed to both verify previously found and identify novel epigenetic changes in PD, in particular isolating different brain cell types and analyzing cell type-specific epigenetic changes. Secondly, future work should also investigate the interplay between genetics and epigenetics in PD. In particular, multiomic approaches might help shed light on the pathogenesis, for example, by identifying biologically relevant genes and cell types as well as potential therapeutic targets. Additionally, the concurrent screening of brain samples for genetic and epigenetic alterations could help determine whether the presence of epigenetic modifications combined with genetic mutations increases disease risk. Furthermore, future work should involve stratifying individuals by genetics to compare epigenetic profiles between different genetic subtypes of the disease. Thirdly, human postmortem brain studies are the most direct method for studying epigenetic modifications in neurodegenerative diseases. However, in postmortem tissue, it is difficult to establish whether epigenetic changes play a causal role in the disease. Studies comparing epigenetic alterations in PD samples of different Braak stages, including very early-stage individuals, may help address this question. In addition, this could allow the study of epigenetic dysregulation over the course of disease progression. Finally, continued experiments in animal models and iPSC-derived cells are also crucial, the latter of which more closely resemble affected individuals. Therefore, functional studies in cell and animal models should help investigate (1) whether epigenetic dysregulation is a cause or consequence of PD pathology and (2) whether epigenetic alterations can be targeted for novel therapeutic approaches. Furthermore, CRISPR/Cas9-based epigenome editing technologies could be utilized to introduce or remove a particular epigenetic modification at specific sites in the genome in order to evaluate its exact biological consequences.

Funding: This research was supported by the Wellcome Trust Seed Award in Science, grant number 217561/Z/19/Z. AK is funded by the GW4 BioMed MRC Doctoral Training Partnership.

Acknowledgments: The authors thank Jonathan Mill (University of Exeter) and Eilis Hannon (University of Exeter) for their scientific input. For the purpose of open access, the author has applied a CC BY public copyright license to any Author Accepted Manuscript version arising from this submission. This study did not generate any new data.

Conflicts of Interest: The authors declare no conflicts of interest.

References

1. Parkinson's Foundation Statistics | Parkinson's Foundation. Available online: https://www.parkinson.org/Understanding-Parkinsons/Statistics (accessed on 12 March 2024).
2. Parkinson's UK. *The Incidence and Prevalence of Parkinson's in the UK*; Parkinson's UK: London, UK, 2017.
3. Dorsey, E.R.; Sherer, T.; Okun, M.S.; Bloemd, B.R. The Emerging Evidence of the Parkinson Pandemic. *J. Park. Dis.* **2018**, *8*, S3–S8. [CrossRef]
4. Yang, W.; Hamilton, J.L.; Kopil, C.; Beck, J.C.; Tanner, C.M.; Albin, R.L.; Ray Dorsey, E.; Dahodwala, N.; Cintina, I.; Hogan, P.; et al. Current and Projected Future Economic Burden of Parkinson's Disease in the U.S. *NPJ Park. Dis.* **2020**, *6*, 15. [CrossRef]
5. Fahn, S. Description of Parkinson's Disease as a Clinical Syndrome. In Proceedings of the Annals of the New York Academy of Sciences. *N. Y. Acad. Sci.* **2003**, *991*, 1–14. [CrossRef]
6. Pont-Sunyer, C.; Hotter, A.; Gaig, C.; Seppi, K.; Compta, Y.; Katzenschlager, R.; Mas, N.; Hofeneder, D.; Brücke, T.; Bayés, A.; et al. The Onset of Nonmotor Symptoms in Parkinson's Disease (the Onset Pd Study). *Mov. Disord.* **2015**, *30*, 229–237. [CrossRef]
7. Poewe, W. Non-Motor Symptoms in Parkinson's Disease. *Eur. J. Neurol.* **2008**, *15*, 14–20. [CrossRef]
8. Hobson, P.; Meara, J. Mortality and Quality of Death Certification in a Cohort of Patients with Parkinson's Disease and Matched Controls in North Wales, UK at 18 Years: A Community-Based Cohort Study. *BMJ Open* **2018**, *8*, 18969. [CrossRef]
9. Dickson, D.W. Parkinson's Disease and Parkinsonism: Neuropathology. *Cold Spring Harb. Perspect. Med.* **2012**, *2*, a009258. [CrossRef]
10. Yasuda, T.; Nakata, Y.; Mochizuki, H. α-Synuclein and Neuronal Cell Death. *Mol. Neurobiol.* **2013**, *47*, 466–483. [CrossRef]
11. Luk, K.C.; Lee, V.M.Y. Modeling Lewy Pathology Propagation in Parkinson's Disease. *Park. Relat. Disord.* **2014**, *20*, S85–S87. [CrossRef]
12. Ma, J.; Gao, J.; Wang, J.; Xie, A. Prion-like Mechanisms in Parkinson's Disease. *Front. Neurosci.* **2019**, *13*, 552. [CrossRef]
13. Stefanoni, G.; Sala, G.; Tremolizzo, L.; Brighina, L.; Ferrarese, C. Role of Autophagy in Parkinson's Disease. In *Autophagy: Principles, Regulation and Roles in Disease;* Nova Science Publishers, Inc.: Hauppauge, NY, USA, 2012; Volume 26, pp. 243–264, ISBN 9781619422667.
14. Puspita, L.; Chung, S.Y.; Shim, J.W. Oxidative Stress and Cellular Pathologies in Parkinson's Disease. *Mol. Brain* **2017**, *10*, 53. [CrossRef]
15. Bose, A.; Beal, M.F. Mitochondrial Dysfunction in Parkinson's Disease. *J. Neurochem.* **2016**, *139* (Suppl. 1), 216–231. [CrossRef] [PubMed]
16. De Virgilio, A.; Greco, A.; Fabbrini, G.; Inghilleri, M.; Rizzo, M.I.; Gallo, A.; Conte, M.; Rosato, C.; Ciniglio Appiani, M.; de Vincentiis, M. Parkinson's Disease: Autoimmunity and Neuroinflammation. *Autoimmun. Rev.* **2016**, *15*, 1005–1011. [CrossRef]
17. Goldman, S.M. Environmental Toxins and Parkinson's Disease. *Annu. Rev. Pharmacol. Toxicol.* **2014**, *54*, 141–164. [CrossRef]
18. Raza, C.; Anjum, R.; Shakeel, N.U.A. Parkinson's Disease: Mechanisms, Translational Models and Management Strategies. *Life Sci.* **2019**, *226*, 77–90. [CrossRef]
19. Nardin, A.; Schrepfer, E.; Ziviani, E. Counteracting PINK/Parkin Deficiency in the Activation of Mitophagy: A Potential Therapeutic Intervention for Parkinson's Disease. *Curr. Neuropharmacol.* **2016**, *14*, 250–259. [CrossRef] [PubMed]
20. Verstraeten, A.; Theuns, J.; Van Broeckhoven, C. Progress in Unraveling the Genetic Etiology of Parkinson Disease in a Genomic Era. *Trends Genet.* **2015**, *31*, 140–149. [CrossRef] [PubMed]
21. Migdalska-Richards, A.; Schapira, A.H.V. The Relationship between Glucocerebrosidase Mutations and Parkinson Disease. *J. Neurochem.* **2016**, *139*, 77–90. [CrossRef]

22. Nalls, M.A.; Blauwendraat, C.; Vallerga, C.L.; Heilbron, K.; Bandres-Ciga, S.; Chang, D.; Tan, M.; Kia, D.A.; Noyce, A.J.; Xue, A.; et al. Identification of Novel Risk Loci, Causal Insights, and Heritable Risk for Parkinson's Disease: A Meta-Analysis of Genome-Wide Association Studies. *Lancet Neurol.* **2019**, *18*, 1091–1102. [CrossRef]
23. Smith, L.; Schapira, A.H.V. GBA Variants and Parkinson Disease: Mechanisms and Treatments. *Cells* **2022**, *11*, 1261. [CrossRef]
24. Balestrino, R.; Tunesi, S.; Tesei, S.; Lopiano, L.; Zecchinelli, A.L.; Goldwurm, S. Penetrance of Glucocerebrosidase (GBA) Mutations in Parkinson's Disease: A Kin Cohort Study. *Mov. Disord.* **2020**, *35*, 2111–2114. [CrossRef]
25. Anheim, M.; Elbaz, A.; Lesage, S.; Durr, A.; Condroyer, C.; Viallet, F.; Pollak, P.; Bonaïti, B.; Bonaïti-Pellié, C.; Brice, A.; et al. Penetrance of PD in Glucocerebrosidase Gene Mutation Carriers. *Neurology* **2012**, *79*, 106–107. [CrossRef]
26. McNeill, A.; Duran, R.; Hughes, D.A.; Mehta, A.; Schapira, A.H.V. A Clinical and Family History Study of Parkinson's Disease in Heterozygous Glucocerebrosidase Mutation Carriers. *J. Neurol. Neurosurg. Psychiatry* **2012**, *83*, 853–854. [CrossRef]
27. Menozzi, E.; Schapira, A.H.V. Exploring the Genotype–Phenotype Correlation in GBA-Parkinson Disease: Clinical Aspects, Biomarkers, and Potential Modifiers. *Front. Neurol.* **2021**, *12*, 694764. [CrossRef] [PubMed]
28. Kim, J.J.; Vitale, D.; Otani, D.V.; Lian, M.M.; Heilbron, K.; Aslibekyan, S.; Auton, A.; Babalola, E.; Bell, R.K.; Bielenberg, J.; et al. Multi-Ancestry Genome-Wide Association Meta-Analysis of Parkinson's Disease. *Nat. Genet.* **2023**, *56*, 27–36. [CrossRef]
29. Maurano, M.T.; Humbert, R.; Rynes, E.; Thurman, R.E.; Haugen, E.; Wang, H.; Reynolds, A.P.; Sandstrom, R.; Qu, H.; Brody, J.; et al. Systematic Localization of Common Disease-Associated Variation in Regulatory DNA. *Science* **2012**, *337*, 1190. [CrossRef] [PubMed]
30. Goldman, S.M.; Marek, K.; Ottman, R.; Meng, C.; Comyns, K.; Chan, P.; Ma, J.; Marras, C.; Langston, J.W.; Ross, G.W.; et al. Concordance for Parkinson's Disease in Twins: A 20-Year Update. *Ann. Neurol.* **2019**, *85*, 600–605. [CrossRef] [PubMed]
31. Li, Y. Modern Epigenetics Methods in Biological Research. *Methods* **2021**, *187*, 104. [CrossRef]
32. Zhang, L.; Lu, Q.; Chang, C. Epigenetics in Health and Disease. *Adv. Exp. Med. Biol.* **2020**, *1253*, 3–55. [CrossRef]
33. Cavalli, G.; Heard, E. Advances in Epigenetics Link Genetics to the Environment and Disease. *Nature* **2019**, *571*, 489–499. [CrossRef]
34. Hwang, J.Y.; Aromolaran, K.A.; Zukin, R.S. The Emerging Field of Epigenetics in Neurodegeneration and Neuroprotection. *Nat. Rev. Neurosci.* **2017**, *18*, 347–361. [CrossRef]
35. Kim, S.; Kaang, B.K. Epigenetic Regulation and Chromatin Remodeling in Learning and Memory. *Exp. Mol. Med.* **2017**, *49*, e281. [CrossRef]
36. Nebbioso, A.; Tambaro, F.P.; Dell'Aversana, C.; Altucci, L. Cancer Epigenetics: Moving Forward. *PLoS Genet.* **2018**, *14*, e1007362. [CrossRef]
37. Smith, R.G.; Pishva, E.; Shireby, G.; Smith, A.R.; Roubroeks, J.A.Y.; Hannon, E.; Wheildon, G.; Mastroeni, D.; Gasparoni, G.; Riemenschneider, M.; et al. A Meta-Analysis of Epigenome-Wide Association Studies in Alzheimer's Disease Highlights Novel Differentially Methylated Loci across Cortex. *Nat. Commun.* **2021**, *12*, 3517. [CrossRef]
38. Pihlstrøm, L.; Berge, V.; Rengmark, A.; Toft, M. Parkinson's Disease Correlates with Promoter Methylation in the α-Synuclein Gene. *Mov. Disord.* **2015**, *30*, 577–580. [CrossRef]
39. Lee, A.J.; Kim, C.; Park, S.; Joo, J.; Choi, B.; Yang, D.; Jun, K.; Eom, J.; Lee, S.J.; Chung, S.J.; et al. Characterization of Altered Molecular Mechanisms in Parkinson's Disease through Cell Type–Resolved Multiomics Analyses. *Sci. Adv.* **2023**, *9*, 15–18. [CrossRef]
40. Sharma, A.; Osato, N.; Liu, H.; Asthana, S.; Dakal, T.C.; Ambrosini, G.; Bucher, P.; Schmitt, I.; Wüllner, U. Common Genetic Variants Associated with Parkinson's Disease Display Widespread Signature of Epigenetic Plasticity. *Sci. Rep.* **2019**, *9*, 18464. [CrossRef]
41. Vermunt, M.W.; Reinink, P.; Korving, J.; de Bruijn, E.; Creyghton, P.M.; Basak, O.; Geeven, G.; Toonen, P.W.; Lansu, N.; Meunier, C.; et al. Large-Scale Identification of Coregulated Enhancer Networks in the Adult Human Brain. *Cell Rep.* **2014**, *9*, 767–779. [CrossRef]
42. Jaenisch, R.; Bird, A. Epigenetic Regulation of Gene Expression: How the Genome Integrates Intrinsic and Environmental Signals. *Nat. Genet.* **2003**, *33*, 245–254. [CrossRef]
43. Feil, R.; Fraga, M.F. Epigenetics and the Environment: Emerging Patterns and Implications. *Nat. Rev. Genet.* **2012**, *13*, 97–109. [CrossRef]
44. Babenko, O.; Golubov, A.; Ilnytskyy, Y.; Kovalchuk, I.; Metz, G.A. Genomic and Epigenomic Responses to Chronic Stress Involve MiRNA-Mediated Programming. *PLoS ONE* **2012**, *7*, e29441. [CrossRef]
45. Thomas, E.A. DNA Methylation in Huntington's Disease: Implications for Transgenerational Effects. *Neurosci. Lett.* **2016**, *625*, 34–39. [CrossRef]
46. Park, C.; Rosenblat, J.D.; Brietzke, E.; Pan, Z.; Lee, Y.; Cao, B.; Zuckerman, H.; Kalantarova, A.; McIntyre, R.S. Stress, Epigenetics and Depression: A Systematic Review. *Neurosci. Biobehav. Rev.* **2019**, *102*, 139–152. [CrossRef]
47. Bellou, V.; Belbasis, L.; Tzoulaki, I.; Evangelou, E.; Ioannidis, J.P.A. Environmental Risk Factors and Parkinson's Disease: An Umbrella Review of Meta-Analyses. *Park. Relat. Disord.* **2016**, *23*, 1–9. [CrossRef]
48. Fraga, M.F.; Ballestar, E.; Paz, M.F.; Ropero, S.; Setien, F.; Ballestar, M.L.; Heine-Suñer, D.; Cigudosa, J.C.; Urioste, M.; Benitez, J.; et al. Epigenetic Differences Arise during the Lifetime of Monozygotic Twins. *Proc. Natl. Acad. Sci. USA* **2005**, *102*, 10604–10609. [CrossRef]

49. Talens, R.P.; Christensen, K.; Putter, H.; Willemsen, G.; Christiansen, L.; Kremer, D.; Suchiman, H.E.D.; Slagboom, P.E.; Boomsma, D.I.; Heijmans, B.T. Epigenetic Variation during the Adult Lifespan: Cross-Sectional and Longitudinal Data on Monozygotic Twin Pairs. *Aging Cell* **2012**, *11*, 694. [CrossRef]
50. Reynolds, C.A.; Tan, Q.; Munoz, E.; Jylhävä, J.; Hjelmborg, J.; Christiansen, L.; Hägg, S.; Pedersen, N.L. A Decade of Epigenetic Change in Aging Twins: Genetic and Environmental Contributions to Longitudinal DNA Methylation. *Aging Cell* **2020**, *19*, e13197. [CrossRef]
51. Oertel, W.; Schulz, J.B. Current and Experimental Treatments of Parkinson Disease: A Guide for Neuroscientists. *J. Neurochem.* **2016**, *139* (Suppl. 1), 325–337. [CrossRef]
52. Poewe, W.; Antonini, A.; Zijlmans, J.C.; Burkhard, P.R.; Vingerhoets, F. Levodopa in the Treatment of Parkinson's Disease: An Old Drug Still Going Strong. *Clin. Interv. Aging* **2010**, *5*, 229–238.
53. Laird, P.W. Principles and Challenges of Genome-Wide DNA Methylation Analysis. *Nat. Rev. Genet.* **2010**, *11*, 191–203. [CrossRef]
54. Mattei, A.L.; Bailly, N.; Meissner, A. DNA Methylation: A Historical Perspective. *Trends Genet.* **2022**, *38*, 676–707. [CrossRef]
55. Tate, P.H.; Bird, A.P. Effects of DNA Methylation on DNA-Binding Proteins and Gene Expression. *Curr. Opin. Genet. Dev.* **1993**, *3*, 226–231. [CrossRef]
56. Nan, X.; Campoy, F.J.; Bird, A. MeCP2 Is a Transcriptional Repressor with Abundant Binding Sites in Genomic Chromatin. *Cell* **1997**, *88*, 471–481. [CrossRef]
57. Watt, F.; Molloy, P.L. Cytosine Methylation Prevents Binding to DNA of a HeLa Cell Transcription Factor Required for Optimal Expression of the Adenovirus Major Late Promoter. *Genes. Dev.* **1988**, *2*, 1136–1143. [CrossRef]
58. Yang, X.; Han, H.; DeCarvalho, D.D.; Lay, F.D.; Jones, P.A.; Liang, G. Gene Body Methylation Can Alter Gene Expression and Is a Therapeutic Target in Cancer. *Cancer Cell* **2014**, *26*, 577–590. [CrossRef]
59. Wang, Q.; Xiong, F.; Wu, G.; Liu, W.; Chen, J.; Wang, B.; Chen, Y. Gene Body Methylation in Cancer: Molecular Mechanisms and Clinical Applications. *Clin. Epigenetics* **2022**, *14*, 154. [CrossRef]
60. Ball, M.P.; Li, J.B.; Gao, Y.; Lee, J.H.; Leproust, E.M.; Park, I.H.; Xie, B.; Daley, G.Q.; Church, G.M. Targeted and Genome-Scale Strategies Reveal Gene-Body Methylation Signatures in Human Cells. *Nat. Biotechnol.* **2009**, *27*, 361–368. [CrossRef]
61. Tost, J.; Gut, I.G. DNA Methylation Analysis by Pyrosequencing. *Nat. Protoc.* **2007**, *2*, 2265–2275. [CrossRef]
62. Polymeropoulos, M.H.; Lavedan, C.; Leroy, E.; Ide, S.E.; Dehejia, A.; Dutra, A.; Pike, B.; Root, H.; Rubenstein, J.; Boyer, R.; et al. Mutation in the α-Synuclein Gene Identified in Families with Parkinson's Disease. *Science* **1997**, *276*, 2045–2047. [CrossRef]
63. Krüger, R.; Kuhn, W.; Müller, T.; Woitalla, D.; Graeber, M.; Kösel, S.; Przuntek, H.; Epplen, J.T.; Schöls, L.; Riess, O. Ala30Pro Mutation in the Gene Encoding α-Synuclein in Parkinson's Disease. *Nat. Genet.* **1998**, *18*, 106–108. [CrossRef]
64. Zarranz, J.J.; Alegre, J.; Gómez-Esteban, J.C.; Lezcano, E.; Ros, R.; Ampuero, I.; Vidal, L.; Hoenicka, J.; Rodriguez, O.; Atarés, B.; et al. The New Mutation, E46K, of α-Synuclein Causes Parkinson and Lewy Body Dementia. *Ann. Neurol.* **2004**, *55*, 164–173. [CrossRef] [PubMed]
65. Chartier-Harlin, M.C.; Kachergus, J.; Roumier, C.; Mouroux, V.; Douay, X.; Lincoln, S.; Levecque, C.; Larvor, L.; Andrieux, J.; Hulihan, M.; et al. α-Synuclein Locus Duplication as a Cause of Familial Parkinson's Disease. *Lancet* **2004**, *364*, 1167–1169. [CrossRef] [PubMed]
66. Singleton, A.B.; Farrer, M.; Johnson, J.; Singleton, A.; Hague, S.; Kachergus, J.; Hulihan, M.; Peuralinna, T.; Dutra, A.; Nussbaum, R.; et al. α-Synuclein Locus Triplication Causes Parkinson's Disease. *Science* **2003**, *302*, 841. [CrossRef] [PubMed]
67. Jowaed, A.; Schmitt, I.; Kaut, O.; Wüllner, U. Methylation Regulates Alpha-Synuclein Expression and Is Decreased in Parkinson's Disease Patients' Brains. *J. Neurosci.* **2010**, *30*, 6355–6359. [CrossRef] [PubMed]
68. Matsumoto, L.; Takuma, H.; Tamaoka, A.; Kurisaki, H.; Date, H.; Tsuji, S.; Iwata, A. CpG Demethylation Enhances Alpha-Synuclein Expression and Affects the Pathogenesis of Parkinson's Disease. *PLoS ONE* **2010**, *5*, e15522. [CrossRef] [PubMed]
69. Desplats, P.; Spencer, B.; Coffee, E.; Patel, P.; Michael, S.; Patrick, C.; Adame, A.; Rockenstein, E.; Masliah, E. α-Synuclein Sequesters Dnmt1 from the Nucleus: A Novel Mechanism for Epigenetic Alterations in Lewy Body Diseases. *J. Biol. Chem.* **2011**, *286*, 9031–9037. [CrossRef] [PubMed]
70. Schmitt, I.; Evert, B.O.; Sharma, A.; Khazneh, H.; Murgatroyd, C.; Wüllner, U. The Alpha-Synuclein Gene (SNCA) Is a Genomic Target of Methyl-CpG Binding Protein 2 (MeCP2)—Implications for Parkinson's Disease and Rett Syndrome. *Mol. Neurobiol.* 2024, *online ahead of print*. [CrossRef]
71. Guhathakurta, S.; Evangelista, B.A.; Ghosh, S.; Basu, S.; Kim, Y.S. Hypomethylation of Intron1 of α-Synuclein Gene Does Not Correlate with Parkinson's Disease. *Mol. Brain* **2017**, *10*, 6. [CrossRef] [PubMed]
72. Gu, J.; Barrera, J.; Yun, Y.; Murphy, S.K.; Beach, T.G.; Woltjer, R.L.; Serrano, G.E.; Kantor, B.; Chiba-Falek, O. Cell-Type Specific Changes in DNA Methylation of SNCA Intron 1 in Synucleinopathy Brains. *Front. Neurosci.* **2021**, *15*, 652226. [CrossRef] [PubMed]
73. Smith, A.R.; Richards, D.M.; Lunnon, K.; Schapira, A.H.V.; Migdalska-Richards, A. DNA Methylation of α-Synuclein Intron 1 Is Significantly Decreased in the Frontal Cortex of Parkinson's Individuals with GBA1 Mutations. *Int. J. Mol. Sci.* **2023**, *24*, 2687. [CrossRef]
74. De Mena, L.; Cardo, L.F.; Coto, E.; Alvarez, V. No Differential DNA Methylation of PARK2 in Brain of Parkinson's Disease Patients and Healthy Controls. *Mov. Disord.* **2013**, *28*, 2032–2033. [CrossRef]
75. Navarro-Sánchez, L.; Águeda-Gómez, B.; Aparicio, S.; Pérez-Tur, J. Epigenetic Study in Parkinson's Disease: A Pilot Analysis of DNA Methylation in Candidate Genes in Brain. *Cells* **2018**, *7*, 150. [CrossRef] [PubMed]

76. Barrachina, M.; Ferrer, I. DNA Methylation of Alzheimer Disease and Tauopathy-Related Genes in Postmortem Brain. *J. Neuropathol. Exp. Neurol.* **2009**, *68*, 880–891. [CrossRef] [PubMed]
77. Coupland, K.G.; Mellick, G.D.; Silburn, P.A.; Mather, K.; Armstrong, N.J.; Sachdev, P.S.; Brodaty, H.; Huang, Y.; Halliday, G.M.; Hallupp, M.; et al. DNA Methylation of the *MAPT* Gene in Parkinson's Disease Cohorts and Modulation by Vitamin E *In Vitro*. *Mov. Disord.* **2014**, *29*, 1606–1614. [CrossRef] [PubMed]
78. Su, X.; Chu, Y.; Kordower, J.H.; Li, B.; Cao, H.; Huang, L.; Nishida, M.; Song, L.; Wang, D.; Federoff, H.J. PGC-1α Promoter Methylation in Parkinson's Disease. *PLoS ONE* **2015**, *10*, e0134087. [CrossRef] [PubMed]
79. Zheng, L.; Bernard-Marissal, N.; Moullan, N.; D'Amico, D.; Auwerx, J.; Moore, D.J.; Knott, G.; Aebischer, P.; Schneider, B.L. Parkin Functionally Interacts with PGC-1α to Preserve Mitochondria and Protect Dopaminergic Neurons. *Hum. Mol. Genet.* **2017**, *26*, 582–598. [CrossRef] [PubMed]
80. St-Pierre, J.; Drori, S.; Uldry, M.; Silvaggi, J.M.; Rhee, J.; Jäger, S.; Handschin, C.; Zheng, K.; Lin, J.; Yang, W.; et al. Suppression of Reactive Oxygen Species and Neurodegeneration by the PGC-1 Transcriptional Coactivators. *Cell* **2006**, *127*, 397–408. [CrossRef] [PubMed]
81. Mastronardi, F.G.; Noor, A.; Wood, D.D.; Paton, T.; Moscarello, M.A. Peptidyl Argininedeiminase 2 CpG Island in Multiple Sclerosis White Matter Is Hypomethylated. *J. Neurosci. Res.* **2007**, *85*, 2006–2016. [CrossRef] [PubMed]
82. Pieper, H.C.; Evert, B.O.; Kaut, O.; Riederer, P.F.; Waha, A.; Wüllner, U. Different Methylation of the TNF-Alpha Promoter in Cortex and Substantia Nigra: Implications for Selective Neuronal Vulnerability. *Neurobiol. Dis.* **2008**, *32*, 521–527. [CrossRef]
83. Villar-Menéndez, I.; Porta, S.; Buira, S.P.; Pereira-Veiga, T.; Díaz-Sánchez, S.; Albasanz, J.L.; Ferrer, I.; Martín, M.; Barrachina, M. Increased Striatal Adenosine A2A Receptor Levels Is an Early Event in Parkinson's Disease-Related Pathology and It Is Potentially Regulated by MiR-34b. *Neurobiol. Dis.* **2014**, *69*, 206–214. [CrossRef]
84. Blanch, M.; Mosquera, J.L.; Ansoleaga, B.; Ferrer, I.; Barrachina, M. Altered Mitochondrial DNA Methylation Pattern in Alzheimer Disease-Related Pathology and in Parkinson Disease. *Am. J. Pathol.* **2016**, *186*, 385–397. [CrossRef]
85. Kaut, O.; Schmitt, I.; Stahl, F.; Fröhlich, H.; Hoffmann, P.; Gonzalez, F.J.; Wüllner, U. Epigenome-Wide Analysis of DNA Methylation in Parkinson's Disease Cortex. *Life* **2022**, *12*, 502. [CrossRef] [PubMed]
86. Kaut, O.; Schmitt, I.; Wüllner, U. Genome-Scale Methylation Analysis of Parkinson's Disease Patients' Brains Reveals DNA Hypomethylation and Increased MRNA Expression of Cytochrome P450 2E1. *Neurogenetics* **2012**, *13*, 87–91. [CrossRef]
87. Masliah, E.; Dumaop, W.; Galasko, D.; Desplats, P. Distinctive Patterns of DNA Methylation Associated with Parkinson Disease: Identification of Concordant Epigenetic Changes in Brain and Peripheral Blood Leukocytes. *Epigenetics* **2013**, *8*, 1030–1038. [CrossRef] [PubMed]
88. Go, R.C.P.; Corley, M.J.; Ross, G.W.; Petrovitch, H.; Masaki, K.H.; Maunakea, A.K.; He, Q.; Tiirikainen, M.I. Genome-Wide Epigenetic Analyses in Japanese Immigrant Plantation Workers with Parkinson's Disease and Exposure to Organochlorines Reveal Possible Involvement of Glial Genes and Pathways Involved in Neurotoxicity. *BMC Neurosci.* **2020**, *21*, 31. [CrossRef] [PubMed]
89. Costello, S.; Cockburn, M.; Bronstein, J.; Zhang, X.; Ritz, B. Parkinson's Disease and Residential Exposure to Maneb and Paraquat From Agricultural Applications in the Central Valley of California. *Am. J. Epidemiol.* **2009**, *169*, 919. [CrossRef]
90. Sinha, D.; D'Silva, P. Chaperoning Mitochondrial Permeability Transition: Regulation of Transition Pore Complex by a J-Protein, DnaJC15. *Cell Death Dis.* **2014**, *5*, e1101. [CrossRef] [PubMed]
91. Kochmanski, J.; Kuhn, N.C.; Bernstein, A.I. Parkinson's Disease-Associated, Sex-Specific Changes in DNA Methylation at PARK7 (DJ-1), SLC17A6 (VGLUT2), PTPRN2 (IA-2β), and NR4A2 (NURR1)in Cortical Neurons. *NPJ Park. Dis.* **2022**, *8*, 120. [CrossRef] [PubMed]
92. Shendelman, S.; Jonason, A.; Martinat, C.; Leete, T.; Abeliovich, A. DJ-1 Is a Redox-Dependent Molecular Chaperone That Inhibits α-Synuclein Aggregate Formation. *PLoS Biol.* **2004**, *2*, 308. [CrossRef] [PubMed]
93. Marshall, L.L.; Killinger, B.A.; Ensink, E.; Li, P.; Li, K.X.; Cui, W.; Lubben, N.; Weiland, M.; Wang, X.; Gordevicius, J.; et al. Epigenomic Analysis of Parkinson's Disease Neurons Identifies Tet2 Loss as Neuroprotective. *Nat. Neurosci.* **2020**, *23*, 1203–1214. [CrossRef]
94. Li, P.; Ensink, E.; Lang, S.; Marshall, L.; Schilthuis, M.; Lamp, J.; Vega, I.; Labrie, V. Hemispheric Asymmetry in the Human Brain and in Parkinson's Disease Is Linked to Divergent Epigenetic Patterns in Neurons. *Genome Biol.* **2020**, *21*, 61. [CrossRef]
95. Gordevicius, J.; Li, P.; Marshall, L.L.; Killinger, B.A.; Lang, S.; Ensink, E.; Kuhn, N.C.; Cui, W.; Maroof, N.; Lauria, R.; et al. Epigenetic Inactivation of the Autophagy–Lysosomal System in Appendix in Parkinson's Disease. *Nat. Commun.* **2021**, *12*, 5134. [CrossRef] [PubMed]
96. Dashtipour, K.; Tafreshi, A.; Adler, C.; Beach, T.; Chen, X.; Serrano, G.; Tashiro, S.; Wang, C. Hypermethylation of Synphilin-1, Alpha-Synuclein-Interacting Protein (SNCAIP) Gene in the Cerebral Cortex of Patients with Sporadic Parkinson's Disease. *Brain Sci.* **2017**, *7*, 74. [CrossRef] [PubMed]
97. Alvarez-Castelao, B.; Castaño, J.G. Synphilin-1 Inhibits Alpha-Synuclein Degradation by the Proteasome. *Cell. Mol. Life Sci.* **2011**, *68*, 2643–2654. [CrossRef] [PubMed]
98. Young, J.I.; Sivasankaran, S.K.; Wang, L.; Ali, A.; Mehta, A.; Davis, D.A.; Dykxhoorn, D.M.; Petito, C.K.; Beecham, G.W.; Martin, E.R.; et al. Genome-Wide Brain DNA Methylation Analysis Suggests Epigenetic Reprogramming in Parkinson Disease. *Neurol. Genet.* **2019**, *5*, e342. [CrossRef] [PubMed]

99. Stafa, K.; Trancikova, A.; Webber, P.J.; Glauser, L.; West, A.B.; Moore, D.J. GTPase Activity and Neuronal Toxicity of Parkinson's Disease-Associated LRRK2 Is Regulated by ArfGAP1. *PLoS Genet.* **2012**, *8*, e1002526. [CrossRef] [PubMed]
100. Xiong, Y.; Yuan, C.; Chen, R.; Dawson, T.M.; Dawson, V.L. ArfGAP1 Is a GTPase Activating Protein for LRRK2: Reciprocal Regulation of ArfGAP1 by LRRK2. *J. Neurosci.* **2012**, *32*, 3877–3886. [CrossRef]
101. Vishweswaraiah, S.; Akyol, S.; Yilmaz, A.; Ugur, Z.; Gordevičius, J.; Oh, K.J.; Brundin, P.; Radhakrishna, U.; Labrie, V.; Graham, S.F. Methylated Cytochrome P450 and the Solute Carrier Family of Genes Correlate With Perturbations in Bile Acid Metabolism in Parkinson's Disease. *Front. Neurosci.* **2022**, *16*, 804261. [CrossRef] [PubMed]
102. Mansell, G.; Gorrie-Stone, T.J.; Bao, Y.; Kumari, M.; Schalkwyk, L.S.; Mill, J.; Hannon, E. Guidance for DNA Methylation Studies: Statistical Insights from the Illumina EPIC Array. *BMC Genom.* **2019**, *20*, 366.
103. International Parkinson's Disease Genomics Consortium; Wellcome Trust Case Control Consortium 2. A Two-Stage Meta-Analysis Identifies Several New Loci for Parkinson's Disease. *PLoS Genet.* **2011**, *7*, e1002142. [CrossRef]
104. Rawlik, K.; Rowlatt, A.; Tenesa, A. Imputation of DNA Methylation Levels in the Brain Implicates a Risk Factor for Parkinson's Disease. *Genetics* **2016**, *204*, 771–781. [CrossRef]
105. Guitton, R.; Dölle, C.; Alves, G.; Ole-Bjørn, T.; Nido, G.S.; Tzoulis, C. Ultra-Deep Whole Genome Bisulfite Sequencing Reveals a Single Methylation Hotspot in Human Brain Mitochondrial DNA. *Epigenetics* **2022**, *17*, 906–921. [CrossRef] [PubMed]
106. Xie, A.; Ensink, E.; Li, P.; Gordevi, J.; Marshall, L.L.; George, S.; Pospisilik, J.A.; Aho, V.T.E.; Houser, M.C.; Pereira, P.A.B.; et al. Bacterial Butyrate in Parkinson' s Disease Is Linked to Epigenetic Changes and Depressive Symptoms. *Mov. Disord.* **2022**, *37*, 1644–1653. [CrossRef] [PubMed]
107. Fisher, D.W.; Tulloch, J.; Yu, C.-E.; Tsuang, D. A Preliminary Comparison of the Methylome and Transcriptome from the Prefrontal Cortex Across Alzheimer's Disease and Lewy Body Dementia. *J. Alzheimers Dis. Rep.* **2023**, *7*, 279–297. [CrossRef] [PubMed]
108. Shen, L.; Song, C.X.; He, C.; Zhang, Y. Mechanism and Function of Oxidative Reversal of DNA and RNA Methylation. *Annu. Rev. Biochem.* **2014**, *83*, 585–614. [CrossRef] [PubMed]
109. Richa, R.; Sinha, R.P. Hydroxymethylation of DNA: An Epigenetic Marker. *EXCLI J.* **2014**, *13*, 592–610. [PubMed]
110. Wen, L.; Tang, F. Genomic Distribution and Possible Functions of DNA Hydroxymethylation in the Brain. *Genomics* **2014**, *104*, 341–346. [CrossRef] [PubMed]
111. Rustad, S.R.; Papale, L.A.; Alisch, R.S. DNA Methylation and Hydroxymethylation and Behavior. In *Behavioral Neurogenomics. Current Topics in Behavioral Neurosciences*; Binder, E.B., Klengel, T., Eds.; Springer International Publishing: Cham, Switzerland, 2019; pp. 51–82, ISBN 978-3-030-31265-7.
112. Stöger, R.; Scaife, P.J.; Shephard, F.; Chakrabarti, L. Elevated 5hmC Levels Characterize DNA of the Cerebellum in Parkinson's Disease. *NPJ Park. Dis.* **2017**, *3*, 6. [CrossRef] [PubMed]
113. Kaut, O.; Kuchelmeister, K.; Moehl, C.; Wüllner, U. 5-Methylcytosine and 5-Hydroxymethylcytosine in Brains of Patients with Multiple System Atrophy and Patients with Parkinson's Disease. *J. Chem. Neuroanat.* **2019**, *96*, 41–48. [CrossRef] [PubMed]
114. Min, S.; Xu, Q.; Qin, L.; Li, Y.; Li, Z.; Chen, C.; Wu, H.; Han, J.; Zhu, X.; Jin, P.; et al. Altered Hydroxymethylome in the Substantia Nigra of Parkinson's Disease. *Hum. Mol. Genet.* **2022**, *31*, 3494–3503. [CrossRef]
115. Bannister, A.J.; Kouzarides, T. Regulation of Chromatin by Histone Modifications. *Cell Res.* **2011**, *21*, 381–395. [CrossRef]
116. Millán-Zambrano, G.; Burton, A.; Bannister, A.J.; Schneider, R. Histone Post-Translational Modifications—Cause and Consequence of Genome Function. *Nat. Rev. Genet.* **2022**, *23*, 563–580. [CrossRef] [PubMed]
117. Gebremedhin, K.G.; Rademacher, D.J. Histone H3 Acetylation in the Postmortem Parkinson's Disease Primary Motor Cortex. *Neurosci. Lett.* **2016**, *627*, 121–125. [CrossRef] [PubMed]
118. Harrison, I.F.; Smith, A.D.; Dexter, D.T. Pathological Histone Acetylation in Parkinson's Disease: Neuroprotection and Inhibition of Microglial Activation through SIRT 2 Inhibition. *Neurosci. Lett.* **2018**, *666*, 48–57. [CrossRef] [PubMed]
119. Toker, L.; Tran, G.T.; Sundaresan, J.; Tysnes, O.B.; Alves, G.; Haugarvoll, K.; Nido, G.S.; Dölle, C.; Tzoulis, C. Genome-Wide Histone Acetylation Analysis Reveals Altered Transcriptional Regulation in the Parkinson's Disease Brain. *Mol. Neurodegener.* **2021**, *16*, 31. [CrossRef] [PubMed]
120. Guhathakurta, S.; Kim, J.; Adams, L.; Basu, S.; Song, M.K.; Adler, E.; Je, G.; Fiadeiro, M.B.; Kim, Y. Targeted Attenuation of Elevated Histone Marks at SNCA Alleviates Α-synuclein in Parkinson's Disease. *EMBO Mol. Med.* **2021**, *13*, e12188. [CrossRef]
121. Park, G.; Tan, J.; Garcia, G.; Kang, Y.; Salvesen, G.; Zhang, Z. Regulation of Histone Acetylation by Autophagy in Parkinson Disease. *J. Biol. Chem.* **2016**, *291*, 3531–3540. [CrossRef] [PubMed]
122. Esteller, M. Non-Coding RNAs in Human Disease. *Nat. Rev. Genet.* **2011**, *12*, 861–874. [CrossRef] [PubMed]
123. Zhang, P.; Wu, W.; Chen, Q.; Chen, M. Non-Coding RNAs and Their Integrated Networks. *J. Integr. Bioinform.* **2019**, *16*, 20190027. [CrossRef] [PubMed]
124. Liu, B.; Li, J.; Cairns, M.J. Identifying MiRNAs, Targets and Functions. *Brief. Bioinform.* **2014**, *15*, 1–19. [CrossRef]
125. Shang, R.; Lee, S.; Senavirathne, G.; Lai, E.C. MicroRNAs in Action: Biogenesis, Function and Regulation. *Nat. Rev. Genet.* **2023**, *24*, 816–833. [CrossRef]
126. Kim, J.; Inoue, K.; Ishii, J.; Vanti, W.B.; Voronov, S.V.; Murchison, E.; Hannon, G.; Abeliovich, A. A MicroRNA Feedback Circuit in Midbrain Dopamine Neurons. *Science* **2007**, *317*, 1220–1224. [CrossRef] [PubMed]
127. Kim, W.; Lee, Y.; McKenna, N.D.; Yi, M.; Simunovic, F.; Wang, Y.; Kong, B.; Rooney, R.J.; Seo, H.; Stephens, R.M.; et al. MiR-126 Contributes to Parkinson's Disease by Dysregulating the Insulin-like Growth Factor/Phosphoinositide 3-Kinase Signaling. *Neurobiol. Aging* **2014**, *35*, 1712–1721. [CrossRef] [PubMed]

128. Alvarez-Erviti, L.; Seow, Y.; Schapira, A.H.V.; Rodriguez-Oroz, M.C.; Obeso, J.A.; Cooper, J.M. Influence of MicroRNA Deregulation on Chaperone-Mediated Autophagy and α-Synuclein Pathology in Parkinson's Disease. *Cell Death Dis.* **2013**, *4*, e545. [CrossRef] [PubMed]
129. Miñones-Moyano, E.; Porta, S.; Escaramís, G.; Rabionet, R.; Iraola, S.; Kagerbauer, B.; Espinosa-Parrilla, Y.; Ferrer, I.; Estivill, X.; Martí, E. MicroRNA Profiling of Parkinson's Disease Brains Identifies Early Downregulation of MiR-34b/c Which Modulate Mitochondrial Function. *Hum. Mol. Genet.* **2011**, *20*, 3067–3078. [CrossRef] [PubMed]
130. Kabaria, S.; Choi, D.C.; Chaudhuri, A.D.; Mouradian, M.M.; Junn, E. Inhibition of MiR-34b and MiR-34c Enhances α-Synuclein Expression in Parkinson's Disease. *FEBS Lett.* **2015**, *589*, 319–325. [CrossRef] [PubMed]
131. Jin Cho, H.; Liu, G.; Min Jin, S.; Parisiadou, L.; Xie, C.; Yu, J.; Sun, L.; Ma, B.; Ding, J.; Vancraenenbroeck, R.; et al. MicroRNA-205 Regulates the Expression of Parkinson's Disease-Related Leucine-Rich Repeat Kinase 2 Protein. *Hum. Mol. Genet.* **2013**, *22*, 608–620. [CrossRef] [PubMed]
132. Hoss, A.G.; Labadorf, A.; Beach, T.G.; Latourelle, J.C.; Myers, R.H. MicroRNA Profiles in Parkinson's Disease Prefrontal Cortex. *Front. Aging Neurosci.* **2016**, *8*, 36. [CrossRef] [PubMed]
133. Siebert, M.; Westbroek, W.; Chen, Y.C.; Moaven, N.; Li, Y.; Velayati, A.; Luiza Saraiva-Pereira, M.; Martin, S.E.; Sidransky, E. Identification of MiRNAs That Modulate Glucocerebrosidase Activity in Gaucher Disease Cells. *RNA Biol.* **2014**, *11*, 1291–1300. [CrossRef] [PubMed]
134. Briggs, C.E.; Wang, Y.; Kong, B.; Woo, T.U.W.; Iyer, L.K.; Sonntag, K.C. Midbrain Dopamine Neurons in Parkinson's Disease Exhibit a Dysregulated MiRNA and Target-Gene Network. *Brain Res.* **2015**, *1618*, 111–121. [CrossRef]
135. Tatura, R.; Kraus, T.; Giese, A.; Arzberger, T.; Buchholz, M.; Höglinger, G.; Müller, U. Parkinson's Disease: SNCA-, PARK2-, and LRRK2- Targeting MicroRNAs Elevated in Cingulate Gyrus. *Park. Relat. Disord.* **2016**, *33*, 115–121. [CrossRef]
136. Chatterjee, P.; Roy, D. Comparative Analysis of RNA-Seq Data from Brain and Blood Samples of Parkinson's Disease. *Biochem. Biophys. Res. Commun.* **2017**, *484*, 557–564. [CrossRef] [PubMed]
137. Cardo, L.F.; Coto, E.; Ribacoba, R.; Menéndez, M.; Moris, G.; Suárez, E.; Alvarez, V. MiRNA Profile in the Substantia Nigra of Parkinson's Disease and Healthy Subjects. *J. Mol. Neurosci.* **2014**, *54*, 830–836. [CrossRef] [PubMed]
138. Nair, V.D.; Ge, Y. Alterations of MiRNAs Reveal a Dysregulated Molecular Regulatory Network in Parkinson's Disease Striatum. *Neurosci. Lett.* **2016**, *629*, 99–104. [CrossRef] [PubMed]
139. Gui, Y.X.; Liu, H.; Zhang, L.S.; Lv, W.; Hu, X.Y. Altered MicroRNA Profiles in Cerebrospinal Fluid Exosome in Parkinson Disease and Alzheimer Disease. *Oncotarget* **2015**, *6*, 37043–37053. [CrossRef] [PubMed]
140. Burgos, K.; Malenica, I.; Metpally, R.; Courtright, A.; Rakela, B.; Beach, T.; Shill, H.; Adler, C.; Sabbagh, M.; Villa, S.; et al. Profiles of Extracellular MiRNA in Cerebrospinal Fluid and Serum from Patients with Alzheimer's and Parkinson's Diseases Correlate with Disease Status and Features of Pathology. *PLoS ONE* **2014**, *9*, e94839. [CrossRef] [PubMed]
141. Martins, M.; Rosa, A.; Guedes, L.C.; Fonseca, B.V.; Gotovac, K.; Violante, S.; Mestre, T.; Coelho, M.; Rosa, M.M.; Martin, E.R.; et al. Convergence of MiRNA Expression Profiling, α-Synuclein Interacton and GWAS in Parkinson's Disease. *PLoS ONE* **2011**, *6*, e25443. [CrossRef] [PubMed]
142. Chen, L.; Yang, J.; Lü, J.; Cao, S.; Zhao, Q.; Yu, Z. Identification of Aberrant Circulating MiRNAs in Parkinson's Disease Plasma Samples. *Brain Behav.* **2018**, *8*, e00941. [CrossRef] [PubMed]
143. Botta-Orfila, T.; Morató, X.; Compta, Y.; Lozano, J.J.; Falgàs, N.; Valldeoriola, F.; Pont-Sunyer, C.; Vilas, D.; Mengual, L.; Fernández, M.; et al. Identification of Blood Serum Micro-RNAs Associated with Idiopathic and *LRRK2* Parkinson's Disease. *J. Neurosci. Res.* **2014**, *92*, 1071–1077. [CrossRef] [PubMed]
144. Serafin, A.; Foco, L.; Zanigni, S.; Blankenburg, H.; Picard, A.; Zanon, A.; Giannini, G.; Pichler, I.; Facheris, M.F.; Cortelli, P.; et al. Overexpression of Blood MicroRNAs 103a, 30b, and 29a in L-Dopa-Treated Patients with PD. *Neurology* **2015**, *84*, 645–653. [CrossRef]
145. Cao, X.Y.; Lu, J.M.; Zhao, Z.Q.; Li, M.C.; Lu, T.; An, X.S.; Xue, L.J. MicroRNA Biomarkers of Parkinson's Disease in Serum Exosome-like Microvesicles. *Neurosci. Lett.* **2017**, *644*, 94–99. [CrossRef]
146. Schlaudraff, F.; Gründemann, J.; Fauler, M.; Dragicevic, E.; Hardy, J.; Liss, B. Orchestrated Increase of Dopamine and PARK MRNAs but Not MiR-133b in Dopamine Neurons in Parkinson's Disease. *Neurobiol. Aging* **2014**, *35*, 2302–2315. [CrossRef]
147. Zhao, N.; Jin, L.; Fei, G.; Zheng, Z.; Zhong, C. Serum MicroRNA-133b Is Associated with Low Ceruloplasmin Levels in Parkinson's Disease. *Park. Relat. Disord.* **2014**, *20*, 1177–1180. [CrossRef]
148. Schulz, J.; Takousis, P.; Wohlers, I.; Itua, I.O.G.; Dobricic, V.; Rücker, G.; Binder, H.; Middleton, L.; Ioannidis, J.P.A.; Perneczky, R.; et al. Meta-analyses Identify Differentially Expressed MicroRNAs in Parkinson's Disease. *Ann. Neurol.* **2019**, *85*, 835–851. [CrossRef] [PubMed]
149. Xing, R.; Li, L.; Liu, X.; Tian, B.; Cheng, Y. Down Regulation of MiR-218, MiR-124, and MiR-144 Relates to Parkinson's Disease via Activating NF-κB Signaling. *Kaohsiung J. Med. Sci.* **2020**, *36*, 786–792. [CrossRef]
150. Scheper, M.; Iyer, A.; Anink, J.J.; Mesarosova, L.; Mills, J.D.; Aronica, E. Dysregulation of MiR-543 in Parkinson's Disease: Impact on the Neuroprotective Gene SIRT1. *Neuropathol. Appl. Neurobiol.* **2022**, *19*, e12864. [CrossRef] [PubMed]
151. Ravanidis, S.; Bougea, A.; Papagiannakis, N.; Maniati, M.; Koros, C.; Simitsi, A.; Bozi, M.; Pachi, I.; Stamelou, M.; Paraskevas, G.P.; et al. Circulating Brain-Enriched MicroRNAs for Detection and Discrimination of Idiopathic and Genetic Parkinson's Disease. *Mov. Disord.* **2020**, *35*, 457–467. [CrossRef]

152. Quinn, J.J.; Chang, H.Y. Unique Features of Long Non-Coding RNA Biogenesis and Function. *Nat. Rev. Genet.* **2016**, *17*, 47–62. [CrossRef]
153. Ni, Y.; Huang, H.; Chen, Y.; Cao, M.; Zhou, H.; Zhang, Y. Investigation of Long Non-Coding RNA Expression Profiles in the Substantia Nigra of Parkinson's Disease. *Cell Mol. Neurobiol.* **2017**, *37*, 329–338. [CrossRef] [PubMed]
154. Elkouris, M.; Kouroupi, G.; Vourvoukelis, A.; Papagiannakis, N.; Kaltezioti, V.; Matsas, R.; Stefanis, L.; Xilouri, M.; Politis, P.K. Long Non-Coding RNAs Associated with Neurodegeneration-Linked Genes Are Reduced in Parkinson's Disease Patients. *Front. Cell Neurosci.* **2019**, *13*, 58. [CrossRef]
155. Pantano, L.; Friedländer, M.R.; Escaramís, G.; Lizano, E.; Pallarès-Albanell, J.; Ferrer, I.; Estivill, X.; Martí, E. Specific Small-RNA Signatures in the Amygdala at Premotor and Motor Stages of Parkinson's Disease Revealed by Deep Sequencing Analysis. *Bioinformatics* **2016**, *32*, 673–681. [CrossRef]
156. Wake, C.; Labadorf, A.; Dumitriu, A.; Hoss, A.G.; Bregu, J.; Albrecht, K.H.; DeStefano, A.L.; Myers, R.H. Novel MicroRNA Discovery Using Small RNA Sequencing in Post-Mortem Human Brain. *BMC Genom.* **2016**, *17*, 776. [CrossRef] [PubMed]
157. McMillan, K.J.; Murray, T.K.; Bengoa-Vergniory, N.; Cordero-Llana, O.; Cooper, J.; Buckley, A.; Wade-Martins, R.; Uney, J.B.; O'Neill, M.J.; Wong, L.F.; et al. Loss of MicroRNA-7 Regulation Leads to α-Synuclein Accumulation and Dopaminergic Neuronal Loss In Vivo. *Mol. Ther.* **2017**, *25*, 2404–2414. [CrossRef] [PubMed]
158. Schulze, M.; Sommer, A.; Plötz, S.; Farrell, M.; Winner, B.; Grosch, J.; Winkler, J.; Riemenschneider, M.J. Sporadic Parkinson's Disease Derived Neuronal Cells Show Disease-Specific MRNA and Small RNA Signatures with Abundant Deregulation of PiRNAs. *Acta Neuropathol. Commun.* **2018**, *6*, 58. [CrossRef] [PubMed]
159. Hu, Y.-B.; Zhang, Y.-F.; Wang, H.; Ren, R.-J.; Cui, H.-L.; Huang, W.-Y.; Cheng, Q.; Chen, H.-Z.; Wang, G. MiR-425 Deficiency Promotes Necroptosis and Dopaminergic Neurodegeneration in Parkinson's Disease. *Cell Death Dis.* **2019**, *10*, 589. [CrossRef] [PubMed]
160. Caldi Gomes, L.; Galhoz, A.; Jain, G.; Roser, A.E.; Maass, F.; Carboni, E.; Barski, E.; Lenz, C.; Lohmann, K.; Klein, C.; et al. Multi-Omic Landscaping of Human Midbrains Identifies Disease-Relevant Molecular Targets and Pathways in Advanced-Stage Parkinson's Disease. *Clin. Transl. Med.* **2022**, *12*, e692. [CrossRef] [PubMed]
161. Dobricic, V.; Schilling, M.; Farkas, I.; Gveric, D.O.; Ohlei, O.; Schulz, J.; Middleton, L.; Gentleman, S.M.; Parkkinen, L.; Bertram, L.; et al. Common Signatures of Differential MicroRNA Expression in Parkinson's and Alzheimer's Disease Brains. *Brain Commun.* **2022**, *4*, fcac274. [CrossRef] [PubMed]
162. Gong, X.; Huang, M.; Chen, L. Mechanism of MiR-132-3p Promoting Neuroinflammation and Dopaminergic Neurodegeneration in Parkinson's Disease. *eNeuro* **2022**, *9*. [CrossRef] [PubMed]
163. Soreq, L.; Guffanti, A.; Salomonis, N.; Simchovitz, A.; Israel, Z.; Bergman, H.; Soreq, H. Long Non-Coding RNA and Alternative Splicing Modulations in Parkinson's Leukocytes Identified by RNA Sequencing. *PLoS Comput. Biol.* **2014**, *10*, 1003517. [CrossRef] [PubMed]
164. Kraus, T.F.J.; Haider, M.; Spanner, J.; Steinmaurer, M.; Dietinger, V.; Kretzschmar, H.A. Altered Long Noncoding RNA Expression Precedes the Course of Parkinson's Disease—A Preliminary Report. *Mol. Neurobiol.* **2017**, *54*, 2869–2877. [CrossRef]
165. Simchovitz, A.; Hanan, M.; Niederhoffer, N.; Madrer, N.; Yayon, N.; Bennett, E.R.; Greenberg, D.S.; Kadener, S.; Soreq, H. NEAT1 Is Overexpressed in Parkinson's Disease Substantia Nigra and Confers Drug-Inducible Neuroprotection from Oxidative Stress. *FASEB J.* **2019**, *33*, 11223–11234. [CrossRef]
166. Cappelletti, C.; Henriksen, S.P.; Geut, H.; Rozemuller, A.J.M.; van de Berg, W.D.J.; Pihlstrøm, L.; Toft, M. Transcriptomic Profiling of Parkinson's Disease Brains Reveals Disease Stage Specific Gene Expression Changes. *Acta Neuropathol.* **2023**, *146*, 227–244. [CrossRef]
167. Asad Samani, L.; Ghaedi, K.; Majd, A.; Peymani, M.; Etemadifar, M. Coordinated Modification in Expression Levels of HSPA1A/B, DGKH, and NOTCH2 in Parkinson's Patients' Blood and Substantia Nigra as a Diagnostic Sign: The Transcriptomes' Relationship. *Neurol. Sci.* **2023**, *44*, 2753–2761. [CrossRef]
168. di Domenico, A.; Carola, G.; Calatayud, C.; Pons-Espinal, M.; Muñoz, J.P.; Richaud-Patin, Y.; Fernandez-Carasa, I.; Gut, M.; Faella, A.; Parameswaran, J.; et al. Patient-Specific IPSC-Derived Astrocytes Contribute to Non-Cell-Autonomous Neurodegeneration in Parkinson's Disease. *Stem Cell Rep.* **2019**, *12*, 213–229. [CrossRef]
169. Agarwal, D.; Sandor, C.; Volpato, V.; Caffrey, T.M.; Monzón-Sandoval, J.; Bowden, R.; Alegre-Abarrategui, J.; Wade-Martins, R.; Webber, C. A Single-Cell Atlas of the Human Substantia Nigra Reveals Cell-Specific Pathways Associated with Neurological Disorders. *Nat. Commun.* **2020**, *11*, 4183. [CrossRef] [PubMed]
170. Bryois, J.; Skene, N.G.; Hansen, T.F.; Kogelman, L.J.A.; Watson, H.J.; Liu, Z.; Adan, R.; Alfredsson, L.; Ando, T.; Andreassen, O.; et al. Genetic Identification of Cell Types Underlying Brain Complex Traits Yields Insights into the Etiology of Parkinson's Disease. *Nat. Genet.* **2020**, *52*, 482–493. [CrossRef]
171. Imamura, K.; Hishikawa, N.; Sawada, M.; Nagatsu, T.; Yoshida, M.; Hashizume, Y. Distribution of Major Histocompatibility Complex Class II-Positive Microglia and Cytokine Profile of Parkinson's Disease Brains. *Acta Neuropathol.* **2003**, *106*, 518–526. [CrossRef]
172. Croisier, E.; Moran, L.B.; Dexter, D.T.; Pearce, R.K.B.; Graeber, M.B. Microglial Inflammation in the Parkinsonian Substantia Nigra: Relationship to Alpha-Synuclein Deposition. *J. Neuroinflamm.* **2005**, *2*, 14. [CrossRef] [PubMed]

173. Mastroeni, D.; Nolz, J.; Sekar, S.; Delvaux, E.; Serrano, G.; Cuyugan, L.; Liang, W.S.; Beach, T.G.; Rogers, J.; Coleman, P.D. Laser-Captured Microglia in the Alzheimer's and Parkinson's Brain Reveal Unique Regional Expression Profiles and Suggest a Potential Role for Hepatitis B in the Alzheimer's Brain. *Neurobiol. Aging* **2018**, *63*, 12–21. [CrossRef] [PubMed]
174. Smajic, S.; Prada-Medina, C.A.; Landoulsi, Z.; Ghelfi, J.; Delcambre, S.; Dietrich, C.; Jarazo, J.; Henck, J.; Balachandran, S.; Pachchek, S.; et al. Single-Cell Sequencing of Human Midbrain Reveals Glial Activation and a Parkinson-Specific Neuronal State. *Brain* **2022**, *145*, 964–978. [CrossRef] [PubMed]
175. Braak, H.; Sastre, M.; Del Tredici, K. Development of α-Synuclein Immunoreactive Astrocytes in the Forebrain Parallels Stages of Intraneuronal Pathology in Sporadic Parkinson's Disease. *Acta Neuropathol.* **2007**, *114*, 231–241. [CrossRef]
176. Rostami, J.; Fotaki, G.; Sirois, J.; Mzezewa, R.; Bergström, J.; Essand, M.; Healy, L.; Erlandsson, A. Astrocytes Have the Capacity to Act as Antigen-Presenting Cells in the Parkinson's Disease Brain. *J. Neuroinflamm.* **2020**, *17*, 119. [CrossRef]
177. Iwamoto, K.; Bundo, M.; Ueda, J.; Oldham, M.C.; Ukai, W.; Hashimoto, E.; Saito, T.; Geschwind, D.H.; Kato, T. Neurons Show Distinctive DNA Methylation Profile and Higher Interindividual Variations Compared with Non-Neurons. *Genome Res.* **2011**, *21*, 688–696. [CrossRef]
178. Jeffries, A.R.; Mill, J. Profiling Regulatory Variation in the Brain: Methods for Exploring the Neuronal Epigenome. *Biol. Psychiatry* **2017**, *81*, 90–91. [CrossRef]
179. Nott, A.; Schlachetzki, J.C.M.; Fixsen, B.R.; Glass, C.K. Nuclei Isolation of Multiple Brain Cell Types for Omics Interrogation. *Nat. Protoc.* **2021**, *16*, 1629–1646. [CrossRef]
180. Sakib, M.S.; Sokpor, G.; Nguyen, H.P.; Fischer, A.; Tuoc, T. Intranuclear Immunostaining-Based FACS Protocol from Embryonic Cortical Tissue. *STAR Protoc.* **2021**, *2*, 100318. [CrossRef]
181. Policicchio, S.S.; Davies, J.P.; Chioza, B.; Burrage, J.; Mill, J.; Dempster, E.L.; Policicchio, S. *Fluorescence-Activated Nuclei Sorting (FANS) on Human Post-Mortem Cortex Tissue Enabling the Isolation of Distinct Neural Cell Populations for Multiple Omic Profiling*; Neurodegeneration Method Development Community Complex Disease Epigenetics Group: Exeter, UK, 2020. [CrossRef]
182. Hannon, E.; Gorrie-Stone, T.J.; Smart, M.C.; Burrage, J.; Hughes, A.; Bao, Y.; Kumari, M.; Schalkwyk, L.C.; Mill, J. Leveraging DNA-Methylation Quantitative-Trait Loci to Characterize the Relationship between Methylomic Variation, Gene Expression, and Complex Traits. *Am. J. Hum. Genet.* **2018**, *103*, 654–665. [CrossRef]
183. Scherzer, C.R.; Grass, J.A.; Liao, Z.; Pepivani, I.; Zheng, B.; Eklund, A.C.; Ney, P.A.; Ng, J.; McGoldrick, M.; Mollenhauer, B.; et al. GATA Transcription Factors Directly Regulate the Parkinson's Disease-Linked Gene α-Synuclein. *Proc. Natl. Acad. Sci. USA* **2008**, *105*, 10907–10912. [CrossRef]
184. Zhu, Z.; Zhang, F.; Hu, H.; Bakshi, A.; Robinson, M.R.; Powell, J.E.; Montgomery, G.W.; Goddard, M.E.; Wray, N.R.; Visscher, P.M.; et al. Integration of Summary Data from GWAS and EQTL Studies Predicts Complex Trait Gene Targets. *Nat. Genet.* **2016**, *48*, 481–487. [CrossRef]
185. Jacobs, B.M.; Taylor, T.; Awad, A.; Baker, D.; Giovanonni, G.; Noyce, A.J.; Dobson, R. Summary-Data-Based Mendelian Randomization Prioritizes Potential Druggable Targets for Multiple Sclerosis. *Brain Commun.* **2020**, *2*, fcaa119. [CrossRef]
186. Topper, M.J.; Vaz, M.; Marrone, K.A.; Brahmer, J.R.; Baylin, S.B. The Emerging Role of Epigenetic Therapeutics in Immuno-Oncology. *Nat. Rev. Clin. Oncol.* **2020**, *17*, 75–90. [CrossRef]
187. Peng, G.S.; Li, G.; Tzeng, N.S.; Chen, P.S.; Chuang, D.M.; Hsu, Y.D.; Yang, S.; Hong, J.S. Valproate Pretreatment Protects Dopaminergic Neurons from LPS-Induced Neurotoxicity in Rat Primary Midbrain Cultures: Role of Microglia. *Mol. Brain Res.* **2005**, *134*, 162–169. [CrossRef]
188. Kidd, S.K.; Schneider, J.S. Protection of Dopaminergic Cells from MPP+-Mediated Toxicity by Histone Deacetylase Inhibition. *Brain Res.* **2010**, *1354*, 172–178. [CrossRef]
189. Kidd, S.K.; Schneider, J.S. Protective Effects of Valproic Acid on the Nigrostriatal Dopamine System in a 1-Methyl-4-Phenyl-1,2,3,6-Tetrahydropyridine Mouse Model of Parkinson's Disease. *Neuroscience* **2011**, *194*, 189–194. [CrossRef]
190. St. Laurent, R.; O'Brien, L.M.; Ahmad, S.T. Sodium Butyrate Improves Locomotor Impairment and Early Mortality in a Rotenone-Induced Drosophila Model of Parkinson's Disease. *Neuroscience* **2013**, *246*, 382–390. [CrossRef]
191. Sharma, S.; Taliyan, R.; Singh, S. Beneficial Effects of Sodium Butyrate in 6-OHDA Induced Neurotoxicity and Behavioral Abnormalities: Modulation of Histone Deacetylase Activity. *Behav. Brain Res.* **2015**, *291*, 306–314. [CrossRef]
192. Liu, J., Wang, F., Liu, S., Du, J.; Hu, X.; Xiong, J.; Fang, R.; Chen, W.; Sun, J. Sodium Butyrate Exerts Protective Effect against Parkinson's Disease in Mice via Stimulation of Glucagon like Peptide-1. *J. Neurol. Sci.* **2017**, *381*, 176–181. [CrossRef]
193. Li, B.; Yang, Y.; Wang, Y.; Zhang, J.; Ding, J.; Liu, X.; Jin, Y.; Lian, B.; Ling, Y.; Sun, C. Acetylation of NDUFV1 Induced by a Newly Synthesized HDAC6 Inhibitor HGC Rescues Dopaminergic Neuron Loss in Parkinson Models. *iScience* **2021**, *24*, 102302. [CrossRef]
194. Mazzocchi, M.; Goulding, S.R.; Wyatt, S.L.; Collins, L.M.; Sullivan, A.M.; O'Keeffe, G.W. LMK235, a Small Molecule Inhibitor of HDAC4/5, Protects Dopaminergic Neurons against Neurotoxin- and α-Synuclein-Induced Degeneration in Cellular Models of Parkinson's Disease. *Mol. Cell. Neurosci.* **2021**, *115*, 103642. [CrossRef]
195. Mazzocchi, M.; Goulding, S.R.; Morales-Prieto, N.; Foley, T.; Collins, L.M.; Sullivan, A.M.; O'Keeffe, G.W. Peripheral Administration of the Class-IIa HDAC Inhibitor MC1568 Partially Protects against Nigrostriatal Neurodegeneration in the Striatal 6-OHDA Rat Model of Parkinson's Disease. *Brain Behav. Immun.* **2022**, *102*, 151–160. [CrossRef]

196. El-Saiy, K.A.; Sayed, R.H.; El-Sahar, A.E.; Kandil, E.A. Modulation of Histone Deacetylase, the Ubiquitin Proteasome System, and Autophagy Underlies the Neuroprotective Effects of Venlafaxine in a Rotenone-Induced Parkinson's Disease Model in Rats. *Chem. Biol. Interact.* **2022**, *354*, 109841. [CrossRef]
197. Wang, L.; Liu, L.; Han, C.; Jiang, H.; Ma, K.; Guo, S.; Xia, Y.; Wan, F.; Huang, J.; Xiong, N.; et al. Histone Deacetylase 4 Inhibition Reduces Rotenone-Induced Alpha-Synuclein Accumulation via Autophagy in SH-SY5Y Cells. *Brain Sci.* **2023**, *13*, 670. [CrossRef]
198. Meka, S.T.; Bojja, S.L.; Kumar, G.; Birangal, S.R.; Rao, C.M. Novel HDAC Inhibitors Provide Neuroprotection in MPTP-Induced Parkinson's Disease Model of Rats. *Eur. J. Pharmacol.* **2023**, *959*, 176067. [CrossRef]
199. Izco, M.; Blesa, J.; Schleef, M.; Schmeer, M.; Porcari, R.; Al-Shawi, R.; Ellmerich, S.; de Toro, M.; Gardiner, C.; Seow, Y.; et al. Systemic Exosomal Delivery of ShRNA Minicircles Prevents Parkinsonian Pathology. *Mol. Ther.* **2019**, *27*, 2111–2122. [CrossRef]
200. Sun, Z.; Kantor, B.; Chiba-Falek, O. Neuronal-Type-Specific Epigenome Editing to Decrease SNCA Expression: Implications for Precision Medicine in Synucleinopathies. *Mol. Ther. Nucleic Acids* **2024**, *35*, 102084. [CrossRef]
201. Zhou, J.; Zhao, Y.; Li, Z.; Zhu, M.; Wang, Z.; Li, Y.; Xu, T.; Feng, D.; Zhang, S.; Tang, F.; et al. MiR-103a-3p Regulates Mitophagy in Parkinson's Disease through Parkin/Ambra1 Signaling. *Pharmacol. Res.* **2020**, *160*, 105197. [CrossRef]
202. Leggio, L.; Vivarelli, S.; L'Episcopo, F.; Tirolo, C.; Caniglia, S.; Testa, N.; Marchetti, B.; Iraci, N. MicroRNAs in Parkinson's Disease: From Pathogenesis to Novel Diagnostic and Therapeutic Approaches. *Int. J. Mol. Sci.* **2017**, *18*, 2698. [CrossRef] [PubMed]

International Journal of
Molecular Sciences

MDPI

Review

Extracellular Vesicles as Biomarkers for Parkinson's Disease: How Far from Clinical Translation?

Alice Gualerzi [1], Silvia Picciolini [1], Marzia Bedoni [1,*], Franca Rosa Guerini [1,*], Mario Clerici [1,2] and Cristina Agliardi [1]

[1] IRCCS Fondazione Don Gnocchi Onlus, 20148 Milan, Italy; agualerzi@dongnocchi.it (A.G.); spicciolini@dongnocchi.it (S.P.); mclerici@dongnocchi.it (M.C.); cagliardi@dongnocchi.it (C.A.)
[2] Department of Pathophysiology and Transplantation, University of Milan, 20122 Milan, Italy
* Correspondence: mbedoni@dongnocchi.it (M.B.); fguerini@dongnocchi.it (F.R.G.); Tel.: +39-02-4030-8874 (M.B.); +39-02-4030-8376 (F.R.G.)

Abstract: Parkinson's disease (PD) is a neurodegenerative disorder affecting about 10 million people worldwide with a prevalence of about 2% in the over-80 population. The disease brings in also a huge annual economic burden, recently estimated by the Michael J Fox Foundation for Parkinson's Research to be USD 52 billion in the United States alone. Currently, no effective cure exists, but available PD medical treatments are based on symptomatic prescriptions that include drugs, surgical approaches and rehabilitation treatment. Due to the complex biology of a PD brain, the design of clinical trials and the personalization of treatment strategies require the identification of accessible and measurable biomarkers to monitor the events induced by treatment and disease progression and to predict patients' responsiveness. In the present review, we strive to briefly summarize current knowledge about PD biomarkers, focusing on the role of extracellular vesicles as active or involuntary carriers of disease-associated proteins, with particular attention to those research works that envision possible clinical applications.

Keywords: Parkinson's disease (PD); extracellular vesicles; exosomes; biomarkers; α-synuclein

1. Parkinson Disease

First described by James Parkinson in 1817, Parkinson's disease (PD) is today the most common neurodegenerative disease after Alzheimer's disease [1]. According to the World Health Organization (WHO) definition, PD is "a degenerative condition of the brain associated with motor symptoms (slow movement, tremor, rigidity, walking and imbalance) and a wide variety of non-motor complications (cognitive impairment, mental health disorders, sleep disorders and pain and other sensory disturbances)". Disability and death due to PD in recent years are increasing faster than they are for any other neurological condition, and the Global Burden of Disease Study 2015 estimates that there may be nearly 13 million people with PD (pwPD) by 2040 [2], suggesting an increasing social and economic burden on modern societies that has to be faced.

1.1. PD Etiology and Subtypes

The causes of PD are still largely unknown; if previously the disease was believed to be caused mainly by environmental factors, nowadays research reveals that the pathology is caused by a complex interplay of genetic factors and environmental triggers. Histologically, PD is characterized by the prominent death of dopaminergic neurons, especially in the substantia nigra pars compacta (SNpc) area of the brain, and by the presence of Lewy bodies and Lewy neurites, which are inclusions of abnormal aggregates of α-synuclein (α-syn) in neuronal cell bodies and neuronal cell processes, respectively [3]. The study of familial PD has led to the identification of rare mutations in different genes responsible for the monogenic forms of PD, which account for <10% of total PD cases [4]. These genes are SNCA (α-syn), LRRK2 (leucine-rich repeat kinase 2), PARK2 (parkin), PINK1 (PTEN-induced putative kinase 1), VPS-35 (vacuolar protein sorting-associated protein 35), PARK7 (DJ1) and GBA (glucocerebrosidase) [5]. The remaining 90% of patients are considered to suffer from "idiopathic" PD. These cases are sporadic and for them a monogenic inheritance pattern cannot be identified; these cases of PD are considered the result of the interaction between environmental and genetic factors. The genetic background of idiopathic PD is complex, similar to many other common multifactorial conditions, and is characterized by the presence of variants that include single-nucleotide polymorphisms (SNP) and structural variants (microsatellites, minisatellites, insertions, and deletions) that together determine an individual's susceptibility to disease. Meta-analyses of several different genome-wide association studies (GWAS) have identified 41 PD risk loci [6,7] that are common genetic variants conferring an increased risk of developing the disease. Among environmental triggers, the following deserve to be mentioned: exposure to the heroin side product 1-methyl-4-phenyl-1,2,3,6-tetrahydropyridine (MPTP) [8] or to pesticides such as rotenone, paraquat, organophosphates, and pyrethroids [9], head trauma [10], and central nervous system (CNS) infections [11]. Notably, moderate nicotine and caffeine consumption, and use the of non-steroid anti-inflammatory drugs have been reported to be protective against PD [12].

1.2. PD Diagnosis

The clinical diagnosis of PD, especially in the early phases, is still challenging because of the unavailability of specific diagnostic tests. Diagnosis primarily depends on clinical symptoms, along with detecting SNpc neurodegeneration and observing Lewy pathology. Typically, a confirmation of these findings takes place during a post-mortem pathological examination [13]. Studies report that in only 80–90% of PD cases is the diagnosis confirmed at autopsy [14]. The first step of diagnosis is the clinical establishment that the patient is affected by "parkinsonism". This relies on the three key elements, bradykinesia, tremor and rigidity, with bradykinesia being the essential element, accompanied by the presence of at least one of the other two [15]. Besides these evaluations, of critical importance are also exclusion criteria including the absence of a response to high-dose levodopa despite moderate disease severity, normal "DAT scan" functional imaging, and cerebellar abnormalities. Supportive criteria are a clear beneficial response to dopaminergic therapy, levodopa-induced dyskinesia, rest tremor of a limb, and the presence of olfactory loss [16]. Non-motor features present in the prodromal phase of the pathology also contribute to the diagnosis; these are depression, anosmia, constipation, and REM sleep behavior disorder [17]. Several disorders mimic idiopathic PD, in particular dementia with Lewy bodies (DLB), multiple-system atrophy (MSA), progressive supranuclear palsy (PSP), and corticobasal degeneration (CBD). Moreover, secondary causes can lead to parkinsonism, like drug-induced parkinsonism, as in the case of antipsychotics, toxins, and normal-pressure hydrocephalus, and vascular parkinsonism.

1.3. PD Therapeutic Strategies

Currently, disease-modifying treatments for PD are not available, and the drugs focus on controlling the motor symptoms and not on altering the course of the disease. The selective loss of dopaminergic neurons of the SNpc results in dopamine depletion in the striatum [18], hence the mainstay of PD pharmacological treatment is dopaminergic drugs that replace the action of dopamine by activating dopamine receptors, the provision of a precursor that is metabolized to dopamine (e.g., levodopa), or the prevention of the breakdown of endogenous dopamine (e.g., monoamine oxidase-B (MAO-B) inhibitors and catechol-O-methyl transferase inhibitors (COMT) [19,20]. Treatments are tailored to each patient considering the benefits and the side effects, mostly relying on the personal expertise of the neurologist and a trial and error process.

Some novel and promising approaches are currently under investigation including stem cell-based treatments [21], gene therapy [22] and therapies with the aim of contrasting α-syn toxic effects by inhibiting α-syn production, increasing the degradation of α-syn aggregates, and reducing the uptake of extracellular α-syn by neighboring cells [23]. Targeted rehabilitative programs for motor function, swallowing difficulty, and speech disorders are assuming an important role in improving the quality of life of pwPD [24].

As previously explained, there are still many challenges to be overcome regarding PD from the fine understanding of the biological processes and molecular pathways at the basis of the disease, and the discovery of solid and reliable diagnostic and prognostic biomarkers—to be used especially in the prodromal and early phases of the disease—to the development of disease-modifying treatments. We will now review the attractive role of extracellular vesicles (EVs), the state of the art, the need to define laboratory technique guidelines, and future opportunities and perspectives.

2. Extracellular Vesicles

EVs is a general term that refers to a complex multitude of "particles that are released from cells, are delimited by a lipid bilayer, and cannot replicate on their own" [25] that can be detected in the interstitial fluid as well as in all biofluids, including blood, saliva, urine, milk, and intraperitoneal fluids [26]. EVs were first described in the 1960s and since their discovery, they have undergone a transformation in the definitions of their structure and role. Initially, they were assumed to be the result of membrane turnover and a waste removal pathway. Later, their release mechanisms were investigated, shedding new light on their function and role with the description of the formation of intraluminal vesicles (ILVs) in late endosomes by the inward budding of the endosomal limiting membrane, followed by the fusion of so-called multivesicular bodies (MVBs). In 1996, Raposo et al. reported the immune-modulating activity of B cell-derived EVs, suggesting new biological implications of these vesicles, far beyond a mere recycling circle [27]. Since then, the investigation of the mechanisms underlying EV release, their classification and nomenclature, as well as their involvement in physiological and pathological cell functions has dramatically increased. The technological evolution that has been accompanying the study of EVs has progressively increased our knowledge about EV heterogeneity in size, structure, and functions in physiological and pathological conditions [28]. This led to an exponential increase in EV-related scientific publications, patents, and clinical trials in the first two decades of this century (Figure 1) as well as to the appearance of EV-focused journals and scientific communities of national and international interest.

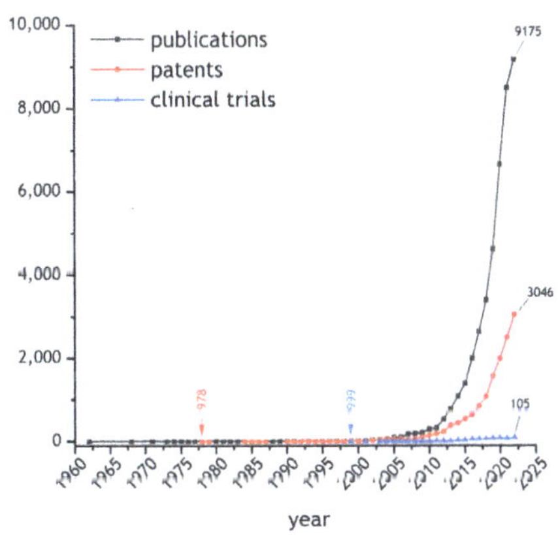

Figure 1. Schematic representation of the exponential growth of the number of publications (grey line) and patents (red line) reported on the Scopus database (https://www.scopus.com/; accessed on 31 December 2022). The blue line indicates the number of registered trials on clinicaltrials.gov. The search for publications patents and trials involved using the following keywords: "extracellular vesicle" OR "exosome" OR "microvesicle".

In the current review, our aim is to emphasize the evidence concerning EVs in PD research. For this reason, we will broadly refer to EVs as a wide family, without going deeply into the numerous distinctions related to their biogenesis. However, it is important to acknowledge that while this overarching term encompasses a diverse range of vesicles, it might obscure variations in their origins, physical and chemical properties, and functions. These distinctions, unraveled over time thanks to innovation and technological advancements, could also underlie certain controversies and conflicting evidence that will be later described.

3. Extracellular Vesicles in Parkinson's Disease

3.1. Pathophysiology: Spread of Pathology

As mentioned, one of the hallmarks of PD is the aggregation and spread of cytotoxic forms of α-syn. Starting from this premise, several studies have tried to understand if EVs might have a role in either the sequestration and removal or in the transport across the body of misfolded and aggregated α-syn. The reasons for the toxicity of α-syn in PD are complex and involve multiple mechanisms, including the loss of function of aggregated proteins, mitochondrial dysfunction, interference with axonal transport, proteasomal inhibition, synaptic toxicity, and endoplasmic reticulum stress. Mounting evidence suggests that EVs have a role in the propagation of aggregated proteins, in particular in the prion-like propagation of aggregated α-syn, and predominantly in the intercellular spread of aggregates.

One of the first observations proposing EV-mediated transport as one (but not a unique) form of transport for α-syn was made by Emmanouilidou and colleagues [29], who observed that in vitro-cultured α-syn-expressing cells could release the α-syn monomeric and oligomeric forms both when free in the medium and within exosomes, in a calcium-dependent manner. Subsequently, it was shown that lysosomal dysfunction—a PD relevant stress condition—increases the release of α-syn-containing vesicles that cause toxicity in the recipient cells [30,31]. Further investigations on the mechanisms of the vesicle-mediated release of α-syn in the extracellular space proposed ubiquitination [32] and sumoylation [33] as possible mechanisms involved in the inclusion of α-syn into multivesicular bodies and, thus, in exosomes. It was also found that cells take advantage of the exosomal secretion of α-syn to discard α-syn in specific conditions, i.e., when autophagy is impaired [34]. Nonetheless, criticisms were raised about the impact of EV transport on α-syn release as some in vitro works hypothesized that less than 2% of α-syn is found in association with exosomes [35]. Indeed, even though the percentage of α-syn released by brain and non-brain cells might be small, the specific composition of EVs seems to have a critical role in the pathogenic mechanisms underlying PD.

Although the mechanisms of EV α-syn loading remain to be elucidated, it was reported that the α-syn–membrane interaction has a role in the conformational changes of α-syn that might alter the protein function and/or drive aggregation in PD development. At the same time, the targeting of α-syn in its different conformations (especially the most neurotoxic ones, i.e., oligomers) at the membrane surface can alter lipid composition because of its ability to disrupt biological membranes [36]. Studies focusing on the chemo-physical interaction of α-syn and membrane components demonstrated that the protein adopts an α-helix conformation when interacting with the cell membrane, in particular with some phospholipids, including phosphatidylserine and phosphatidyl inositol [37]. The α-helix remains entirely buried within the depth of the membrane, whereas the rest of the protein segments present lower membrane penetration and higher flexibility [36], influencing α-syn's role in vesicle biogenesis and recycling. Data suggest that monomer binding to the membrane is accompanied by the further clustering of α-syn, which can perturb lipid membranes, in particular nanosized vesicles like EVs [38]. The aggregated and toxic form of the protein, oligomeric α-syn, was found to interact with lipids causing structural changes in lipid membranes, leading to membrane disruption, membrane thinning, pore formation, and lipid clustering. Membrane-associated oxidative stress further promotes α-syn aggregation inside the cells, with the consequent release of toxic α-syn species in EVs, both inside and on the surface of the vesicles, propagating the neurodegenerative process to adjacent neurons [39].

EVs derived from pwPD were proven to be enriched in α-syn and to be able to induce the oligomerization of α-syn in a dose-dependent manner [40]. Exogenous α-syn strains can also seed the assembly of endogenous α-syn and its propagation, leading to long-term functional effects [40]. Furthermore, α-syn strains display differential seeding capacities and elongation processes and induce a strain-specific pathology and specific neurotoxic phenotypes, possibly explaining the heterogeneity in the spread and in the symptoms of the pathology [40].

What is of particular interest in the investigation of the role of EVs in PD progression is the observation that α-syn assemblies cross the blood–brain barrier (BBB); intravenous α-syn injection results in its dissemination in the CNS [38]. This can be mediated by EV transport, which is facilitated by the leakage of the BBB, as seen in pwPD [41]. Microglia are also known to play a major role in the neuroinflammation that accompanies the neurodegenerative process. In pwPD, microglia were shown to fail to clear and/or to promote the release of EVs that contained toxic forms of α-syn or other pathogenic factors, possibly potentiating neurodegeneration [42].

Although α-syn is the most studied player in the pathogenic mechanisms of PD, other genes are mutated in familial PD, and other proteins were found to be associated with EVs in pwPD. For example, mutant LRRK2 acts at multiple points within the endosomal pathway and thus has the potential to modulate exosome biogenesis. It was shown that LRRK2 is released within EVs and regulates their biogenesis [43], with LRRK2-loaded EVs being also found in the urine of pwPD [44]. Similarly, the microtubule-stabilizing protein Tau is known to become hyperphosphorylated and aggregation-prone in so-called taupathies, including PD. Different mechanisms, including EV transport, allow brain tau to cross the BBB and circulate in the bloodstream, where it can associate to exosomes, and in particular with neuron-derived exosomes expressing the L1CAM protein marker [45].

Finally, the oxidative stress of neurons is crucial in PD progression, contributing to the cellular dysfunction of neurons. In this regard, studies performed on preclinical models of PD have demonstrated that EV-associated microRNAs have an impact on brain tissue. In particular, the inhibition of exosomal miR-137 relieved oxidative stress, promoted neuron viability, and inhibited the apoptosis of neurons by up-regulating oxidation resistance 1 (OXR1), which inhibits oxidative DNA damage [46].

3.2. *Biomarkers for Diagnosis and Prognosis*

Over the years, accumulating evidence has demonstrated that EVs are crucial for intercellular communication within the brain, and it is commonly accepted that EVs can move from the bloodstream to the CNS crossing the BBB, even though the mechanisms of this transfer are largely unclear. The possibility to monitor and observe the processes occurring within the brain taking advantage of complex multifaceted nanovesicles, i.e., EVs, has inspired a huge body of literature about brain disorders including neurodegenerative diseases where the disruption of the BBB facilitates the transfer of EVs from the systemic circulation to the brain and vice versa [47]. EVs and their constituents, either proteins, nucleic acids, lipids or small metabolites, have been the object of several studies that over the years have tried to identify molecular biomarkers for the diagnosis of PD.

One of the first studies performed on the blood of pwPD to verify the actual potential of EVs as biomarkers of PD was attempted by Shi and colleagues, who demonstrated that both plasma exosomal α-syn [48] and tau [45] correlated with PD severity, performing even better than the quantification of free α-syn or tau in cerebrospinal fluid (CSF). First, a large cohort of 267 PD and 215 age- and sex-matched healthy controls were considered and α-syn concentration was assessed in plasma L1CAM-containing (L1CAM+) exosomes (using Luminex assays). No significant difference in the total plasma α-syn concentrations was observed in PD versus controls, but the α-syn concentration in plasma L1CAM+

exosomes was significantly higher in pwPD compared to that in healthy controls, with the plasma exosomal α-syn/total α-syn ratio being also significantly higher in pwPD than in controls [48]. The ROC curve revealed moderate diagnostic performance for α-syn within L1CAM+ EVs (area under the curve/AUC = 0.654) as well as for the exosomal α-syn/total α-syn ratio (AUC = 0.657). Interestingly, significant correlations were found between plasma α-syn in L1CAM+ exosomes and the plasma exosomal α-syn/total α-syn ratio with UPDRS motor scores (UPDRS III) [48]. These data were confirmed by the observation that neuronal (L1CAM+) EVs can predict and differentiate pwPD from atypical parkinsonisms. Using a combined protocol of immunoprecipitation and electrochemiluminescence assays, it was established that an α-synuclein level in L1CAM-immunocaptured exosomes of above 14 pg/mL is a robust biomarker for the differential diagnosis of PD and atypical parkinsonisms (ROC AUC = 0.98) [49]. Similarly, it was reported by Cerri and coauthors [50] that the quantification of plasma exosomal α-syn (not only L1CAM+) normalized for total α-syn (whole plasma) via ELISA was significantly increased in pwPD (39 pwPD and 33 healthy controls) compared to that in controls. The results were also correlated with the measures of disease severity, indexed by the UPDRS III and Hoehn and Yahr (H&Y) scores, as well as with the Glucocerebrosidase (GCase) activity measured in lymphocytes [50]. This study is in line with other data that demonstrated the abundance of α-syn not only in neuronal circulating EVs, but also in non-brain EVs [51]. In particular, it was reported that the quantification of plasma EV-associated α-syn (without prior enrichment for brain-derived EVs) was effective in distinguishing pwPD (n = 94) from people with DLB (n = 48) and PSP (n = 49). ROC analysis showed an AUC of 0.804 for PD and DLB and an AUC of 0.815 between PD and PSP, whereas the ROC AUC was 0.769, comparing pwPD and control subjects. A significant inverse correlation was also observed with UPDRS III and H&Y scores [51].

Interestingly, EVs from non-brain cells were found to be loaded with α-syn in pwPD, as were brain-derived EVs, for example those from red blood cells [52] and peripheral lymphocytes [53]. This observation can partially explain the better performance of the quantification of α-syn in the entire population of blood-derived EVs compared to that of its quantification in specific neuronal populations. Moreover, L1CAM is known to identify vesicles released by neurons, although its specificity is controversial and not all of the neuronal EVs will be loaded with L1CAM. It was also shown that the blood concentration of EVs derived from neurons (SNAP-25), astrocytes (EAAT1), and oligodendrocytes (OMG) is significantly augmented in pwPD compared to those in other forms of parkinsonisms, with good diagnostic potential reported for these EV subfamilies (AUCs of the ROC curve for plasma neuron-, astrocyte- and oligodendrocyte-derived EVs were 0.82, 0.75, and 0.78, respectively) [54].

A comprehensive analysis of neuron-derived EVs (L1CAM+) in pwPD (n = 32) showed that the toxic oligomeric form of α-syn is significantly increased in L1CAM+ EVs, whereas total α-syn, STX-1A, and VAMP-2, proteins that are known to be involved in synaptic function, are decreased [55]. The analysis of the diagnostic performance of the proposed biomarkers in L1CAM+ EVs showed good performance for oligomeric α-syn (ROC curve AUC = 0.824); the results were even better when two biomarkers were considered in L1CAM+ EVs: the oligomeric α-syn/STX-1A ratio (AUC = 0.871) and oligomeric α-Syn/VAMP-2 ratio (AUC 0.876). Importantly, also in this study, the amount of the oligomeric form of α-syn in L1CAM+ EVs correlated with disease duration and clinical severity (UPDRS III and H&Y Scales) [55]. The same protocol was also applied for the combined quantification of oligomeric α-syn and aggregated tau in L1CAM+ EVs comparing pwPD (n = 70) and atypical parkinsonisms. Even though in this case the diagnostic performance did not include control subjects, the diagnostic performance of the combination between the two biomarkers (oligomeric α-syn/Tau aggregates ratio) showed far better diagnostic performance than did that of the single biomarker, with ROC analysis obtaining AUC = 0.902 for pwPD compared to CBD and AUC = 0.908 for pwPD compared to PSP [56].

Following a similar approach of exosomal quantitation [48] but using a different technology, single-molecule array (SiMoA) assays, Shi and colleagues evaluated the possibility of using tau as a PD biomarker, measuring its concentration both in whole plasma and L1CAM+ plasma exosomes in pwPD, Alzheimer's disease (AD), and healthy controls [45]. Plasma exosomal tau was significantly higher in PD than in healthy controls, but was not significantly different in AD patients; importantly, exosomal tau correlated with CSF tau concentration [45]. The diagnostic performance, as evaluated via ROC analysis, showed that L1CAM+ exosome-associated tau was modestly predictive in distinguishing between PD and healthy controls (AUC = 0.607), and was associated with disease duration but not with other clinical scales in pwPD [45].

Taken together, these data suggest variations in the subpopulations of vesicles circulating in the blood of pwPD that might relate not only to a single protein variation, but possibly to multiple factors associated with EVs. In line with this observation, Tomlinson and colleagues performed a proteomic analysis of serum EVs and found that, despite the absence of significant differences in EV numbers or sizes, 23 EV-associated proteins were differentially abundant in PD [57]. Subpopulations of vesicles that are either differentially regulated or enriched for certain proteins (including but not limited to α-syn) in response to the disease process could explain this observation. An overall biochemical characterization of EVs was proposed as a potential biomarker for PD in 2019 [58]. Taking advantage of the Raman spectroscopy approach, serum-derived EVs from a small cohort of pwPD (n = 22) and healthy controls (n = 18) were analyzed; the results identified a specific biochemical fingerprint for pwPD [58]. Although the complexity of the Raman spectrum did not allow the complete definition of the molecules responsible for the observed differences, the ROC curve demonstrated moderate diagnostic accuracy (AUC = 0.71) and a correlation between the biochemical features of serum EVs and the clinical scales used to profile pwPD (UPDRS III and H&Y scores) [58].

Of course, EVs are carriers of multiple bioactive compounds—other than proteins—that can play a role as PD biomarkers, and among them miRNAs have received increasing attention. The first observation in this topic focused on EV-associated miRNAs isolated from CSF. Gui and colleagues demonstrated that five miRNAs that circulate in CSF in association with EVs could discriminate PD from healthy controls with an accuracy calculated by the ROC curve of higher than 90%, with the highest performance being achieved with miR-409-3p. Notably, the combination of miR-153 and miR-409-3p in CSF-derived EVs could enhance the performance of discrimination significantly (AUC = 0.99) [59]. In blood, miR-331-5p and miR-505 were found to have optimal diagnostic performance (AUC > 0.85) between pwPD and healthy subjects [60]. In a different cohort, the downregulation of miR-19b and upregulation of miR-195 and miR-24 were found to have consistent diagnostic potential with variable accuracy ranging from 70 to 90% compared to that in healthy subjects (n = 40) [61].

Recently, the search for PD biomarkers has expanded towards organic fluids other than blood and CSF. Urine and saliva are intriguing sources of biomarkers for chronic pathologies like PD, and measurements performed in accessible biofluid would allow clinicians to periodically monitor disease progression and therapy effectiveness, without the need for blood and CSF collection. Studies performed on urinary samples of pwPD showed that EVs loaded with elevated levels of autophosphorylated Ser(P)-1292 LRRK2 could be detected in people with idiopathic PD [44,62]. SNAP23 and calbindin were found to be elevated as well in urinary EVs from pwPD compared to healthy subjects, with both markers having moderate diagnostic accuracy after ROC curve calculation (AUC = 0.68 for SNAP23 and AUC = 0.75 for calbindin; AUC = 0.76 for the combination of calbindin and SNAP23 in the same logistic model) [63] (Table 1).

Table 1. Summary of the study on PD biomarkers with reported data on clinical samples. Only studies with reported clinical power calculations are included. Abbreviations: AD: Alzheimer's disease; AUC: area under the curve after ROC curve calculation; CBD: corticobasal degeneration; CSF: cerebrospinal fluid; DLB: dementia with Lewy bodies; ECL: electrochemi-luminescence; HC: healthy control; MSA: multiple-system atrophy; PD: Parkinson's disease; PSP: progressive supranuclear palsy; SEC: size exclusion chromatography.

Biomarker	**Source**	**EV Subtype**	**Isolation Method**	**Considered Population**	**Detection Method**	**Diagnostic Performance**	**Ref.**
				Blood			
α-syn	serum	L1CAM+	Immunoaffinity	PD = 290 MSA = 50 PSP = 116 CBD = 88 HC = 191	ECL	AUC = 0.86 (PD vs. HC); AUC = 0.98 (PD vs. MSA); AUC = 0.94 (PD vs. PSP + CBD)	[49]
	plasma	L1CAM+	Immunoaffinity	PD = 267 HC = 215	Luminex® (fluorimetric)	AUC = 0.654	[48]
	plasma	Total EVs	SEC	PD = 96 DLB = 50 PSP = 50 HC = 42	ECL	AUC = 0.804 (PD vs. DLB); AUC = 0.815 (PD vs. PSP); AUC = 0.769 (PD vs. HC)	[51]
Oligomeric α-syn; STX-1A; VAMP2	serum	L1CAM+	Immunoaffinity	PD = 32 HC = 40	ELISA	oligomeric α-syn: AUC = 0.820; oligomeric α-syn/STX1: AUC = 0.871; oligomeric α-syn/VAMP2: AUC = 0.876	[55]
Oligomeric α-syn; Tau aggregates	serum	L1CAM+	Immunoaffinity	PD = 70 PSP = 21 CBD = 19	ELISA	oligomeric α-syn/Tau aggregates: AUC = 0.902 (PD vs. CBD) Tau aggregates: AUC = 0.908 (PSP vs. PD)	[56]

Table 1. *Cont.*

Biomarker	Source	EV Subtype	Isolation Method	Considered Population	Detection Method	Diagnostic Performance	Ref.
Tau	plasma	L1CAM+	Immunoaffinity	PD = 91 AD = 106 HC = 106	SiMoA	AUC = 0.607 (PD vs. HC)	[45]
EV count	plasma	SNAP25+ EAAT1+ OMG+	Immunoaffinity	PD = 15 MSA = 15 PSP = 7 HC = 15	ELISA	SNAP25+ EVs: AUC = 0.82 (PD vs. HC); EAAT1+ EVs: AUC = 0.75 (PD vs. HC) OMG+ EVs: AUC = 0.78 (PD vs. HC)	[54]
Raman spectrum	serum	Total EVs	SEC and ultracen-trifugation	PD = 22 HC = 19	Raman spec-troscopy	AUC = 0.71	[58]
miR-331-5p miR-505	plasma	Total EVs	Polymer based pre-cipitation and Targeted filtration	PD = 52 HC = 48	RT-qPCR	miR-331-5p: AUC = 0.849; miR-505: AUC = 0.898	[60]

miR-19b
miR-24
miR-195
	serum	Total EVs	Polymer based pre-cipitation	PD = 109
				HC = 40
					RT-qPCR	miR-19b: AUC = 0.753;
						miR-24: AUC = 0.908;
						miR-195: AUC = 0.697
							[61]

				Other biofluids			
miR-409-3p miR-153	CSF	Total EVs	Ultracentrif-ugation	PD = 47 HC = 27	RT-qPCR	miR-409-3p: AUC = 0.90; miR-153 and miR-409-3p: AUC = 0.99	[59]
SNAP23 calbindin	urine	Total EVs	Ultracentrif-ugation	Discovery cohort PD = 28 HC = 22	Mass spec-trom-etry	Discovery Cohort SNAP23: AUC = 0.80 Calbindin: AUC = 0.75	[63]
				Replication cohort PD = 57 HC = 51		Replication cohort SNAP23: AUC = 0.68 Calbindin: AUC = 0.75	

Saliva has received attention in the field of neurodegenerative disease diagnostics as it shares multiple components with plasma, is collected in non-invasive ways, and has the advantage of a reduced interference of lipoproteins. A pilot study performed on pwPD showed an increased level of L1CAM and phosphorylated α-syn in salivary EVs compared to those in healthy controls, suggesting salivary EVs as a source of PD biomarkers [64].

3.3. Therapy

As mentioned above, current therapies for PD are far from optimal, treating the symptoms of the disease but being unable to alter the undergoing neuronal damage. One of the main problems with OD drugs is that they must cross the BBB, a natural dynamic barrier composed of specialized epithelial cells that together with pericytes and astrocytes form a tight barrier that regulates the flow of substances into the brain, including therapeutic agents. Different strategies have been designed in an attempt to bypass this obstacle [65], including the use of nanocarriers that may be of synthetic or biological origin.

EVs, due to their properties, are emerging as a new class of nanocarriers of drugs and therapeutic molecules (proteins—including growth factors—lipids, various DNA and RNA species, etc.) that could be used in neurodegenerative diseases including PD [66]. EVs-based drug design could be divided into two main groups: the first takes advantage of the paracrine effect of EVs as part of the "secretome" of cell therapies (e.g., mesenchymal stem cell (MSC), and the second involves the use of EVs as drug delivery systems to deliver biomolecules and drugs to recipient cells.

In the first case, the intrinsic therapeutic effect of stem cell-derived EVs is exploited: the therapeutic effect of MSCs is mainly mediated by soluble factors, among which EVs play a crucial role [67]. Moreover, the use of MSC-derived EVs is safer than that of MSCs themselves, which presents the risk of malignant transformation [68]. Several promising in vitro and in vivo experiments were carried out: neural stem cell (NSC) EVs attenuated the ROS-induced apoptotic pathway and neuroinflammation in cultured dopaminergic neuroblastoma neurons [69]. The incubation of dopaminergic neuroblastoma neurons exposed to 6-OHDA neurotoxin and umbilical cord MSC-EVs increased cell viability and proliferation, and inhibited apoptosis by inducing autophagy. Furthermore, in vivo, these EVs reduced the loss of dopaminergic neurons in the *substantia nigra* and upregulated dopamine in the striatum in a 6-OHDA-induced PD rat model [70]. EVs derived from human exfoliated deciduous teeth (SHED) were also shown to suppress 6-OHDA-induced dopaminergic neurons apoptosis [71]; the same EVs administrated intranasally in a PD rat model increased tyrosine hydroxylase expression in the *substantia nigra* and improved gait parameters [72].

Biological EVs also represent a potential avenue for drug delivery systems; notably, they can also overcome the drawbacks of synthetic nanovectors: allergic reactions and off-target accumulation in the liver and spleen. In addition, biological EVs can bypass the macrophagic system due to the presence of CD47 surface molecule [73,74], are highly biocompatible and stable, and are less immunogenic than are synthetic nanovectors. Therapeutic molecules can be loaded in EVs, both modifying the parent cells before isolation and altering EVs after isolation. In addition, EVs' protein surface composition can be modified to enhance their ability to cross the BBB [48]. Rabies virus glycoprotein (RVG)-modified EVs loaded with α-syn siRNA were administered into the brain of PD mice; this reduced total and aggregated α-syn in *sustantia nigra* dopaminergic neurons [75]. Izco et al. improved the system by increasing the efficacy of siRNA treatment using short hairpin RNA microcircles, and obtained a decrease in α-syn aggregation, a reduced loss of dopaminergic neurons, and an improvement in clinical symptoms in an α-syn PFF intrastriatally injected PD mouse model [76]. Several small molecules have also been encapsulated into EVs; EVs loaded with curcumin, an anti-inflammatory compound, were administered intranasally in different neurological disease models. The results showed an inhibition of inflammation and an increased apoptosis of IL1β^+ microglia [77]. Finally, dopamine, which cannot cross the

BBB when administered intravenously [78], improved functional recovery in a murine PD mouse model without any toxicity when loaded in blood EVs [79].

EV-based therapies, in general, have drawn a lot of interest; however, this field is still in its infancy and many problems have yet to be solved, including the large-scale production of highly purified EVs, standardized methods of purification and characterization, the efficiency of drug encapsulation, and knowledge of their mechanisms of action [80].

4. Clinical Translation and Future Perspective

From the above-mentioned results, it is apparent that EVs are valuable candidates for the identification of biomarkers to allow the early diagnosis of PD and to differentiate between different tauopathies, as well as having possible usefulness as therapeutics. However, it must be noted that the huge body of evidence that emerged from the literature of the last decades has not led to a clinical translation of the proposed biomarkers, with only a few clinical trials registered (Table 2).

The reasons for the discrepancy between research findings and clinical trials can possibly be found in the lack of standard procedures related to the laboratory activities of EV isolation, characterization, and storage that limit the reproducibility and clinical translation of the procedures. Indeed, several variables are known to impact the downstream results of EV analysis, including, for example, storage conditions that can affect EVs concentration, physical properties, and functionality [81]. The EV scientific community, represented by the International Society for Extracellular Vesicles (ISEV), is active in monitoring the current literature, creating dedicated task forces, and releasing guidelines and recommendations [25,82–85]. No consensus, though, has been reached on standard operating procedures in the isolation and characterization of EVs from liquid biopsies, because of a number of problems including the heterogeneity of EV themselves, the variable EV sources, and the plethora of techniques used to detect EV related molecules like protein markers and nucleic acids [86]. Moreover, the increasing knowledge of EV biogenesis and function led to changes in the design of EV analyses, which can now focus on studying single vesicles and pure preparations, or can even concentrate on the biomolecular corona that can be co-isolated with EVs. Technical limitations and a small sample size, together with a lack of replicates and standard operating procedures, are often the basis of the lack of translation of the research results to clinics.

From a methodological point of view, difficulties in the quantification of biomarkers like α-syn in EVs can be encountered in several steps of the analytical procedure. As an example, some concerns were raised on the specificity of one of the most widely used neuronal biomarkers, L1CAM [87], although several studies could successfully enrich brain EVs using this protein, confirming its specificity [88]. In the future it will be essential to identify and standardize robust biomarkers and methods to identify the exact cellular origin of EVs. Similarly, criticisms were made for some commercial kits for the preparation of EV suspensions that might lead to the co-isolation of non-specific molecules [89]. Indeed, the standardization of isolation and characterization methods will be crucial to improving the reproducibility of the results and their actual translation to clinics.

Clinically, little information is available on EVs, including their half-life in biological fluids and whether or not biological fluids can contain levels of disease-associated EVs that are suitable for the detection range of current technologies [90].

Table 2. List of the current clinical trials registered on clinicaltrial.gov on the use of EVs as a biomarker or as a treatment for PD.

NCT Number	Study Title/Funding	Study Type	Intervention	Single (S) or Multicenter (M)	Locations	Link
NCT05320250	Saliva and Extracellular Vesicles for Parkinson's Disease (RaSPiD)	Observational	PROCEDURE: Saliva collection	M	Italy	https://clinicaltrials.gov/study/NCT05320250 *
NCT05871359	Transcranial Direct Current Stimulation and Dual Tasks (Tdcs&DT)	Interventional	DEVICE: Transcranial Direct Current Stimulation PROCEDURE: Dual task	S	Italy	https://clinicaltrials.gov/study/NCT05871359 *
NCT05452655	Intensive Multidisciplinary Rehabilitation and Biomarkers in Parkinson's Disease	Interventional	BEHAVIORAL: Multidisciplinary Intensive Rehabilitation vs. Muscle-stretching and active mobilization exercises	S	Italy	https://clinicaltrials.gov/study/NCT05452655 *
NCT03775447	Fox BioNet Project: ECV-003	Observational	PROCEDURE: Lumbar Puncture	M	Canada, United States of America	https://clinicaltrials.gov/study/NCT03775447 *
NCT05902065	Effect of a Progressive Treadmill Training Protocol for Parkinson's Disease	Interventional	DEVICE: AVR Treadmill training with C-Mill vs. Conventional Treadmill training with C-Mill	S	Italy	https://clinicaltrials.gov/study/NCT05902065 *
NCT04603326	FoxBioNet: ECV (Extracellular Vesicle) 004	Observational	PROCEDURE: Lumbar Puncture	M	Canada, United States of America	https://clinicaltrials.gov/study/NCT04603326 *
NCT05807581	Clinical, Molecular and Electrophysiological Profiling of Parkinson's Disease: the Role of Non-pharmacological Therapies	Interventional	OTHER: physical activity vs. iTBS	M	Italy	https://clinicaltrials.gov/study/NCT05807581 *

Table 2. *Cont.*

NCT Number	Study Title/Funding	Study Type	Intervention	Single (S) or Multicenter (M)	Locations	Link
NCT04350177	A Study to Assess Single and Multiple Doses of IkT-148009 in Healthy Elderly Participants and Parkinson's Patients	Interventional	DRUG: IkT-148009 vs. Placebo	M	United States of America	https://clinicaltrials.gov/study/NCT04350177 *
NCT01860118	LRRK2 and Other Novel Exosome Proteins in Parkinson's Disease	Observational		S	United States of America	https://clinicaltrials.gov/study/NCT01860118 *
NCT03415984	Prevalence of Age Related Macular Degeneration (ARMD) in Parkinson's Patients and Assesment of the Role of L-DOPA (AMD-PARK)	Observational	DIAGNOSTIC_TEST: Color retinograph; Optical coherence tomograph; Fundus autofluorescence imaging	S	France	https://clinicaltrials.gov/study/NCT03415984 *
NCT05815524	Physical Activity in Patients With Parkinson's Disease: a "Disease Modifying" Intervention?	Interventional	OTHER: Physical activity training	S	Italy	https://clinicaltrials.gov/study/NCT05815524 *
NCT05109364	Terazosin and Parkinson's Disease Extension Study	Interventional	DRUG: Terazosin	S	United States of America	https://clinicaltrials.gov/study/NCT05109364 *

* accessed on 30 October 2023.

Among the obstacles to the validation of specific PD biomarkers, the selection of the analyzed cohort of subjects and improvements in clinical study design could significantly improve the translational ability of the studies in this field. As highlighted in Table 1, most works cited in the present review were performed in very small groups of subjects, which is reasonable for pilot studies demonstrating a proof of concept, but were not further validated on larger cohorts. Studies performed in larger groups of patients in different disease stages will nevertheless be needed to validate the findings and to better evaluate potential correlations with motor impairment and disease progression [51], avoiding misleading conclusions. Additionally, authors developing new techniques and assays should be alerted about the need for different control groups in their studies when designing a trial. Indeed, pilot studies are usually performed on disease cohorts, PD in this case, which are compared to healthy sex- and age-matched groups, but ROC results can be an overestimate of the real potential of the diagnostic assay due to the lack of an experimental group that shares the comorbidities of pwPD.

Finally, it has to be noted that the tremendous potential of new techniques and methods should be always cautiously assessed in single-center studies. The lack of EV standardization for isolation and characterization methods limits the reproducibility of results, as was recently reported [84]. For this reason, data should be validated with multicenter studies or studies that take advantage of biobanking procedures as they rely on standard operating procedures, whereas single-center studies mainly rely on a few operators and one single laboratory.

5. Conclusions

Taken together, the data summarized in the present review lead us to two main conclusions: (i) EVs have remarkable potential in PD diagnosis, prognosis and treatment, specifically because of their multifaceted nature and their ability to co-transport multiple biomarkers providing a snapshot of PD progression; (ii) the path towards clinical validation is still tortuous as some pieces of the puzzle (representing a knowledge gap) as well as standardized highly sensitive techniques (representing a technological gap) are still missing. Although both knowledge and technological gaps might require quite a lot of time and efforts to be fulfilled, in our opinion, three main solutions could significantly impact the field in the near future: the accurate selection of a cohort; the design of multicenter studies; the standardization of methods. If the latter cannot be achieved, detailed reporting could be significantly beneficial for translation to clinics.

Author Contributions: Conceptualization, C.A. and A.G.; investigation, C.A. and A.G.; writing—original draft preparation, C.A., and A.G.; writing—review and editing, S.P., M.B., F.R.G. and M.C.; visualization, C.A. and A.G.; supervision, M.B. and M.C. All authors have read and agreed to the published version of the manuscript.

Funding: This research was funded by Italian Ministry of Health Ricerca Corrente 2021-2023, RCR-2022-23682291 Fondo progetti reti EF 2022 and PNRR Project MNESYS (PE00000006–B43D22000650006). It was also partially supported by Fondazione Alessandro e Vincenzo Negroni Prati Morosini e Fondazione Romeo ed Enrica Invernizzi.

Institutional Review Board Statement: Not applicable.

Informed Consent Statement: Not applicable.

Data Availability Statement: Not applicable.

Conflicts of Interest: The authors declare no conflicts of interest.

References

1. Kalia, L.V.; Lang, A.E. Parkinson's Disease. *Lancet* **2015**, *386*, 896–912. [CrossRef]
2. Feigin, V.L.; Abajobir, A.A.; Abate, K.H.; Abd-Allah, F.; Abdulle, A.M.; Abera, S.F.; Abyu, G.Y.; Ahmed, M.B.; Aichour, A.N.; Aichour, I.; et al. Global, Regional, and National Burden of Neurological Disorders during 1990–2015: A Systematic Analysis for the Global Burden of Disease Study 2015. *Lancet Neurol.* **2017**, *16*, 877–897. [CrossRef]

3. Zafar, S.; Yaddanapudi, S.S. Parkinson Disease. In *StatPearls*; StatPearls Publishing: Treasure Island, FL, USA, 2023.
4. Ross, O.A. A Prognostic View on the Application of Individualized Genomics in Parkinson's Disease. *Curr. Genet. Med. Rep.* **2013**, *1*, 52–57. [CrossRef]
5. Deng, H.; Wang, P.; Jankovic, J. The Genetics of Parkinson Disease. *Ageing Res. Rev.* **2018**, *42*, 72–85. [CrossRef]
6. International Parkinson's Disease Genomics Consortium; 23andMe Research Team; Chang, D.; Nalls, M.A.; Hallgrímsdóttir, I.B.; Hunkapiller, J.; Van Der Brug, M.; Cai, F.; Kerchner, G.A.; Ayalon, G.; et al. A Meta-Analysis of Genome-Wide Association Studies Identifies 17 New Parkinson's Disease Risk Loci. *Nat. Genet.* **2017**, *49*, 1511–1516. [CrossRef]
7. International Parkinson's Disease Genomics Consortium (IPDGC); Parkinson's Study Group (PSG) Parkinson's Research: The Organized GENetics Initiative (PROGENI); 23andMe; GenePD; NeuroGenetics Research Consortium (NGRC); Hussman Institute of Human Genomics (HIHG); The Ashkenazi Jewish Dataset Investigator; Cohorts for Health and Aging Research in Genetic Epidemiology (CHARGE); North American Brain Expression Consortium (NABEC); United Kingdom Brain Expression Consortium (UKBEC); et al. Large-Scale Meta-Analysis of Genome-Wide Association Data Identifies Six New Risk Loci for Parkinson's Disease. *Nat. Genet.* **2014**, *46*, 989–993. [CrossRef]
8. Langston, J.W.; Langston, E.B.; Irwin, I. MPTP-Induced Parkinsonism in Human and Non-Human Primates—Clinical and Experimental Aspects. *Acta Neurol. Scand. Suppl.* **1984**, *100*, 49–54.
9. Nandipati, S.; Litvan, I. Environmental Exposures and Parkinson's Disease. *Int. J. Environ. Res. Public. Health* **2016**, *13*, 881. [CrossRef]
10. Goldman, S.M.; Kamel, F.; Ross, G.W.; Bhudhikanok, G.S.; Hoppin, J.A.; Korell, M.; Marras, C.; Meng, C.; Umbach, D.M.; Kasten, M.; et al. Genetic Modification of the Association of Paraquat and Parkinson's Disease. *Mov. Disord.* **2012**, *27*, 1652–1658. [CrossRef]
11. Fang, F.; Wirdefeldt, K.; Jacks, A.; Kamel, F.; Ye, W.; Chen, H. CNS Infections, Sepsis and Risk of Parkinson's Disease. *Int. J. Epidemiol.* **2012**, *41*, 1042–1049. [CrossRef]
12. Kieburtz, K.; Wunderle, K.B. Parkinson's Disease: Evidence for Environmental Risk Factors: Evidence for Environmental Risk Factors. *Mov. Disord.* **2013**, *28*, 8–13. [CrossRef]
13. Braak, H.; Del Tredici, K.; Rüb, U.; de Vos, R.A.I.; Jansen Steur, E.N.H.; Braak, E. Staging of Brain Pathology Related to Sporadic Parkinson's Disease. *Neurobiol. Aging* **2003**, *24*, 197–211. [CrossRef]
14. Litvan, I.; Bhatia, K.P.; Burn, D.J.; Goetz, C.G.; Lang, A.E.; McKeith, I.; Quinn, N.; Sethi, K.D.; Shults, C.; Wenning, G.K. SIC Task Force Appraisal of Clinical Diagnostic Criteria for Parkinsonian Disorders. *Mov. Disord.* **2003**, *18*, 467–486. [CrossRef]
15. Postuma, R.B.; Berg, D. Advances in Markers of Prodromal Parkinson Disease. *Nat. Rev. Neurol.* **2016**, *12*, 622–634. [CrossRef]
16. Jimenez-Ferrer, I.; Swanberg, M. Immunogenetics of Parkinson's Disease. In *Parkinson's Disease: Pathogenesis and Clinical Aspects*; Stoker, T.B., Greenland, J.C., Eds.; Codon Publications: Brisbane, AU, USA, 2018; ISBN 978-0-9944381-6-4.
17. Tolosa, E.; Gaig, C.; Santamaria, J.; Compta, Y. Diagnosis and the Premotor Phase of Parkinson Disease. *Neurology* **2009**, *72*, S12–S20. [CrossRef]
18. Samii, A.; Nutt, J.G.; Ransom, B.R. Parkinson's Disease. *Lancet* **2004**, *363*, 1783–1793. [CrossRef]
19. National Collaborating Centre for Chronic Conditions (UK). *Parkinson's Disease: National Clinical Guideline for Diagnosis and Management in Primary and Secondary Care*; National Institute for Health and Clinical Excellence: Guidance; Royal College of Physicians (UK): London, UK, 2006; ISBN 978-1-86016-283-1.
20. Connolly, B.S.; Lang, A.E. Pharmacological Treatment of Parkinson Disease: A Review. *JAMA* **2014**, *311*, 1670. [CrossRef]
21. Barker, R.A.; Björklund, A. Restorative Cell and Gene Therapies for Parkinson's Disease. In *Handbook of Clinical Neurology*; Elsevier: Amsterdam, The Netherlands, 2023; Volume 193, pp. 211–226; ISBN 978-0-323-85555-6.
22. Palfi, S.; Gurruchaga, J.M.; Lepetit, H.; Howard, K.; Ralph, G.S.; Mason, S.; Gouello, G.; Domenech, P.; Buttery, P.C.; Hantraye, P.; et al. Long-Term Follow-Up of a Phase I/II Study of ProSavin, a Lentiviral Vector Gene Therapy for Parkinson's Disease. *Hum. Gene Ther. Clin. Dev.* **2018**, *29*, 148–155. [CrossRef]
23. Brundin, P.; Dave, K.D.; Kordower, J.H. Therapeutic Approaches to Target Alpha-Synuclein Pathology. *Exp. Neurol.* **2017**, *298*, 225–235. [CrossRef]
24. Yoon, S.Y. Update on Parkinson's Disease Rehabilitation. *Brain Neurorehabilit.* **2022**, *15*, e15. [CrossRef]
25. Welsh, J.A.; Goberdhan, D.C.I.; O'Driscoll, L.; Buzas, E.I.; Blenkiron, C.; Bussolati, B. Minimal Information for Studies of Extracellular Vesicles (MISEV2023): From Basic to Advanced Approaches. *J. Extracell. Vesicles* 2024, *in press*.
26. Yáñez-Mó, M.; Siljander, P.R.-M.; Andreu, Z.; Zavec, A.B.; Borràs, F.E.; Buzas, E.I.; Buzas, K.; Casal, E.; Cappello, F.; Carvalho, J.; et al. Biological Properties of Extracellular Vesicles and Their Physiological Functions. *J. Extracell. Vesicles* **2015**, *4*, 27066. [CrossRef]
27. Raposo, G.; Nijman, H.W.; Stoorvogel, W.; Liejendekker, R.; Harding, C.V.; Melief, C.J.; Geuze, H.J. B Lymphocytes Secrete Antigen-Presenting Vesicles. *J. Exp. Med.* **1996**, *183*, 1161–1172. [CrossRef]
28. Van Niel, G.; Carter, D.R.F.; Clayton, A.; Lambert, D.W.; Raposo, G.; Vader, P. Challenges and Directions in Studying Cell–Cell Communication by Extracellular Vesicles. *Nat. Rev. Mol. Cell Biol.* **2022**, *23*, 369–382. [CrossRef]
29. Emmanouilidou, E.; Melachroinou, K.; Roumeliotis, T.; Garbis, S.D.; Ntzouni, M.; Margaritis, L.H.; Stefanis, L.; Vekrellis, K. Cell-Produced α-Synuclein Is Secreted in a Calcium-Dependent Manner by Exosomes and Impacts Neuronal Survival. *J. Neurosci.* **2010**, *30*, 6838–6851. [CrossRef]

30. Alvarez-Erviti, L.; Seow, Y.; Schapira, A.H.; Gardiner, C.; Sargent, I.L.; Wood, M.J.A.; Cooper, J.M. Lysosomal Dysfunction Increases Exosome-Mediated Alpha-Synuclein Release and Transmission. *Neurobiol. Dis.* **2011**, *42*, 360–367. [CrossRef]
31. Danzer, K.M.; Kranich, L.R.; Ruf, W.P.; Cagsal-Getkin, O.; Winslow, A.R.; Zhu, L.; Vanderburg, C.R.; McLean, P.J. Exosomal Cell-to-Cell Transmission of Alpha Synuclein Oligomers. *Mol. Neurodegener.* **2012**, *7*, 42. [CrossRef]
32. Tofaris, G.K.; Kim, H.T.; Hourez, R.; Jung, J.-W.; Kim, K.P.; Goldberg, A.L. Ubiquitin Ligase Nedd4 Promotes α-Synuclein Degradation by the Endosomal–Lysosomal Pathway. *Proc. Natl. Acad. Sci. USA* **2011**, *108*, 17004–17009. [CrossRef]
33. Kunadt, M.; Eckermann, K.; Stuendl, A.; Gong, J.; Russo, B.; Strauss, K.; Rai, S.; Kügler, S.; Falomir Lockhart, L.; Schwalbe, M.; et al. Extracellular Vesicle Sorting of α-Synuclein Is Regulated by Sumoylation. *Acta Neuropathol.* **2015**, *129*, 695–713. [CrossRef]
34. Fussi, N.; Höllerhage, M.; Chakroun, T.; Nykänen, N.-P.; Rösler, T.W.; Koeglsperger, T.; Wurst, W.; Behrends, C.; Höglinger, G.U. Exosomal Secretion of α-Synuclein as Protective Mechanism after Upstream Blockage of Macroautophagy. *Cell Death Dis.* **2018**, *9*, 757. [CrossRef]
35. Tofaris, G.K. A Critical Assessment of Exosomes in the Pathogenesis and Stratification of Parkinson's Disease. *J. Park. Dis.* **2017**, *7*, 569–576. [CrossRef]
36. Alza, N.P.; Iglesias González, P.A.; Conde, M.A.; Uranga, R.M.; Salvador, G.A. Lipids at the Crossroad of α-Synuclein Function and Dysfunction: Biological and Pathological Implications. *Front. Cell. Neurosci.* **2019**, *13*, 175. [CrossRef]
37. Shvadchak, V.V.; Falomir-Lockhart, L.J.; Yushchenko, D.A.; Jovin, T.M. Specificity and Kinetics of α-Synuclein Binding to Model Membranes Determined with Fluorescent Excited State Intramolecular Proton Transfer (ESIPT) Probe. *J. Biol. Chem.* **2011**, *286*, 13023–13032. [CrossRef]
38. Hannestad, J.K.; Rocha, S.; Agnarsson, B.; Zhdanov, V.P.; Wittung-Stafshede, P.; Höök, F. Single-Vesicle Imaging Reveals Lipid-Selective and Stepwise Membrane Disruption by Monomeric α-Synuclein. *Proc. Natl. Acad. Sci. USA* **2020**, *117*, 14178–14186. [CrossRef]
39. Zhang, S.; Eitan, E.; Wu, T.-Y.; Mattson, M.P. Intercellular Transfer of Pathogenic α-Synuclein by Extracellular Vesicles Is Induced by the Lipid Peroxidation Product 4-Hydroxynonenal. *Neurobiol. Aging* **2018**, *61*, 52–65. [CrossRef]
40. Peelaerts, W.; Bousset, L.; Van Der Perren, A.; Moskalyuk, A.; Pulizzi, R.; Giugliano, M.; Van Den Haute, C.; Melki, R.; Baekelandt, V. α-Synuclein Strains Cause Distinct Synucleinopathies after Local and Systemic Administration. *Nature* **2015**, *522*, 340–344. [CrossRef]
41. Hill, A.F. Extracellular Vesicles and Neurodegenerative Diseases. *J. Neurosci.* **2019**, *39*, 9269–9273. [CrossRef]
42. Bliederhaeuser, C.; Grozdanov, V.; Speidel, A.; Zondler, L.; Ruf, W.P.; Bayer, H.; Kiechle, M.; Feiler, M.S.; Freischmidt, A.; Brenner, D.; et al. Age-Dependent Defects of Alpha-Synuclein Oligomer Uptake in Microglia and Monocytes. *Acta Neuropathol.* **2016**, *131*, 379–391. [CrossRef]
43. Fraser, K.B.; Moehle, M.S.; Daher, J.P.L.; Webber, P.J.; Williams, J.Y.; Stewart, C.A.; Yacoubian, T.A.; Cowell, R.M.; Dokland, T.; Ye, T.; et al. LRRK2 Secretion in Exosomes Is Regulated by 14-3-3. *Hum. Mol. Genet.* **2013**, *22*, 4988–5000. [CrossRef]
44. Fraser, K.B.; Rawlins, A.B.; Clark, R.G.; Alcalay, R.N.; Standaert, D.G.; Liu, N.; Parkinson's Disease Biomarker Program Consortium; West, A.B. Ser(P)-1292 LRRK2 in Urinary Exosomes Is Elevated in Idiopathic Parkinson's Disease: P-LRRK2 Predicts Severity of PD. *Mov. Disord.* **2016**, *31*, 1543–1550. [CrossRef]
45. Shi, M.; Kovac, A.; Korff, A.; Cook, T.J.; Ginghina, C.; Bullock, K.M.; Yang, L.; Stewart, T.; Zheng, D.; Aro, P.; et al. CNS Tau Efflux via Exosomes Is Likely Increased in Parkinson's Disease but Not in Alzheimer's Disease. *Alzheimers Dement.* **2016**, *12*, 1125–1131. [CrossRef]
46. Jiang, Y.; Liu, J.; Chen, L.; Jin, Y.; Zhang, G.; Lin, Z.; Du, S.; Fu, Z.; Chen, T.; Qin, Y.; et al. Serum Secreted miR-137-Containing Exosomes Affects Oxidative Stress of Neurons by Regulating OXR1 in Parkinson's Disease. *Brain Res.* **2019**, *1722*, 146331. [CrossRef]
47. Thompson, A.G.; Gray, E.; Heman-Ackah, S.M.; Mäger, I.; Talbot, K.; Andaloussi, S.E.; Wood, M.J.; Turner, M.R. Extracellular Vesicles in Neurodegenerative Disease—Pathogenesis to Biomarkers. *Nat. Rev. Neurol.* **2016**, *12*, 346–357. [CrossRef]
48. Shi, M.; Liu, C.; Cook, T.J.; Bullock, K.M.; Zhao, Y.; Ginghina, C.; Li, Y.; Aro, P.; Dator, R.; He, C.; et al. Plasma Exosomal α-Synuclein Is Likely CNS-Derived and Increased in Parkinson's Disease. *Acta Neuropathol.* **2014**, *128*, 639–650. [CrossRef]
49. Jiang, C.; Hopfner, F.; Berg, D.; Hu, M.T.; Pilotto, A.; Borroni, B.; Davis, J.J.; Tofaris, G.K. Validation of α-Synuclein in L1CAM -Immunocaptured Exosomes as a Biomarker for the Stratification of Parkinsonian Syndromes. *Mov. Disord.* **2021**, *36*, 2663–2669. [CrossRef]
50. Cerri, S.; Ghezzi, C.; Sampieri, M.; Siani, F.; Avenali, M.; Dornini, G.; Zangaglia, R.; Minafra, B.; Blandini, F. The Exosomal/Total α-Synuclein Ratio in Plasma Is Associated With Glucocerebrosidase Activity and Correlates With Measures of Disease Severity in PD Patients. *Front. Cell. Neurosci.* **2018**, *12*, 125. [CrossRef] [PubMed]
51. Stuendl, A.; Kraus, T.; Chatterjee, M.; Zapke, B.; Sadowski, B.; Moebius, W.; Hobert, M.A.; Deuschle, C.; Brockmann, K.; Maetzler, W.; et al. α-Synuclein in Plasma-Derived Extracellular Vesicles Is a Potential Biomarker of Parkinson's Disease. *Mov. Disord. Off. J. Mov. Disord. Soc.* **2021**, *36*, 2508–2518. [CrossRef]
52. Matsumoto, J.; Stewart, T.; Sheng, L.; Li, N.; Bullock, K.; Song, N.; Shi, M.; Banks, W.A.; Zhang, J. Transmission of α-Synuclein-Containing Erythrocyte-Derived Extracellular Vesicles across the Blood-Brain Barrier via Adsorptive Mediated Transcytosis: Another Mechanism for Initiation and Progression of Parkinson's Disease? *Acta Neuropathol. Commun.* **2017**, *5*, 71. [CrossRef] [PubMed]

53. Kim, S.; Jeon, B.S.; Heo, C.; Im, P.S.; Ahn, T.; Seo, J.; Kim, H.; Park, C.H.; Choi, S.H.; Cho, S.; et al. α-Synuclein Induces Apoptosis by Altered Expression in Human Peripheral Lymphocytes in Parkinson's Disease. *FASEB J.* **2004**, *18*, 1615–1617. [CrossRef]
54. Ohmichi, T.; Mitsuhashi, M.; Tatebe, H.; Kasai, T.; El-Agnaf, O.M.A.; Tokuda, T. Quantification of Brain-Derived Extracellular Vesicles in Plasma as a Biomarker to Diagnose Parkinson's and Related Diseases. *Park. Relat. Disord.* **2019**, *61*, 82–87. [CrossRef]
55. Agliardi, C.; Meloni, M.; Guerini, F.R.; Zanzottera, M.; Bolognesi, E.; Baglio, F.; Clerici, M. Oligomeric α-Syn and SNARE Complex Proteins in Peripheral Extracellular Vesicles of Neural Origin Are Biomarkers for Parkinson's Disease. *Neurobiol. Dis.* **2021**, *148*, 105185. [CrossRef]
56. Meloni, M.; Agliardi, C.; Guerini, F.R.; Zanzottera, M.; Bolognesi, E.; Picciolini, S.; Marano, M.; Magliozzi, A.; Di Fonzo, A.; Arighi, A.; et al. Oligomeric α-Synuclein and Tau Aggregates in NDEVs Differentiate Parkinson's Disease from Atypical Parkinsonisms. *Neurobiol. Dis.* **2023**, *176*, 105947. [CrossRef] [PubMed]
57. Tomlinson, P.R.; Zheng, Y.; Fischer, R.; Heidasch, R.; Gardiner, C.; Evetts, S.; Hu, M.; Wade-Martins, R.; Turner, M.R.; Morris, J.; et al. Identification of Distinct Circulating Exosomes in Parkinson's Disease. *Ann. Clin. Transl. Neurol.* **2015**, *2*, 353–361. [CrossRef]
58. Gualerzi, A.; Picciolini, S.; Carlomagno, C.; Terenzi, F.; Ramat, S.; Sorbi, S.; Bedoni, M. Raman Profiling of Circulating Extracellular Vesicles for the Stratification of Parkinson's Patients. *Nanomed. Nanotechnol. Biol. Med.* **2019**, *22*, 102097. [CrossRef]
59. Gui, Y.; Liu, H.; Zhang, L.; Lv, W.; Hu, X. Altered microRNA Profiles in Cerebrospinal Fluid Exosome in Parkinson Disease and Alzheimer Disease. *Oncotarget* **2015**, *6*, 37043–37053. [CrossRef] [PubMed]
60. Yao, Y.-F.; Qu, M.-W.; Li, G.-C.; Zhang, F.-B.; Rui, H.-C. Circulating Exosomal miRNAs as Diagnostic Biomarkers in Parkinson's Disease. *Eur. Rev. Med. Pharmacol. Sci.* **2018**, *22*, 5278–5283. [CrossRef] [PubMed]
61. Cao, X.-Y.; Lu, J.-M.; Zhao, Z.-Q.; Li, M.-C.; Lu, T.; An, X.-S.; Xue, L.-J. MicroRNA Biomarkers of Parkinson's Disease in Serum Exosome-like Microvesicles. *Neurosci. Lett.* **2017**, *644*, 94–99. [CrossRef]
62. Wang, S.; Liu, Z.; Ye, T.; Mabrouk, O.S.; Maltbie, T.; Aasly, J.; West, A.B. Elevated LRRK2 Autophosphorylation in Brain-Derived and Peripheral Exosomes in LRRK2 Mutation Carriers. *Acta Neuropathol. Commun.* **2017**, *5*, 86. [CrossRef] [PubMed]
63. Wang, S.; Kojima, K.; Mobley, J.A.; West, A.B. Proteomic Analysis of Urinary Extracellular Vesicles Reveal Biomarkers for Neurologic Disease. *EBioMedicine* **2019**, *45*, 351–361. [CrossRef]
64. Rani, K.; Mukherjee, R.; Singh, E.; Kumar, S.; Sharma, V.; Vishwakarma, P.; Bharti, P.S.; Nikolajeff, F.; Dinda, A.K.; Goyal, V.; et al. Neuronal Exosomes in Saliva of Parkinson's Disease Patients: A Pilot Study. *Parkinsonism Relat. Disord.* **2019**, *67*, 21–23. [CrossRef]
65. Daneman, R.; Prat, A. The Blood–Brain Barrier. *Cold Spring Harb. Perspect. Biol.* **2015**, *7*, a020412. [CrossRef]
66. Anand, S.; Samuel, M.; Kumar, S.; Mathivanan, S. Ticket to a Bubble Ride: Cargo Sorting into Exosomes and Extracellular Vesicles. *Biochim. Biophys. Acta BBA Proteins Proteom.* **2019**, *1867*, 140203. [CrossRef] [PubMed]
67. Prockop, D.J.; Brenner, M.; Fibbe, W.E.; Horwitz, E.; Le Blanc, K.; Phinney, D.G.; Simmons, P.J.; Sensebe, L.; Keating, A. Defining the Risks of Mesenchymal Stromal Cell Therapy. *Cytotherapy* **2010**, *12*, 576–578. [CrossRef] [PubMed]
68. Yu, B.; Zhang, X.; Li, X. Exosomes Derived from Mesenchymal Stem Cells. *Int. J. Mol. Sci.* **2014**, *15*, 4142–4157. [CrossRef]
69. Lee, E.J.; Choi, Y.; Lee, H.J.; Hwang, D.W.; Lee, D.S. Human Neural Stem Cell-Derived Extracellular Vesicles Protect against Parkinson's Disease Pathologies. *J. Nanobiotechnol.* **2022**, *20*, 198. [CrossRef] [PubMed]
70. Chen, H.-X.; Liang, F.-C.; Gu, P.; Xu, B.-L.; Xu, H.-J.; Wang, W.-T.; Hou, J.-Y.; Xie, D.-X.; Chai, X.-Q.; An, S.-J. Exosomes Derived from Mesenchymal Stem Cells Repair a Parkinson's Disease Model by Inducing Autophagy. *Cell Death Dis.* **2020**, *11*, 288. [CrossRef] [PubMed]
71. Jarmalavičiūtė, A.; Tunaitis, V.; Pivoraitė, U.; Venalis, A.; Pivoriūnas, A. Exosomes from Dental Pulp Stem Cells Rescue Human Dopaminergic Neurons from 6-Hydroxy-Dopamine–Induced Apoptosis. *Cytotherapy* **2015**, *17*, 932–939. [CrossRef]
72. Narbute, K.; Pilipenko, V.; Pupure, J.; Dzirkale, Z.; Jonavičė, U.; Tunaitis, V.; Kriaučiūnaitė, K.; Jarmalavičiūtė, A.; Jansone, B.; Kluša, V.; et al. Intranasal Administration of Extracellular Vesicles Derived from Human Teeth Stem Cells Improves Motor Symptoms and Normalizes Tyrosine Hydroxylase Expression in the Substantia Nigra and Striatum of the 6-Hydroxydopamine-Treated Rats. *Stem Cells Transl. Med.* **2019**, *8*, 490–499. [CrossRef]
73. Elsharkasy, O.M.; Nordin, J.Z.; Hagey, D.W.; De Jong, O.G.; Schiffelers, R.M.; Andaloussi, S.E.; Vader, P. Extracellular Vesicles as Drug Delivery Systems: Why and How? *Adv. Drug Deliv. Rev.* **2020**, *159*, 332–343. [CrossRef] [PubMed]
74. Escudé Martinez De Castilla, P.; Tong, L.; Huang, C.; Sofias, A.M.; Pastorin, G.; Chen, X.; Storm, G.; Schiffelers, R.M.; Wang, J.-W. Extracellular Vesicles as a Drug Delivery System: A Systematic Review of Preclinical Studies. *Adv. Drug Deliv. Rev.* **2021**, *175*, 113801. [CrossRef] [PubMed]
75. Cooper, J.M.; Wiklander, P.B.O.; Nordin, J.Z.; Al-Shawi, R.; Wood, M.J.; Vithlani, M.; Schapira, A.H.V.; Simons, J.P.; El-Andaloussi, S.; Alvarez-Erviti, L. Systemic Exosomal siRNA Delivery Reduced Alpha-Synuclein Aggregates in Brains of Transgenic Mice: siRNA-EXOSOME DECREASED ALPHA-SYNUCLEIN AGGREGATES. *Mov. Disord.* **2014**, *29*, 1476–1485. [CrossRef]
76. Izco, M.; Blesa, J.; Schleef, M.; Schmeer, M.; Porcari, R.; Al-Shawi, R.; Ellmerich, S.; De Toro, M.; Gardiner, C.; Seow, Y.; et al. Systemic Exosomal Delivery of shRNA Minicircles Prevents Parkinsonian Pathology. *Mol. Ther.* **2019**, *27*, 2111–2122. [CrossRef]
77. Zhuang, X.; Xiang, X.; Grizzle, W.; Sun, D.; Zhang, S.; Axtell, R.C.; Ju, S.; Mu, J.; Zhang, L.; Steinman, L.; et al. Treatment of Brain Inflammatory Diseases by Delivering Exosome Encapsulated Anti-Inflammatory Drugs From the Nasal Region to the Brain. *Mol. Ther.* **2011**, *19*, 1769–1779. [CrossRef] [PubMed]
78. Bernheimer, H.; Birkmayer, W.; Hornykiewicz, O.; Jellinger, K.; Seitelberger, F. Brain Dopamine and the Syndromes of Parkinson and Huntington Clinical, Morphological and Neurochemical Correlations. *J. Neurol. Sci.* **1973**, *20*, 415–455. [CrossRef] [PubMed]

79. Qu, M.; Lin, Q.; Huang, L.; Fu, Y.; Wang, L.; He, S.; Fu, Y.; Yang, S.; Zhang, Z.; Zhang, L.; et al. Dopamine-Loaded Blood Exosomes Targeted to Brain for Better Treatment of Parkinson's Disease. *J. Control. Release* **2018**, *287*, 156–166. [CrossRef] [PubMed]
80. Gratpain, V.; Mwema, A.; Labrak, Y.; Muccioli, G.G.; Van Pesch, V.; Des Rieux, A. Extracellular Vesicles for the Treatment of Central Nervous System Diseases. *Adv. Drug Deliv. Rev.* **2021**, *174*, 535–552. [CrossRef]
81. Gelibter, S.; Marostica, G.; Mandelli, A.; Siciliani, S.; Podini, P.; Finardi, A.; Furlan, R. The Impact of Storage on Extracellular Vesicles: A Systematic Study. *J. Extracell. Vesicles* **2022**, *11*, e12162. [CrossRef] [PubMed]
82. Théry, C.; Witwer, K.W.; Aikawa, E.; Alcaraz, M.J.; Anderson, J.D.; Andriantsitohaina, R.; Antoniou, A.; Arab, T.; Archer, F.; Atkin-Smith, G.K.; et al. Minimal Information for Studies of Extracellular Vesicles 2018 (MISEV2018): A Position Statement of the International Society for Extracellular Vesicles and Update of the MISEV2014 Guidelines. *J. Extracell. Vesicles* **2018**, *7*, 1535750. [CrossRef]
83. Clayton, A.; Boilard, E.; Buzas, E.I.; Cheng, L.; Falcón-Perez, J.M.; Gardiner, C.; Gustafson, D.; Gualerzi, A.; Hendrix, A.; Hoffman, A.; et al. Considerations towards a Roadmap for Collection, Handling and Storage of Blood Extracellular Vesicles. *J. Extracell. Vesicles* **2019**, *8*, 1647027. [CrossRef]
84. Lucien, F.; Gustafson, D.; Lenassi, M.; Li, B.; Teske, J.J.; Boilard, E.; Von Hohenberg, K.C.; Falcón-Perez, J.M.; Gualerzi, A.; Reale, A.; et al. MIBlood-EV: Minimal Information to Enhance the Quality and Reproducibility of Blood Extracellular Vesicle Research. *J. Extracell. Vesicles* **2023**, *12*, 12385. [CrossRef] [PubMed]
85. Roux, Q.; Deun, J.V.; Dedeyne, S.; Hendrix, A. The EV-TRACK Summary Add-on: Integration of Experimental Information in Databases to Ensure Comprehensive Interpretation of Biological Knowledge on Extracellular Vesicles. *J. Extracell. Vesicles* **2019**, *9*, 1699367. [CrossRef] [PubMed]
86. Margolis, L.; Sadovsky, Y. The Biology of Extracellular Vesicles: The Known Unknowns. *PLoS Biol.* **2019**, *17*, e3000363. [CrossRef]
87. Norman, M.; Ter-Ovanesyan, D.; Trieu, W.; Lazarovits, R.; Kowal, E.J.K.; Lee, J.H.; Chen-Plotkin, A.S.; Regev, A.; Church, G.M.; Walt, D.R. L1CAM Is Not Associated with Extracellular Vesicles in Human Cerebrospinal Fluid or Plasma. *Nat. Methods* **2021**, *18*, 631–634. [CrossRef] [PubMed]
88. Gomes, D.E.; Witwer, K.W. L1CAM-associated Extracellular Vesicles: A Systematic Review of Nomenclature, Sources, Separation, and Characterization. *J. Extracell. Biol.* **2022**, *1*, e35. [CrossRef] [PubMed]
89. Yan, S.; Jiang, C.; Davis, J.J.; Tofaris, G.K. Methodological Considerations in Neuronal Extracellular Vesicle Isolation for α-Synuclein Biomarkers. *Brain* **2023**, *146*, awad169. [CrossRef] [PubMed]
90. Möller, A.; Lobb, R.J. The Evolving Translational Potential of Small Extracellular Vesicles in Cancer. *Nat. Rev. Cancer* **2020**, *20*, 697–709. [CrossRef] [PubMed]

MDPI AG
Grosspeteranlage 5
4052 Basel
Switzerland
Tel.: +41 61 683 77 34

International Journal of Molecular Sciences Editorial Office
E-mail: ijms@mdpi.com
www.mdpi.com/journal/ijms

www.ingramcontent.com/pod-product-compliance
Lightning Source LLC
Chambersburg PA
CBHW041949260926
51850CB00019B/2143
9783725862740